WORKBOOK/LAB MANUAL
to Accompany

Residential Construction Academy

House Wiring

Third Edition

Greg Fletcher

DELMAR
CENGAGE Learning™

Australia • Brazil • Japan • Korea • Mexico • Singapore • Spain • United Kingdom • United States

DELMAR
CENGAGE Learning™

Workbook/Lab Manual to Accompany Residential Construction Academy: House Wiring, Third Edition
Greg Fletcher

Vice President, Career and Professional Editorial: Dave Garza

Director of Learning Solutions: Sandy Clark

Senior Acquisitions Editor: Jim DeVoe

Managing Editor: Larry Main

Product Manager: Brooke Wilson, Ohlinger Publishing Services

Editorial Assistant: Cris Savino

Vice President, Career and Professional Marketing: Jennifer Baker

Marketing Director: Deborah Yarnell

Marketing Manager: Jillian Borden

Senior Production Director: Wendy Troeger

Production Manager: Mark Bernard

Senior Content Project Manager: Michael Tubbert

Senior Art Director: Casey Kirchmayer

For product information and technology assistance, contact us at
Cengage Learning Customer & Sales Support, 1-800-354-9706

For permission to use material from this text or product,
submit all requests online at **www.cengage.com/permissions.**
Further permissions questions can be e-mailed to
permissionrequest@cengage.com

ISBN-13: 978-1-111-30624-3

ISBN-10: 1-111-30624-9

Delmar
5 Maxwell Drive
Clifton Park, NY 12065-2919
USA

Cengage Learning is a leading provider of customized learning solutions with office locations around the globe, including Singapore, the United Kingdom, Australia, Mexico, Brazil, and Japan. Locate your local office at: **international.cengage.com/region**

Cengage Learning products are represented in Canada by Nelson Education, Ltd.

To learn more about Delmar, visit **www.cengage.com/delmar.**

Purchase any of our products at your local college store or at our preferred online store **www.cengagebrain.com**

Printed in the United States of America
1 2 3 4 5 6 7 15 14 13 12 11

Table of Contents

PART I Workbook

CHAPTER 1 Residential Workplace Safety 3

CHAPTER 2 Hardware and Materials Used in Residential Wiring 13

CHAPTER 3 Tools Used in Residential Wiring 25

CHAPTER 8   Service Entrance Equipment and Installation

CHAPTER 9   General Requirements for Rough-In Wiring

CHAPTER 10   Electrical Box Installation

CHAPTER 11   Cable Installation

CHAPTER 2 Hardware and Materials Used in Residential Wiring 267

CHAPTER 3 Tools Used in Residential Wiring 277

CHAPTER **12** **Raceway Installation** 359

CHAPTER **13** **Switching Circuit Installation** 373

Preface

About This Book

This workbook/lab manual is written to supplement the third edition of *Residential Construction Academy: House Wiring*. It is designed as an instructional resource for secondary, post-secondary, and apprentice residential wiring programs. Various types of learning activities are included in the workbook section, and the lab section includes more than fifty hands-on lab exercises covering wiring practices that are commonly used in residential wiring. Both general safety and electrical safety are stressed throughout the workbook/lab manual.

The exercises in this workbook/lab manual will reinforce the information an electrician needs to know to install residential electrical systems in a practical, hands-on manner. It focuses on basic hands-on wiring skills such as the proper usage of hand and power tools, the proper use of measuring instruments, connecting wires together properly, attaching electrical boxes to framing members, and running cable or raceway to the boxes. It also focuses on service entrance installation; bending conduit; installing voice, video, and data wiring; trimming out a loadcenter; and troubleshooting circuits that do not work properly. In addition, this edition focuses on green wiring practices and the installation of solar photovoltaic systems and small wind turbine systems.

In the case of most secondary, post-secondary, and apprentice electrical programs, students have little or no residential electrical wiring experience before they enter the program and need to be taught all of the basic hands-on skills that are required to perform residential electrical installations. This workbook/lab manual allows them to develop the necessary hands-on skills to become successful residential electricians.

Organization

This workbook/lab manual is organized in the same way that a typical residential wiring project unfolds. The chapters of the workbook section coincide with the chapters in the third edition of *House Wiring* and are designed to be used at the same time as that topic is being covered in the main textbook. Chapters can be covered in the order they are presented or instructors can pick and choose the chapters to meet their teaching plans.

The lab manual part of this supplement includes several hands-on labs and applications. Most of the lab exercises require work to be done on a wiring mock-up. Some of the lab exercises can be completed at a table or desk. The lab exercises coincide with the material covered in the corresponding chapter of the third edition of *House Wiring*. Each lab exercise has its objective clearly stated at the beginning. An introduction section will briefly cover the information necessary for a student to understand the focus of the lab exercise. A material/equipment section lists what is required to complete the lab exercise. The procedure section tells students what to do in a step-by-step format as they proceed through the lab exercise. Finally, there is a review section that asks the students to answer some questions about the lab exercise or to fill out a complete materials list of everything used to complete the lab exercise.

Wiring Mock-up

Most of the hands-on lab exercises are designed to be done on a wiring mock-up that has a 4-foot × 8-foot × ¾-inch plywood wall and a studded wall with four studs (see Figure P-1). However, if the wiring mock-ups in your lab area are not built the same way, the lab exercises can be adapted for other wiring mock-up configurations.

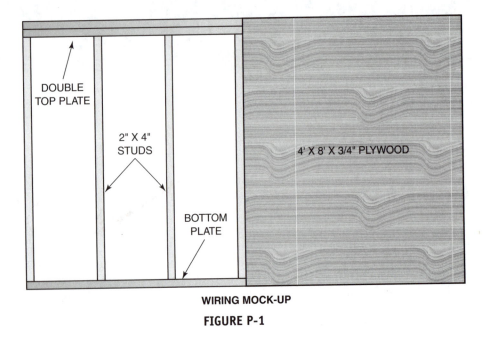

WIRING MOCK-UP

FIGURE P-1

A small circuit breaker panel that can be easily locked out should be mounted on the wiring mock-up or close by. Some lab exercises require "homeruns" from a loadcenter to power up the lab project. A 120-volt, GFCI receptacle also needs to be close to the wiring mock-up. A suggested power panel setup is shown in Figure P-2. This power panel configuration can be cord-and-plug connected or hardwired to a 120/240-volt power source. It can be mounted on the plywood section of the wiring mock-up when lab exercises are being done on the stud section of the mock-up. When lab exercises need to be done on the plywood section, the power panel setup can easily be moved to the stud section.

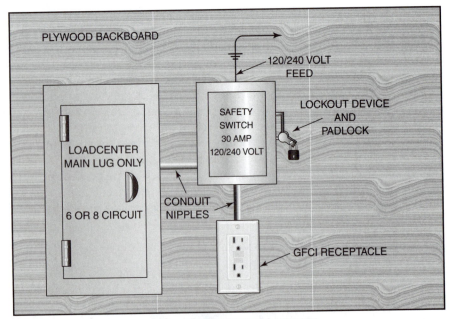

WIRING MOCK-UP POWER SOURCE

FIGURE P-2

Features

The workbook section contains several features that will help a student electrician better understand the presented material. The workbook features include:

- A detailed residential electrician job description. The purpose of the job description is to familiarize students with the skills and training necessary for the electrical field, employment opportunities, and job outlook.

- A detailed outline organized by the chapter objectives. The outline is organized by the objectives for that chapter and correlating chapter material.

- Quizzes on special topics. The quizzes contain items on safety, the *National Electrical Code®*, and tools as appropriate to the chapter. The question style varies among multiple choice, fill-ins, and matching.

- Chapter quizzes. There is a chapter quiz that will cover information from throughout the entire chapter. The quizzes are twenty questions long.

- Troubleshooting/quality exercises. This section uses a variety of question styles. The focus is primarily derived from the From Experience boxes in each chapter of the textbook and includes discussion questions, hands-on applications, and real-world scenarios. The scenarios highlight communication/etiquette issues when dealing with customers and residential situations. They also focus on teamwork/team-building issues for fields where working with others and on construction sites is a factor.

The lab manual section contains several features that reinforce the material covered in the textbook.

- Exercises are written in a consistent manner throughout the entire workbook/lab manual.

- Students are reminded to follow safety guidelines throughout the lab exercises.

- Exercises are designed to get students to recognize common hardware and materials used in the residential electrical field.

- Exercises provide students with the practice they need to become proficient with using common hand and power tools.

- Each exercise requires students to take a look at what they did in the exercise and answer a few thought-provoking questions about the lab.

- Many lab exercises require wiring diagrams to be filled in by the student.

- Instructors must look at the work and verify the work quality before a student can go on.

- Instructors are given a scoring rubric for assigning a score for each lab exercise.

About the Author

The author of this textbook, Greg Fletcher, has over 30 years of experience in the electrical field as both a practicing electrician and as an electrical instructor. His practical experience has been primarily in the residential and commercial electrical construction field. He has been a licensed electrician since 1976; first as a Journeyman Electrician and then as a Master Electrician. He has taught electrical wiring practices at both the secondary level and the post-secondary level. He has taught apprenticeship electrical courses and has facilitated many workshops on topics such as Using the National Electrical Code, Fiber Optics for Electricians, Understanding Electrical Calculations, and Introduction to Photovoltaics. The knowledge gained over those years, specifically on what works and what does not work to effectively teach electrical wiring practices, was used as a guide to help determine the focus of this textbook.

Since 1988 he has been Department Chairman of the Trades and Technology Department and an Instructor in the Electrical Technology program at Kennebec Valley Community College in Fairfield, Maine. He holds an Associate of Applied Science Degree in Electrical Construction and Maintenance, a Bachelor of Science Degree in Applied Technical Education, and a Master of Science Degree in Industrial Education. Mr. Fletcher is a member of the International Association of Electrical Inspectors and The National Fire Protection Association. He lives in Waterville, Maine with his wife and daughter. When not teaching or writing textbooks he enjoys reading, golfing, motorcycling, and spending time with his family.

Job Description: Residential Electrician

Skills and Training

Residential electricians install, maintain, and troubleshoot electrical wiring systems in a house. There are many ways to become a residential electrician. Most people complete an electrical program of study at a technical high school, a community college, a workforce development center, or an apprentice program. A residential electrician's job involves a variety of the following skills, knowledge, and attitudes.

- Electricians must understand and follow both general and electrical safety procedures.
- Electricians must know the common hardware and materials used in a residential electrical system.
- Electricians must know how to read and understand building plans, use hand tools, use power tools, and use testing equipment.
- Electricians must follow the *National Electrical Code®* and comply with state and local codes when installing electrical wiring.
- Electricians install service entrances so that electrical power can be provided from the electric utility to the building's electrical system.
- Electricians install branch and feeder circuits with various types of cable and conduit.
- Electricians fasten metallic or nonmetallic electrical boxes to the framing members of a house.
- Electricians connect conductors to circuit breakers, lighting fixtures, paddle fans, electric motors, and other electrical equipment.
- Electricians install video, voice, and data wiring and equipment.
- Electricians may rewire a home or replace an old fuse box with a new circuit breaker loadcenter.
- Electricians may go on service calls to identify and then fix electrical problems.

Employment

Because of the widespread need for electrical services, jobs for residential electricians are found in all parts of the country. Most states and localities require electricians to be licensed. Although licensing requirements differ from area to area, residential electricians generally have to pass an examination that tests their knowledge of electrical theory, the *NEC®*, local electrical codes, and the various types of wiring methods and materials. Experienced electricians can advance to jobs as estimators, supervisors, project managers, or even electrical inspectors. They may also decide to start their own electrical contracting company.

- A residential electrician's work is sometimes strenuous and may require standing or kneeling for long periods of time.
- A residential electrician may be subject to inclement weather conditions when working outdoors and the work may require traveling long distances to jobsites.

- Being a residential electrician involves frequently working on ladders and scaffolding and includes the risk of injury from electrical shock, burns, falls, and cuts.

- Residential electricians should be in good health and have at least average physical strength.

- Residential electricians should have good agility and manual dexterity as well as have good color vision because they often must identify electrical wires by color.

- Residential electricians should have good interpersonal and communication skills.

Job Outlook

Employment of residential electricians is expected to increase by about 12% though the year 2018. As the population and the economy grow, more residential electricians will be needed to install and maintain electrical equipment in homes.

- Because of their training and relatively high earnings, a smaller proportion of electricians than other trade workers leave their occupation each year.

- In addition to jobs created by the increased demand for electrical work, many openings are expected to occur over the next decade as a large number of electricians are expected to retire.

- The average hourly wage for all electricians based on the latest data available is $22.32. The middle 50% earn between $17.00 and $29.88. The lowest 10% earn less than $13.54, and the highest 10% earn more than $38.18. Residential electricians may earn more or less per hour, depending on a number of factors such as experience, licenses held, and the demand for electricians in a specific geographical area.

- Beginning electricians usually start at between 30% and 50% of the rate paid to fully licensed and trained electricians.

The information presented above is based on information from the Bureau of Labor Statistics, U.S. Department of Labor, *Occupational Outlook Handbook, 2010-11 Edition* ("Electricians") section, and the Web site at http://www.bls.gov/oco/pdf/ocos206.pdf.

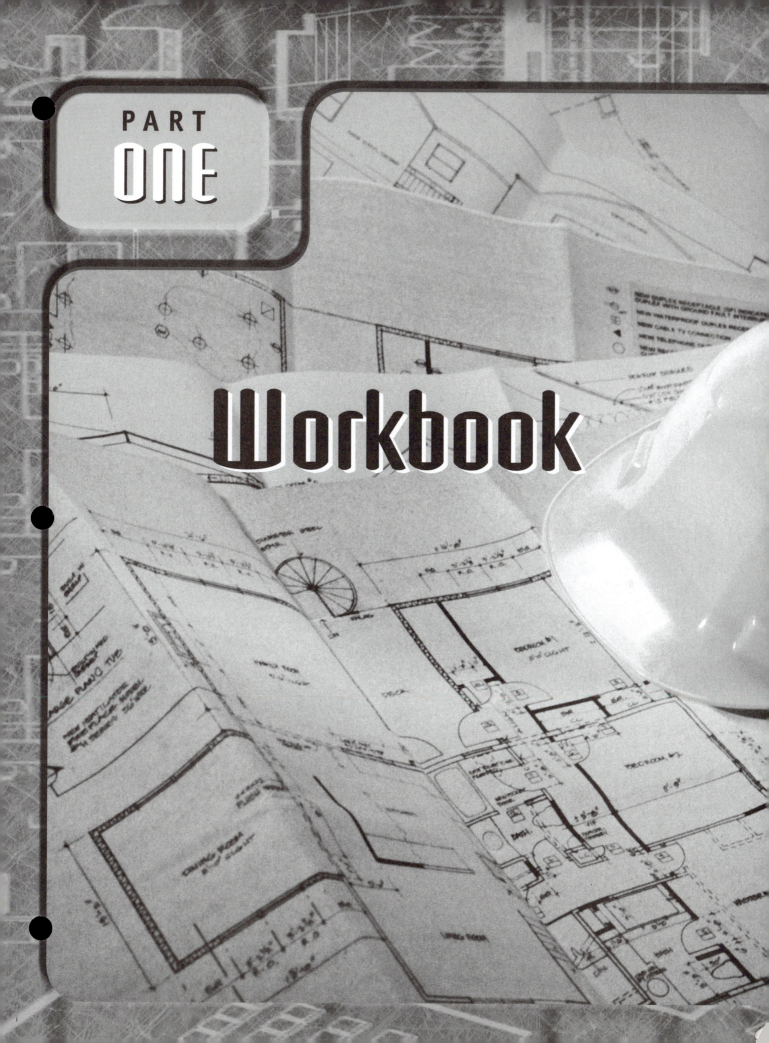

PART

ONE

Workbook

Chapter 1 — Residential Workplace Safety

Outline

Objective: Demonstrate an understanding of the electrical hazards associated with electrical work.

1) Understanding the shock hazard

Key terms: voltage, conductor, insulator, circuit, power source, current, resistance, Ohm's law, amperes, volts, ohms, arc blast

- Electrical shock is considered the biggest safety hazard when doing electrical work.
- Electricity is the flow of electrons through a material.
- The force that causes electrons to move is called voltage. It is measured in volts.
- The amount of electron flow is called current. It is measured in amperes.
- The opposition to electron flow in a material is called resistance. It is measured in ohms.
- Ohm's law is a mathematical representation of the relationship between voltage, current, and resistance in an electrical circuit.
- Assuming that voltage stays the same, less resistance means more current flow; more resistance means less current flow.
- The human body reacts differently to the amount of current flowing through it.
- The longer a person's body is part of an electrical circuit, the greater will be the severity of the injury.
- Burns and arc blasts caused by electricity are other electrical hazards encountered by electricians.
- An arc-flash is a dangerous condition associated with the possible release of energy caused by an electric arc.
- Proper personal protective equipment must be used where the possibility of arcing exists.

Objective: Demonstrate an understanding of the purpose of the *National Electrical Code®*.

2) Understanding the purpose of the *National Electrical Code®*

Key term: *National Electrical Code® (NEC®)*

- The *NEC®* is the guide for safe wiring practices in the electrical field.
- Every three years the *NEC®* is brought up to date to reflect the latest changes and trends in the electrical industry.
- The *NEC®* contains specific rules to help safeguard people and property from the hazards arising from the use of electricity.

Objective: Demonstrate an understanding of the arrangement of the *National Electrical Code*®.

3) Understanding the arrangement of the *National Electrical Code*®

- The *NEC*® is divided into nine chapters.
- Chapters 1–4 consist of rules that apply generally to all electrical installations.
- Chapters 5, 6, and 7 apply to special occupancies and equipment.
- Chapter 8 contains rules for the installation of wiring for communications systems.
- Chapter 9 consists of tables that contain information such as conduit fill and conductor properties.
- The chapters are broken down into a series of articles.
- Most articles are divided into parts.
- Each rule found in the *NEC*® is called a section.
- Tables are used throughout the *NEC*® and contain various amounts and types of information.
- There are a few basic illustrations in the *NEC*® and they are called figures.
- *Exceptions* to *NEC*® rules are listed under the section where they apply and are written in italic text.
- Informational Notes are used to provide explanatory material and are placed after the section or table to which they apply.
- The table of contents lists each article in numerical order.
- Informative Annexes are located after Chapter 9 and contain additional explanatory material that is designed to help people use the *NEC*®.
- The index is located at the very end of the *NEC*®. It contains the major topic areas covered in the *NEC*® and lists them in alphabetical order.

Objective: Cite examples of rules from the *National Electrical Code*® pertaining to common residential electrical safety hazards.

4) Examples of rules from the *National Electrical Code*®

Key term: ground fault circuit interrupter (GFCI)

- Section 210.8(A) specifies where GFCI protection is required in residential wiring applications.
- Section 210.12(A) requires arc fault protection for 120-volt, 15- and 20-amp branch circuits supplying outlets in almost all locations of a house.

Objective: Demonstrate an understanding of the purpose of NFPA 70E Standard for Electrical Safety in the Workplace.

5) Understanding the purpose of NFPA 70E Standard for Electrical Safety in the Workplace

Key Term: NFPA 70E

- The purpose of NFPA 70E is to provide a practical safe working area for employees relative to the hazards arising from the use of electricity.
- NFPA 70E addresses electrical safety concerns in all of the places that the *NEC*® addresses safe installation practices, including residential construction.

Objective: Identify common electrical hazards and how to avoid them on the job.

6) Identify common electrical hazards

- Use GFCI-protected cord sets or plug any extension cord you are using into a GFCI-protected receptacle when installing an electrical system.

- Look for conductor insulation that has deteriorated and fix or replace as necessary.

- Make sure that all noncurrent-carrying metal parts of electrical equipment are grounded properly.

- Never use an electrical plug that is missing the grounding prong.

Objective: Demonstrate an understanding of the purpose of OSHA.

7) Understanding the purpose of OSHA

Key term: Occupational Safety and Health Administration (OSHA)

- The purpose of OSHA is to ensure safe and healthy working conditions for working men and women on the jobsite by authorizing enforcement of safety standards; to assist and encourage the states in their efforts to ensure safe and healthy working conditions; and to provide for research, information, education, and training in the field of occupational safety and health.

- *Code of Federal Regulations (CFR)* Title 29, Part 1910 covers regulations for general industry. Regulations on general electrical safety are covered in *CFR* Title 29, 1910, Subpart S – Electrical.

- *Code of Federal Regulations (CFR)* Title 29, Part 1926 covers regulations for the construction industry. Electrical safety on the construction site is covered in *CFR* Title 29, 1926, Subpart K – Electrical.

Objective: Cite specific OSHA provisions pertaining to various general and electrical safety hazards associated with residential wiring.

8) Examples of OSHA provisions

- OSHA 29 *CFR* 1910.332 requires employees be trained in safety-related work practices.

- OSHA 29 *CFR* 1910.333 and 29 *CFR* 1926.417 outlines the requirements for lock-out/tag-out procedures for the electrical source of the circuit or equipment you may be working on.

- OSHA 29 *CFR* 1926.416 requires that existing conditions be determined by inspection or testing with a meter before any work can be started. Always assume an electrical circuit is energized until you verify otherwise by using a proper test instrument.

Objective: Demonstrate an understanding of the personal protective equipment used by residential electricians.

9) Understanding personal protective equipment requirements

Key terms: hazard, personal protective equipment (PPE)

- Selecting and using the proper PPE is very important.

- PPE is considered to be the last line of defense between you and an injury.

- OSHA requires PPE to be worn. Wearing PPE is not a choice—it is a requirement!

- Safety glasses or face shields must be worn any time a hazard exists that can cause an object to get in your eyes from the front or the sides.

- Head protection (hard hats) must be worn whenever there is a potential for objects to fall from above, bumps to your head from objects fastened in place, or from accidental contact with electrical hazards.

- Wear work shoes or boots with slip-resistant and puncture-proof soles. Safety shoes should be worn to prevent crushed toes when working around heavy equipment or falling objects.

- Wear hand protection (gloves) any time your hands are exposed to a potential hazard.

- Wear hearing protection in high-noise areas.

Objective: List several safety practices pertaining to general and electrical safety.

10) List safety practices pertaining to both general and electrical safety

- General safety rules to follow:
  - Be serious about your work and never fool around on the job.
  - Never use drugs or alcohol on the job.
  - Do not talk unnecessarily when working.
  - Use good judgment in both giving and receiving instructions.
  - Always follow safe practices when handling the materials used to install a residential wiring system.
  - Always follow safe practices when using step ladders, straight or extension ladders, and scaffolding.
  - Follow safe practices when using tools and always use a tool properly.

- Electrical safety rules to follow:
  - Install all wiring according to the *NEC®*.
  - Work with a buddy and avoid working alone.
  - Always turn off the electrical power and lock out and tag out.
  - Never cut off the grounding prong on a three-prong plug.
  - Assume all electrical equipment to be energized until proven otherwise.
  - Do not defeat the purpose of any safety device.
  - Do not open or close switches under load unless absolutely necessary.
  - Clean up all wiring debris and keep the job area clean.

Objective: Demonstrate an understanding of material safety data sheets.

11) Understanding material safety data sheets

Key term: material safety data sheet (MSDS)

- A MSDS is designed to inform an electrician of a material's physical properties as well as the effects on health that make the material dangerous to handle.

- MSDSs must be made available to any worker using any hazardous material.

Objective: Demonstrate an understanding of various classes of fires and the types of extinguishers used on them.

12) Understanding classes of fires and types of extinguishers

- The three components that must be present for a fire to start and sustain itself are fuel, heat, and oxygen.
- If any one of the three components is removed, a fire will be extinguished.
- Class A fires occur in ordinary combustible material like wood or paper.
- Class B fires occur with flammable liquids like gasoline.
- Class C fires occur in or near electrical equipment.
- Class D fires occur with combustible metals like powdered magnesium.
- There are four common types of fire extinguishers: pressurized water, carbon dioxide, dry chemical, and foam.
- The most common type of extinguisher is the dry chemical. It can be used on Class A, B, and C fires.

Quizzes

SAFETY

1. The _____ the body is in the circuit, the greater will be the severity of the injury.
   - a. shorter
   - b. longer
   - c. bigger
2. Never cut off the _____ prong on an electrical plug!
   - a. grounded
   - b. ungrounded
   - c. grounding
3. Always assume an electrical circuit to be _____ unless you verify otherwise.
   - a. energized
   - b. dead
   - c. off

CODE/REGULATIONS/STANDARDS

1. The *NEC*® is published by _____.
   - a. NEMA
   - b. NFPA
   - c. ANSI
2. The *NEC*® is updated every _____ years.
   - a. two
   - b. three
   - c. five

3. Section 210.8(A) of the *NEC*® lists the locations where _____ protection is required in dwelling units.
    a. ground-fault circuit-interrupter
    b. arc-fault circuit-interrupter
    c. fire

TOOLS

1. Use power tools only after you have received _____.
    a. the money
    b. proper instruction
    c. a new tool case
2. If an electrician does not have a hammer available, it is okay to use a pair of pliers to hammer in a nail.
    a. true
    b. false
3. Sharp-edged or pointed tools should be carried in _____.
    a. a tool pouch
    b. your pocket
    c. the truck
4. Any tool used on or around energized electrical equipment must be of the _____ type.
    a. rubber grip
    b. conductive
    c. nonconductive

Chapter Quiz

1. Copper is not a conductor of electricity because only materials that are magnetic can conduct electricity. True or False?
2. The *NEC*® gives minimum safety standards for electrical work and is not intended to be a how-to manual. True or False?
3. One milliampere (mA) equals _____ amperes.
4. The *NEC*® is published by the _____.
5. The *Code of Federal Regulations* is published and administered by _____.
6. Fall protection must be used whenever workers are on a walking or work surface that is _____ feet or higher and has unprotected sides.
7. When placing a straight or extension ladder against a wall, the bottom of the ladder should be placed about _____ of the vertical height from the wall.
8. State the purpose of NFPA 70E Standard for Electrical Safety in the Workplace.
9. Fires that occur with flammable liquids such as gasoline, oil, grease, paints, and thinners are Class _____ fires.
10. Dry chemical fire extinguishers can be used on Class _____ fires.

11. The amount of current that it takes to cause ventricular fibrillation is _____ milliamperes.
    a. 5
    b. 6–30
    c. 50–150
    d. 1000–4300

12. Which of the following types of fire extinguishers should be used on electrical fires?
    a. carbon dioxide
    b. pressurized water
    c. dry chemical
    d. either a or c

13. The *NEC*® chapter that covers methods and materials is Chapter _____.

14. Which of the following items is considered to be personal protective equipment (PPE)?
    a. safety glasses
    b. hard hat
    c. work boots
    d. all of the above

15. Which of the following items is not typically included in a material safety data sheet?
    a. hazardous ingredients/identity information
    b. physical/chemical characteristics
    c. cost of material
    d. fire/explosion data

16. A two-pronged plug that distinguishes between the grounded conductor and the hot conductor by having a grounded conductor prong that is wider than the hot conductor prong is called a(n) _____ plug.

17. All electrical equipment and conductors are to be considered _____ until determined to be otherwise by testing methods.

18. The three components that must be present for a fire to start and sustain itself are _____, _____, and _____.

19. Explain the purpose of material safety data sheets.

    _____

    _____

    _____

    _____

20. Discuss the proper way to lift an object.

    _____

    _____

    _____

    _____

Troubleshooting/Quality Exercises

DISCUSSION QUESTIONS

1. Explain why it is recommended that electricians try to use only one hand when taking measurements or working on live circuits.

   _____

   _____

   _____

   _____

2. Even though it is not correct, a common practice is for electricians to use lineman pliers to hammer in nails and staples. Explain why this is not a safe practice.

   _____

   _____

   _____

   _____

HANDS-ON APPLICATIONS

1. Use Table 1-2 in the textbook to complete the following statements.
   - It takes _____ mA for you to feel a slight shock.
   - It takes _____ mA for you to have heart convulsions and paralysis of breathing.
   - It takes _____ mA for you to possibly feel a tingling sensation.
   - It takes _____ mA for you to receive a painful shock, have your breathing stop, and have severe muscle contractions.

2. Place the letter of the fire extinguisher you would use in front of the following fire types. Some extinguishers may be used on more than one fire type.

   a. pressurized water   b. carbon dioxide   c. dry chemical   d. foam

   _____ Fires that occur in or near electrical equipment

   _____ Fires that occur in materials like wood, rags, and paper

   _____ Fires that occur with combustible metals

   _____ Fires that occur with flammable liquids

3. It is important for an electrician to be able to read a MSDS. Using the textbook as a resource, explain what is covered in each of the eight sections used in the OSHA format.

   Section 1:

   Section 2:

   Section 3:

Section 4:

Section 5:

Section 6:

Section 7:

Section 8:

REAL-WORLD SCENARIOS

1. OSHA outlines the requirements for locking out and then tagging out the electrical source for the circuit or equipment you and possibly your coworkers are working on. Each company is required to establish its own lock-out/tag-out procedure. If a company you work for does not have a procedure established, use the suggested lock-out/tag-out procedure outlined at the end of Chapter 1 in the textbook. List the 10 steps of the suggested lock-out/tag-out procedure.

Step 1:

Step 2:

Step 3:

Step 4:

Step 5:

Step 6:

Step 7:

Step 8:

Step 9:

Step 10:

2. Whenever possible, you should avoid working alone. If you are working with someone and they somehow become part of an electrical circuit and are being shocked, you need to know what to do to help them. List the six steps to follow when a coworker is being shocked. The procedure is outlined at the end of Chapter 1 in the textbook.

   Step 1:

   Step 2:

   Step 3:

   Step 4:

   Step 5:

   Step 6:

Chapter 2 — Hardware and Materials Used in Residential Wiring

Outline

Objective: List several nationally recognized testing laboratories and demonstrate an understanding of the purpose of these labs.

1) List several nationally recognized testing labs (NRTLs) and understand their purpose

- There are many NRTLs, but the three most common are:
  - Underwriters Laboratories (UL), the most recognizable.
  - CSA International (formerly known as the Canadian Standards Association).
  - Intertek Testing Services (formerly known as Electrical Testing Laboratories).
- An NRTL evaluates products against an established standard. If the product meets or exceeds the standard, the NRTL lists it in its directory along with instructions on how it is to be used and puts a label on the product.
- Section 110.3(B) of the *National Electrical Code®* (*NEC®*) requires electricians to install materials and equipment according to the instructions that come with their listing or labeling.
- Electrical inspectors often base their approval of electrical equipment on whether it is listed and labeled by an NRTL.

Objective: Identify common box and enclosure types used in residential wiring.

2) Identify common box and enclosure types

Key terms: device box, outlet box, switch box, spliced, junction box, handy box, new work box, old work box, utility box

- A device box is designed to contain switches or receptacles. A utility box, also called a handy box, is a type of device box.
- Boxes that contain just switches are often referred to as simply switch boxes.
- An outlet box usually has a lighting fixture attached to it but could be used to house switches or receptacles when a mud ring is used. These boxes are either square or octagonal.
- The square box style can also be used when connecting small or large appliances that require a large receptacle for cord and plug connection.
- Junction boxes are designed to contain spliced conductors. They are often referred to as J-boxes. These boxes are usually square or octagonal but device boxes could also contain splices.
- Electrical boxes can be metal or nonmetallic.

- Metal boxes can be ganged together so that more than one switch or receptacle can be installed. Nonmetallic boxes are bought already ganged together.

- New work boxes are used in new construction while old work boxes are typically used in remodel work.

- Boxes used to support a ceiling-suspended paddle fan must be designed for that application and listed as such by an organization like UL. The boxes can be made of metal or nonmetal and are specifically designed and tested to support heavy loads.

Objective: Identify common box covers and raised rings used in residential wiring.

3) Identify common box covers and raised rings

Key terms: raised plaster ring, mud ring

- When an electrical box contains a switch or receptacle, a plate is attached to the device to cover the area around the box and to finish off the installation. Plates are found in both metal and nonmetallic styles.

- When a square or octagon box is used strictly as a junction box, a flat blank cover is used.

- When a square box is installed in a wall or ceiling and will contain a device or have a lighting fixture attached to it, raised plaster rings (sometimes called mud rings) must be used. If a square box is used in a surface mount installation and a device is required, raised covers designed for either receptacles or switches are used.

Objective: Identify common conductor and cable types used in residential wiring.

4) Identify common conductor and cable types

Key terms: cable, sheath, copper-clad aluminum, American Wire Gauge (AWG), wirenut, Romex™

- Electrical conductors used in homes are usually installed as part of a cable assembly. The outer cable covering is referred to as a sheath and can be nonmetallic or metal.

- Conductors used in residential wiring are usually copper but can be aluminum or copper-clad aluminum.

- Conductors are sized using a system called the American Wire Gauge (AWG). In this system, the smaller the number, the larger the conductor size. The larger the number, the smaller the conductor size. For example, a 10-AWG conductor is smaller than a 6-AWG conductor, but is larger than a 14-AWG conductor.

- The ampacity of a conductor refers to the ability of a conductor to carry current. It depends on several factors, including the conductor size, the type of conductor insulation, and how the conductor is being used.

- Table 310.15(B)(16) in the *NEC®* is usually used by residential electricians to determine the ampacity of the conductors they will be using.

- Conductors are color-coded to indicate their function. For example, black is used as a "hot" conductor; red is also used as a "hot" conductor; white is used as the grounded circuit conductor; and bare or green is used to identify the grounding conductor.

- Wirenuts are the most common devices used to splice together two or more conductors.

- The most common cable types used in residential wiring include:
  - Nonmetallic sheathed cable (commonly called Romex™)
  - Underground feeder and branch-circuit cable
  - Armored clad cable
  - Metal clad cable
  - Service entrance cable
  - Manufacturers of nonmetallic sheathed cable are now color-coding their cables. The cable colors are white for 14 AWG Type NM cable, yellow for 12 AWG Type NM cable, orange for 10 AWG Type NM cable, and black for larger sizes.
  - URD (underground residential distribution) cable is an assembly of three or four conductors wrapped around each other with no overall outer sheathing.

Objective: Identify types of cable connectors, conductor terminals, and lugs.

5) Identify types of cable connectors, conductor terminals, and lugs

Key terms: connector, fitting, knockout, pryout

- When a cable is terminated at an electrical box, a connector must be used to secure it to the box. Nonmetallic sheathed cable connectors can be made of metal or a nonmetallic material. Connectors for metal sheathed cables are made of metal.

- A knockout or a pryout is removed from an electrical box so a cable connector can be attached to the box. Some electrical boxes have internal cable clamps and separate cable connectors are not required.

Objective: Identify common raceway types used in residential wiring.

6) Identify common raceway types

Key term: raceway

- Conductors are often installed in raceways when greater protection from physical damage is desired.
- Raceways used in residential wiring are made of metal or polyvinyl chloride (PVC).
- The most common raceway types used in residential wiring include:
  - Rigid metal conduit (RMC)
  - Intermediate metal conduit (IMC)
  - Electrical metallic tubing (EMT)
  - Rigid polyvinyl chloride conduit (PVC)
  - Flexible metal conduit (FMC)
  - Electrical nonmetallic tubing (ENT)
  - Liquid-tight flexible metal conduit (LFMC)
  - Liquid-tight flexible nonmetallic conduit (LFNC)

Objective: Identify common devices used in residential wiring.

7) Identify common devices

Key term: device

- Devices are part of an electrical system that carry, but do not use, electrical energy. Items like receptacles, switches, attachment plugs, and lamp holders are all considered to be devices.

- Receptacles are the most common devices used in a residential wiring system. They provide ready access to the electrical system and are defined as a contact device installed at an outlet for the connection of an attachment plug.

- Receptacles are available in both single and duplex models. Duplex receptacles are the most commonly installed type.

- The *NEC*® requires almost all receptacles in a house to be tamper-resistant.

- Ground-fault circuit-interrupter (GFCI) receptacles can be used to provide GFCI protection in those locations required by the *NEC*®.

- Switches are used to control the various lighting outlets throughout a house.

- Switches are available in a variety of styles, including:

  o Single-pole switches

  o Double-pole switches

  o Three-way switches

  o Four-way switches

  o Dimmer switches

  o Combination devices

Objective: Identify common types of fuses and circuit breakers used in residential wiring.

8) Identify common types of fuses and circuit breakers

Key terms: fuses, circuit breakers

- Circuit breakers are the most common type of overcurrent protection device used in residential wiring.

- Single-pole breakers are used on 120-volt circuits. Double-pole breakers are used on both 240-volt and 120/240-volt circuits.

- Common circuit breaker sizes used in residential applications include 15-, 20-, 30-, 40-, 50-, and 60-amp ratings. Other sizes may be used.

- Circuit breakers are also available in ground-fault circuit-interrupter (GFCI) and arc-fault circuit-interrupter (AFCI) models.

- Fuses are not often used in residential wiring anymore but electricians will run into them from time to time on new installations and especially when doing remodel work.

- Plug fuses are found in two types: the Edison base and the Type S. The Edison base plug fuse can be used only as a replacement for an existing Edison base fuse. All new installations that use a plug fuse require the use of Type S fuses. Type S fuses require the use of an adapter.

- Cartridge fuses are also available and come in two styles: the ferrule cartridge fuse and the blade-type cartridge fuse.

Objective: Describe the operation of a fuse and circuit breaker.

9) Describe the operation of a fuse and circuit breaker

Key terms: overcurrent, overload, short circuit, ground fault, bimetallic strip

- An overload, a short circuit, or a ground fault can cause a circuit breaker to trip or a fuse to blow.

- Circuit breakers used in residential applications are of the thermal/magnetic type and have a time-delay feature built in.

- When a larger current than what the breaker is rated for is flowing, a bimetallic strip heats up and starts to bend. If the larger-than-normal current flow continues, the bimetallic strip continues to bend to a point where the latching mechanism is tripped, causing the breaker contacts to open. The result is no more current flow.

- When a short circuit or ground fault occurs, a very large amount of current flows through the breaker. The high current causes a strong magnetic field that causes the latching mechanism to trip, opening the breaker contacts and stopping current flow.

- When a short circuit or ground fault occurs in a circuit protected by a fuse, the fusible link in the fuse melts away and opens the circuit.

- Fuses are available with a time-delay feature. When an overload occurs, it is allowed to continue for a short length of time. If it continues, the fuse will blow and open the circuit.

Objective: Identify common panelboards, loadcenters, and safety switches used in residential wiring.

10) Identify common panelboards, loadcenters, and safety switches

Key terms: panelboard, cabinet, loadcenter, safety switch, National Electrical Manufacturer's Association (NEMA)

- Panelboards are designed to be installed in a cabinet. The cabinet is placed in or on a wall, partition, or other support. This arrangement is often called a loadcenter in residential wiring. The loadcenter usually contains the main service entrance disconnect and the branch-circuit overcurrent protection devices.

- If located inside a house or garage, the loadcenter enclosure can be a NEMA Type 1 enclosure. If located on the outside, the enclosure must be a NEMA Type 3R. This enclosure type protects the internal wiring from damage due to rain and snow.

- Loadcenters are available with a main breaker or with main lugs. Loadcenters with just main lugs are typically used as subpanels and are referred to as main lug only (MLO) panels.

- Large electrical equipment used in homes such as an outside air-conditioning unit requires a way to disconnect it for servicing. Safety switches are typically used. It is mounted on the surface and usually contains fuses or a circuit breaker. Nonfusible safety switches are also available.

Objective: Identify common types of fasteners, fittings, and supports used in residential wiring.

11) Identify common types of fasteners, fittings, and supports

- When cable is installed in a house, it must be secured and supported at certain intervals as required by the *NEC®*. Staples are the most common items used to secure and support cabling in a residential application. Straps, clamps, or tie-wraps could also be used.

- When a raceway is installed in a house, it must also be secured and supported at certain intervals as required by the *NEC®*. One-hole and two-hole straps are the most common fittings used to secure and support raceways.

- Fasteners are used to assemble and install electrical equipment. There are several different fastener types used in residential wiring.

- Nails are used to mount electrical boxes to framing members, to install running boards to the underside of joists, and for any other fastening job in which a nail would work.

- Anchors are used to mount electrical boxes and enclosures on solid surfaces such as concrete, as well as in hollow wall surfaces such as wallboard. Toggle bolts, wallboard anchors, molly bolts, plastic anchors, lead anchors, and drive studs are examples.

- Screws are used to mount enclosures to wood surfaces. The types typically used include wood screws, sheet metal screws, and wallboard screws.

- When threaded fasteners are encountered, nuts and washers will need to be used. Nuts come in either a square or hexagonal shape and are threaded onto a bolt or machine screw. Washers are used to help distribute the pressure of the bolt or screw over a larger area.

- Tie wraps, often called cable ties, are one of the most often used fasteners in residential wiring. They come in many sizes and styles and are used to secure a bundle of cables or wires together. Some electricians use tie wraps in a loadcenter in an effort to make a neater installation.

Quizzes

SAFETY

1. Eye protection is not always required when you are driving nails. True or False?

2. Special training and certification in the use of powder-actuated fastening systems is required by OSHA before an electrician can use this type of fastening system. True or False?

3. If a piece of electrical equipment has a nationally recognized testing laboratory _____ on it, it means that the item has met specific standards and is suitable for its intended purpose.
   a. marking
   b. label
   c. sticker

CODE/REGULATIONS/STANDARDS

1. The _____ develops electrical equipment standards that all member manufacturers follow when designing and producing electrical equipment.
   a. NEMA
   b. NFPA
   c. ANSI

2. The *NEC®* requires a junction box to be accessible after installation; therefore, it cannot be buried in a wall or ceiling. True or False?

3. The _____ is the system used to size the conductors used in residential wiring.
   a. international wire sizing chart
   b. American Wire Gauge (AWG)
   c. *NEC®* wire table

4. The article in the *NEC®* that covers nonmetallic sheathed cable is Article.
   a. 90
   b. 220
   c. 334

TOOLS

1. Drive studs are installed using a _____.
   a. hammer
   b. powder-actuated fastening system
   c. a screwdriver

2. The most common sizes of machine screws used in residential wiring are 6-32, 10-32, and 12-32. True or False?

3. When ganging metal device boxes together, a _____ is needed to tighten the screws holding the boxes together.
   a. screwdriver
   b. pair of pliers
   c. nut driver

Chapter Quiz

1. The _____ is a scale of specified diameters and cross sections for wire sizing and is the standard wire sizing scale for the United States.
   a. American Wire Gauge
   b. metric system
   c. United States measuring system
   d. none of the above

2. An accidental connection of a "hot" ungrounded conductor and a grounded piece of equipment or the grounded conductor is called a(n) _____.
   a. open circuit
   b. grounding conductor
   c. ground fault
   d. complete circuit

3. It is the _____ who has the job of approving the electrical materials used for a residential electrical installation.
   a. electrician
   b. foreman
   c. architect
   d. authority having jurisdiction

4. When nonmetallic device boxes that are two-gang or more are used, the installed cable must be supported no more than _____ inches from the box.

5. When using aluminum conductors, _____ size conductor must be used to carry the same amount of current as a copper conductor.
   a. a smaller
   b. the same
   c. a larger
   d. the same or larger

6. The _____ colored circuit conductor is used as the grounded conductor and returns current from the load back to the source.
   a. black
   b. white
   c. red
   d. green

7. Type _____ cable is used for underground installations of branch and feeder circuits.

8. Type _____ cable is used for underground service entrance installations.

9. The *NEC*® requires rigid metal conduit to be supported within _____ feet of each box and then no more than every _____ feet thereafter. (General rule)
   a. 3, 8
   b. 5, 10
   c. 2, 10
   d. 3, 10

10. The ungrounded "hot" conductor is terminated to the _____ colored terminal on a receptacle.

11. A switch that is used on 240-volt circuits to control a load from one location is called a _____ switch.
    a. single-pole
    b. double-pole
    c. occupancy
    d. combination

12. Which of the following switches has ON/OFF marked on the switch handle?
    a. three-way
    b. double-pole
    c. four-way
    d. none of the above

13. _____ are the most often used type of overcurrent protection device used in residential wiring.
    a. Circuit breakers
    b. Edison base plug fuses
    c. Plug fuses
    d. Cartridge fuses

14. _____ fuses are considered non-tamperable and are used in new installations requiring fuses.
    a. Edison base plug
    b. Type S plug
    c. Cartridge
    d. Plug

15. A part of an electrical box that is designed to be removed so that a cable or raceway can be connected to the box is called a _____.
    a. knockout
    b. mounting bracket
    c. cable clamp
    d. plaster ear

16. When two or more electrical device boxes are joined together for the purpose of holding more than one device, it is called _____.

17. All _____ boxes must be accessible after they are installed.

18. In residential wiring, the *NEC®* requires a _____ -amp circuit breaker or fuse to be used with 12-AWG copper conductors.

19. A _____ is an overcurrent device that opens a circuit when a fusible link is melted away by the extreme heat caused by an overcurrent.

20. _____ fuses cannot be used in new installations.
    a. Edison base plug
    b. Type S plug
    c. Cartridge
    d. Plug

Troubleshooting/Quality Exercises

DISCUSSION QUESTIONS

1. Section 110.3(B) of the *NEC®* states that electrical equipment must be installed according to any instructions included in the listing and labeling of the item. Discuss what this means if the installation instructions call for a threaded fastener to be torqued to 20 inch-pounds.

_____

_____

_____

_____

2. Sometimes there are locations in residential wiring where two switches are needed but there is not enough room between studs to place a two-gang device box. Discuss what could be done to install two switches in a location where only a single-gang device box can be installed.

_____

_____

_____

_____

HANDS-ON APPLICATIONS

1. Using Table 2-1 in the textbook, write in the names of the conductor types permitted.

Type of Device	Marking	Conductor Type Permitted
15- or 20-amp receptacle	CO/ALR	
15- or 20-amp receptacle	None	
Wire connector	CC/CU	
Wire connector	AL/CU	

2. Using Table 2-2 in the textbook, write in the size of the copper conductor typically used for each of the following applications.

Application	Copper Conductor Size
15-amp lighting branch circuits	
20-amp small-appliance branch circuits	
30-amp electric clothes dyer branch circuit	
40-amp electric range branch circuit	

3. Using Table 2-4 in the textbook, write in the NEMA designation for the following general-purpose non-locking receptacle applications.

Receptacle Application	NEMA Designation
15-amp, 125-volt, two-pole, three-wire, grounding	
20-amp, 125-volt, two-pole, three-wire, grounding	
30-amp, 125/250-volt, three-pole, four-wire, grounding	
40-amp, 125/250-volt, three-pole, four-wire, grounding	

REAL-WORLD SCENARIOS

1. When you are installing aluminum conductors, it is often necessary to use an antioxidant. List the six steps of the procedure to follow when installing aluminum conductors. The procedure is outlined at the end of Chapter 2 in the textbook.

   Step 1:

   Step 2:

Step 3:

Step 4:

Step 5:

Step 6:

2. Conductors are often connected together in residential wiring with a wirenut. List the five steps of the procedure to follow when installing a wirenut. The procedure is outlined at the end of Chapter 2 in the textbook.

Step 1:

Step 2:

Step 3:

Step 4:

Step 5:

Chapter 3 — Tools Used in Residential Wiring

Outline

Objective: Identify common electrical hand tools and their uses in the residential electrical trade.

1) Identify common electrical hand tools

Key terms: strip, crimp, wallboard, tempered

- Screwdrivers

  - Keystone and cabinet tip screwdriver: Used to remove and install slot-head screws and to remove and install slot-head lugs.

  - Phillips tip screwdriver: Used to remove and install Phillips head screws and to tighten or loosen Phillips head lugs.

  - Stubby screwdriver: Used to remove and install screws in a limited working space.

  - Ratcheting screwdriver: Uses multiple tips to allow an electrician to carry just one screwdriver when working on different headed screws.

  - Quick-rotating screwdriver: This type of screwdriver is great for quickly tightening or loosening screws in switches, receptacles, and lighting fixtures.

- Pliers

  - Lineman (side-cutter) pliers: Used to cut cables, conductors, and small screws, form large conductors, and pull and hold conductors.

  - Long-nose (needle-nose) pliers: Used to form small conductors, to cut, and to hold and pull conductors.

  - Diagonal (dike) pliers: Used to cut cables and conductors in limited spaces.

  - Pump (channel-lock) pliers: Used to hold and tighten raceway couplings and connectors, and to hold and turn conduit and tubing.

  - Cable cutter plier: Used to cut cables without the distortion of the cable ends that you can get when cutting a cable with lineman pliers.

  - Crimping plier: Used to crimp bare or insulated terminals and splices.

- Wire strippers

  - T-stripper: Designed to strip insulation from several different wire sizes without having to be adjusted for each size.

  - Multipurpose (six-in-one) tool: Designed to strip insulation, cut conductors, crimp smaller solderless connectors, cut small bolts/screws, and thread small bolts/screws; also used as a wire gauge.

  - Electrician's knife: Strips large conductors and cables, opens cardboard boxes, cuts string, and sharpens pencils.

- Other common hand tools
  - Tap tool: Taps holes for securing equipment to metal, enlarges existing holes, re-taps damaged threads, and determines screw sizes.
  - Adjustable (crescent) wrench: Tightens couplings and connectors, tightens pressure-type wire connectors, and removes and holds nuts and bolts.
  - Awl or "scratch" awl: Starts screw holes, makes pilot holes for drilling, and marks metal.
  - Electrician's hammer: Drives/pulls nails, pries boxes loose, breaks wallboard, and strikes awls and chisels. Should have long, straight claws to simplify removal of electrical equipment.
  - Folding rule or tape measure: Checks measurements on prints, determines box locations on prints, and determines depth and setout of electrical boxes.
  - Tool pouch: Holds, organizes, and carries electrician's hand tools.

Objective: Identify common specialty tools and their uses in the residential electrical trade.

2) Identify common specialty tools

Key terms: knockout punch, torque, level, plumb, die, cutter, bender, hydraulic

- Nut driver set: Tightens and loosens various sizes of nuts and bolts.
- Knockout punch: Cuts holes for installing cable or conduit in metal boxes, equipment, and appliances.
- Keyhole (compass) saw: Cuts holes in wallboard for installing electrical boxes.
- Jab saw: Used for same job as a keyhole saw and has replaceable blades for a variety of jobs and comes with an ergonomic grip for easy handling.
- Hacksaw: Cuts some conduit types and cuts larger conductors and cables.
- Fish tape and reel: Pulls wires or cables through electrical conduit and pulls/pushes cables in wall or ceiling cavities.
- Conduit bender: Bends electrical metallic tubing (up to 1¼") or rigid metal conduit (up to 1").
- PVC heater box or blanket: Softens PVC pipe for bending.
- Torpedo level: Levels or plumbs conduit, equipment, and appliances.
- Plumb bob: Transfers location points from ceiling to floor or floor to ceiling.
- Metal and wood files: Metal files deburr conduit, sharpen tools, and cut and form metal. Wood files enlarge drilled holes in wooden framing members.
- Chisel: Notches wood for boxes and cables, cuts and shapes masonry or metal.
- Cable cutter: Cuts larger size cables and conductors.
- Sledgehammer: Drives ground rods into the ground.
- Hex key set (Allen wrenches): Tightens/loosens countersunk hexagonal setscrews and conductor lugs.
- Fuse puller: Removes cartridge fuses from electrical enclosures.
- Torque screwdrivers and wrenches: Tightens smaller screws, lugs, Allen head, and bolt-type lugs to manufacturer's recommended torque requirements.
- Rotary armored cable cutter: Strips outside sheathing from Type AC and Type MC cable.
- Insulated tools: Used to work on energized circuits.

Objective: Identify common power tools and their uses in the residential electrical trade.

3) Identify common power tools

Key terms: auger, chuck key, chuck, reciprocating

- Whether you are using a corded power tool or a cordless power tool it is important to review the information on the tool's nameplate.

- Power drills: Bores holes for installation of cables, conduits, and other electrical equipment in wood, metal, plastic, or other material.

- Right-angle drill: Drills holes for the installation of cables through wooden framing members.

- Pistol grip drill: An electric drill very popular because it is small, relatively lightweight, and easy to use.

- Hammer drills: Used to drill holes in masonry or concrete walls and floors.

- Rotary hammer drills: Used when larger holes need to be drilled in masonry.

- Cordless drills: Pistol grip drills, right-angle drills, and hammer drills available in cordless models.

- Circular saws: Used to cut wood products such as studs and plywood to desired sizes. Available in a cordless model.

- Reciprocating saws or Sawzalls®: Instead of rotating, the blade moves back and forth to make a cut. Available in a cordless model.

Objective: List several guidelines for the care and safe use of electrical hand tools, specialty tools, and power tools.

4) List several guidelines for the care and safe use of tools

- Use the correct tool for the job.

- Keep all cutting tools sharp. Dull cutting tools require much more force to do their job.

- Keep tools clean and dry.

- Lubricate tools when necessary.

- Inspect tools frequently to make sure they are in good condition.

- Repair damaged tools promptly and dispose of broken or damaged tools that cannot be repaired.

- Store tools properly when not in use.

- Pay attention when using tools.

- Wear eye protection when using tools.

Objective: Demonstrate an understanding of the procedures for using several common hand tools, specialty tools, and power tools.

5) Understand the procedures for using hand tools, specialty tools, and power tools

- Review the following procedures located at the end of Chapter 3 in the textbook.
    - Using a screwdriver
    - Using lineman pliers to cut cable

- ○ Using a wire stripper
- ○ Using a knife to strip the sheathing from a Type NM cable
- ○ Using a knife to strip insulation from large conductors
- ○ Using an adjustable wrench
- ○ Using a knockout punch to cut a hole in a metal box
- ○ Setting up and using a hacksaw
- ○ Using a torque screwdriver
- ○ Using a rotary armored cable cutter
- ○ Using a cordless pistol grip power drill
- ○ Drilling a hole with an auger bit and a corded right-angle drill
- ○ Cutting a hole with a hole saw and a corded right-angle drill
- ○ Drilling a hole in masonry with a corded hammer drill
- ○ Using a corded circular saw
- ○ Using a corded reciprocating saw

Quizzes

SAFETY

1. Eye protection is always required when you are using tools to install a residential electrical wiring system. True or False?
2. The comfort grip handles on screwdrivers are designed for comfort and to provide a secure grip, and not to act as an insulator from electrical shock. True or False?
3. You should never use a power tool before you have received the proper training on how to use it and have read its owner's manual. True or False?

CODE/REGULATIONS/STANDARDS

1. OSHA 1910.335 requires _____ tools to be used whenever an electrician is working on exposed energized circuits or equipment.
   a. clean
   b. insulated
   c. covered
2. Always follow OSHA 1910 S and NFPA 70E safety requirements when working on energized parts. This includes wearing the proper _____.
   a. personal protective equipment (PPE)
   b. T-shirt
   c. tool belt
3. Section 110.3(B) of the *National Electrical Code*® requires that all equipment be installed according to the _____ that comes with the equipment.
   a. instructions
   b. paperwork
   c. warranty

TOOLS

1. Another name for a Skilsaw® is a _____ saw.
   a. circular
   b. reciprocating
   c. jig

2. _____ are used to cut cables, cut conductors, form large conductors, and pull and hold conductors.
   a. Diagonal pliers
   b. Lineman pliers
   c. Long-nose pliers

3. A _____ can be used to strip the outside metal coverings from common sizes of cable, such as Type AC and Type MC.
   a. sharp knife
   b. rotary armored cable cutter
   c. keyhole saw

4. A(n) _____ is usually 9 inches long and is used to level or plumb conduit, equipment, and appliances.
   a. lineman pliers
   b. plumb bob
   c. torpedo level

Chapter Quiz

1. A drill bit with a spiral cutting edge used to bore holes in wood is called a(n) _____ bit.

2. A _____ is used to cut holes in metal electrical boxes for installing cable or conduit.

3. If a run of conduit is plumb, it is _____ to the floor.
   a. parallel
   b. at a right angle
   c. 45°
   d. positioned near

4. When installing a ground rod with a sledgehammer, ensure that at least _____ feet of the rod is in the ground.

5. _____ drills are used to drill holes in masonry blocks or concrete walls and floors.
   a. Hammer
   b. Pistol grip
   c. Auger
   d. Right-angle

6. Another name for a Sawzall® is a _____ saw.

7. _____ pliers are used to cut cables, cut conductors, form large conductors, and pull and hold conductors.

8. _____ pliers are used to form small conductors, cut conductors, and hold and pull conductors.

9. _____ pliers are used to cut cables and conductors in a limited space.

10. _____ pliers are used to strip insulation, crimp solderless connectors, cut and size conductors, and cut and thread small bolts.

11. A _____ is used to transfer location points from ceiling to floor and floor to ceiling and to plumb conduit and equipment.

12. Which of the following is not a commonly used drill chuck size?

    a. ⅛ inch

    b. ¼ inch

    c. ⅜ inch

    d. ½ inch

13. OSHA 1910.335 requires _____ tools to be used whenever an electrician is working on exposed energized circuits or equipment.

    a. clean

    b. insulated

    c. covered

    d. new

14. A _____ is used to deburr conduit, sharpen tools, and cut and form metal.

    a. wood rasp

    b. plumb bob

    c. metal file

    d. chisel

15. A(n) _____ is used to check measurements on prints, determine box location on prints, and determine the depth and setout of electrical boxes.

    a. tape measure

    b. awl

    c. triple-tap tool

    d. torpedo level

16. A(n) _____ is used to pull wire and cables through conduit and to pull cables in insulated walls.

    a. awl

    b. fish tape

    c. plumb bob

    d. auger bit

17. _____ screwdrivers are used to remove and install screws in limited working space and to tighten and loosen lugs in limited working space.

18. The component of a knockout punch that is used in conjunction with the cutter to make a hole in a metal electrical box is called a _____.

19. The styles of screwdrivers used most often in residential electrical work include the keystone tip, the cabinet tip, and the _____ tip.

20. The three parts of a screwdriver include the handle, the tip, and the _____.

Troubleshooting/Quality Exercises

DISCUSSION QUESTIONS

1. Conductors used in residential wiring are available in both solid and stranded configurations. Discuss the differences between solid and stranded conductors. Include information on the types of wire strippers used on these conductors in your discussion.

   _____

   _____

   _____

   _____

2. Electricians use many different tools in the course of wiring a house. Discuss some of the ways that they carry their tools.

   _____

   _____

   _____

   _____

3. Work on energized circuits and equipment should be done only when absolutely necessary and only by qualified electricians. Discuss what safety precautions you would need to follow if you had to work on an energized circuit.

   _____

   _____

   _____

   _____

4. It is common practice to use cordless power tools on the job. Discuss what you need to do to make sure that your cordless power tool is ready to be used at all times when wiring a house.

   _____

   _____

   _____

   _____

HANDS-ON APPLICATIONS

1. Match the following pouch tools to their common residential electrical trades use. Write the correct number for the tool in the blanks.

   1. Channel-lock pliers
   2. Lineman pliers
   3. Adjustable wrench
   4. Wire stripper
   5. Electrician's knife
   6. Tool pouch
   7. Phillips head screwdriver
   8. Electrician's hammer
   9. Tape measure
   10. Triple-tap tool
   11. Long-nose pliers
   12. Six-in-one tool
   13. Stubby screwdriver
   14. Flat blade screwdriver
   15. Awl
   16. Diagonal cutting pliers

_____ a. Used to check measurements on prints, locate boxes, and determine depth and setout

_____ b. Used to remove and install slot head screws

_____ c. Used to remove and install Phillips head screws

_____ d. Used to remove and install screws in a limited working space

_____ e. Used to cut cables, conductors, small screws; strip conductors; form large conductors; and pull and hold conductors

_____ f. Used to form small conductors, strip conductors, cut conductors, and hold and pull conductors

_____ g. Used to cut cables and conductors in limited space

_____ h. Used to hold and tighten couplings and connectors and to hold and turn conduit and tubing

_____ i. Used to strip insulation from conductors

_____ j. Used to strip insulation, crimp solderless connectors, cut and size conductors, and cut and thread small bolts

_____ k. Used to strip large conductors and cables

_____ l. Used to tap holes for securing equipment to metal, enlarge holes and tap for larger screws, re-tap damaged threads, and determine screw sizes

_____ m. Used to tighten couplings and connectors, tighten pressure-type wire connectors, and remove and hold nuts and bolts

_____ n. Used to start screw holes, make pilot holes for drilling, and mark metal

_____ o. Used to drive and pull nails, pry boxes loose, chip wood, break wallboard, and strike chisels

_____ p. Used to hold, organize, and carry electrician's tools

2. Match the following specialty tools and power tools to their common residential electrical trades use. Write the correct number for the tool in the blanks.

1. Portable drill
2. Keyhole saw
3. Plumb bob
4. Hacksaw
5. Cable cutter

6. Torpedo level
7. Fish tape
8. Wood chisel
9. Knockout punch
10. File

_____ a. Used to cut conduit and large conductors and cables

_____ b. Used to bore holes for cables and conduits

_____ c. Used to level or plumb conduit, equipment, and appliances

_____ d. Used to cut holes for cable or conduit fittings in metal boxes, equipment, and appliances

_____ e. Used to cut holes in wallboard for old-work boxes

_____ f. Used to pull wires or cables through conduit and to pull cables up insulated walls

_____ g. Used to transfer location points from ceiling to floor or floor to ceiling and to plumb conduit and equipment

_____ h. Used to deburr conduit and sharpen tools

_____ i. Used to notch wood for boxes and cables

_____ j. Used to cut large size cables

REAL-WORLD SCENARIOS

1. Electricians often use a knife to strip the insulation from larger cables and conductors. List the six steps of the procedure to follow when using a knife to strip insulation from large conductors. The procedure is outlined at the end of Chapter 3 in the textbook.

    Step 1:

    Step 2:

    Step 3:

    Step 4:

    Step 5:

    Step 6:

2. Electricians often have to set up and use a hacksaw when they are wiring a house. List the six steps of the procedure to follow when setting up and using a hacksaw. The procedure is outlined at the end of Chapter 3 in the textbook.

    Step 1:

    Step 2:

    Step 3:

    Step 4:

    Step 5:

    Step 6:

Chapter 4

Test and Measurement Instruments Used in Residential Wiring

Outline

Objective: Demonstrate an understanding of continuity testers and how to properly use them.

1) Understand a continuity tester and know how to use it

Key terms: continuity tester

- A continuity tester is a device used to indicate whether there is a continuous path for current flow in an electrical circuit or device.

- Never attach a continuity tester to a circuit that is energized.

- Review the procedure for using a continuity tester located in the textbook at the end of Chapter 4.

Objective: Demonstrate an understanding of the differences between a voltage tester and a voltmeter.

2) Understand the difference between a voltage tester and a voltmeter

Key terms: voltage tester, Wiggy®, non-contact voltage tester, voltmeter, analog meter, digital meter

- A voltage tester is designed to indicate approximate values of voltage for either direct current (DC) or alternating current (AC) applications.

- It can be used to:

  o Identify the grounded conductor of a circuit.

  o Check for blown fuses.

  o Distinguish between AC and DC currents.

  o Test for whether a voltage is present and provide an approximate value for that voltage.

- Voltage testers come in several different styles:

  o Solenoid type (Wiggy®)

  o Solenoid type with a continuity tester

  o Digital voltage tester

  o Non-contact voltage tester

- Voltmeters are used for the same applications as voltage testers, but are more accurate and give more precise information.

- There are two common types of voltmeters:

  o Analog meters (older style)

  o Digital meters (most commonly used)

Objective: Connect and properly use a voltage tester and a voltmeter.

3) Connect and use a voltage tester and a voltmeter

Key terms: auto-ranging, manual ranging

- Always connect a voltage tester or a voltmeter across (in parallel with) the circuit being tested.

- Do not leave an analog meter connected with the polarity reversed.

- A voltmeter can be used to determine if the voltage delivered to an electrical load is the proper amount. A too high or too low voltage will cause equipment to not work as efficiently as possible, resulting in excessive energy use.

- Review the procedure for using a voltage tester and a voltmeter in the textbook at the end of Chapter 4.

Objective: Demonstrate an understanding of the differences between an in-line ammeter and a clamp-on ammeter.

4) Understand the differences between an in-line ammeter and a clamp-on ammeter

Key terms: multiwire circuits, clamp-on ammeters, in-line ammeters

- Ammeters are designed to measure the amount of current flowing in a circuit.

- They can be used to balance the loads on multiwire circuits and locate electrical component malfunctions.

- An ammeter can be used to determine if an electrical load is drawing more current than it should. Excessive current draw results in lower efficiency and wasted energy.

- There are two styles of ammeters.
  - Clamp-on ammeters:
    - Are easy to use and are the most common type.
  - In-line ammeters:
    - Do not have high enough amperage ratings to make them practical in residential wiring.
    - Are more difficult to use than clamp-on ammeters.

Objective: Connect and properly use a clamp-on ammeter.

5) Connect and use a clamp-on ammeter

- Clamp-on ammeters have a moveable jaw that can be opened, allowing the meter to be placed around a current-carrying conductor. The meter is able to provide a current reading based on the strength of the magnetic field around the conductor.

- Review the procedure for using a clamp-on ammeter in the textbook at the end of Chapter 4.

Objective: Demonstrate an understanding of ohmmeters, megohmmeters, and ground resistance meters.

6) Understand ohmmeters, megohmmeters, and ground resistance meters

Key terms: ohmmeter, short circuit, open circuit, megohmmeter

- Ohmmeters are used to measure the resistance of a circuit or circuit component.
- A megohmmeter is used to measure very high values of resistance.
  - Commonly called a Megger®
- Ground resistance meters are used to determine the amount of resistance to the earth when the resistance to earth through a grounding electrode is unknown

Objective: Demonstrate an understanding of multimeters.

7) Understand multimeters

- Multimeters are designed to measure more than one electrical value.
- The major advantage of this type of meter is that several different types of test and measurement can be taken with only one meter.
- A newer style of multimeter is a clamp meter. It can measure AC amperage, both DC and AC voltage, and resistance.
- Many clamp meters can also measure values of capacitance and frequency

Objective: Connect and properly use a multimeter to test for voltage, current, resistance, and continuity.

8) Connect and use a multimeter

Key terms: multimeter, VOM, DMM

- The basic volt-ohm milliammeter (VOM) can measure:
  - DC and AC voltages
  - DC and AC current
  - Resistance
- Review the procedure for using a digital multimeter in the textbook at the end of Chapter 4.

Objective: Demonstrate an understanding of the uses for a True RMS meter.

9) Understand the uses of a True RMS meter

Key terms: nonlinear loads, harmonics, True RMS meter

- A True RMS meter is used to provide accurate readings in locations that have harmonics present. However, this meter type can also be used on applications where harmonics are not present.
- True RMS meters are more expensive to purchase than regular "average" reading meters.
- Residential wiring applications usually do not require the use of a True RMS meter.

Objective: Demonstrate an understanding of how to read a kilowatt-hour meter.

10) Understand how to read a kilowatt-hour meter

Key terms: kilowatt-hour meter

- Kilowatt-hour meters measure the amount of power consumed over a specific amount of time.
- Electrical power is measured in watts and kilowatts (1000 watts equals 1 kW).
- The kilowatt-hour meter works on the principle of magnetic induction.
- Kilowatt-hour meters use either dial/needle readouts or digital readouts.
- A new type of kilowatt-hour meter, called a smart meter, is now being installed by electric utility companies in many parts of the country.
- Smart meters fit into the same meter enclosures as the "regular" kilowatt-hour meters.
- Review the procedure for reading a kilowatt-hour meter in the textbook at the end of Chapter 4.

Objective: Demonstrate an understanding of safe practices to follow when using test and measurement instruments.

11) Understand safe practices when using test and measurement instruments

- Recalibration is necessary from time to time to bring a meter back to its intended level of accuracy.
- Residential electricians should always follow safe meter practices when using meters of any kind.
- Always wear safety glasses when using test and measurement instruments.
- Wear rubber gloves when testing or measuring energized electrical circuits or equipment.
- Never work on energized circuits unless absolutely necessary.
- If you must take measurements on energized circuits, make sure you have been properly trained to work with energized circuits.
- Do not work alone, especially on energized circuits.
- Keep your clothing, hands, and feet as dry as possible when taking measurements.
- Make sure the meter you are using has a rating equal to or exceeding the highest value of electrical quantity you are measuring.
- Look for two or more labels from independent testing labs like UL (USA), CSA (Canada), or TUV (Europe) to verify that your meters have been tested and certified.
- NFPA 70E requires that you visually inspect test and measurement tools frequently to help detect damage and ensure proper operation.
- NFPA 70E requires the use of IEC (International Electrotechnical Commission) rated meters. Look for 600-volt or 1000-volt CAT III or 600-volt CAT IV ratings on the front of meters and testers. Also look for the "double insulated" symbol or wording on the back of the meter.

Objective: Demonstrate an understanding of the proper care and maintenance of test and measurement instruments.

12) Understand proper care and maintenance of test and measurement instruments

- Keep the meters clean and dry.
- Do not store analog meters next to strong magnets; magnets can cause the meters to become inaccurate.
- Handle all meters with care; they are very fragile.
- Do not expose meters to large temperature changes.
- Make sure you know the type of circuit you are testing (AC or DC).
- Never let the value being measured exceed the range of the meter.
- Multimeters and ohmmeters will need to have their batteries changed from time to time.
- Many meters have fuses to protect them from exposure to excessive voltage or current values.
- Re-calibrate measuring instruments once a year.

Quizzes

SAFETY

1. Never work on energized electrical equipment unless it is absolutely necessary. True or False?

2. When using measuring instruments, be sure to use an instrument that has a rating of less than the highest value of electricity that you expect to be measuring. True or False?

3. Safety glasses are always required when using test and measurement instruments. True or False?

TOOLS

1. The abbreviation for a digital multimeter is _____.

2. A _____ tester is a device used to indicate whether there is a continuous path for current to flow in an electrical circuit or electrical device.

3. A(n) _____ is designed to measure the amount of current flowing in a circuit.

4. A(n) _____ can indicate that voltage is present on a circuit without actually being connected to the circuit.
   a. Wiggy®
   b. ohmmeter
   c. non-contact voltage tester

5. A Wiggy® is an electrical trade name for a _____.
   a. solenoid type voltmeter
   b. solenoid type voltage tester
   c. solenoid type ammeter

6. An instrument used to measure values of resistance is called a(n) _____.
   a. ammeter
   b. ohmmeter
   c. voltmeter

7. A break in an electrical conductor or cable is called a(n) _____ circuit.
   a. short
   b. ground faulted
   c. open

Chapter Quiz

1. An instrument used to measure values of resistance is called a(n) _____.

2. The letters VOM are an abbreviation for volt, ohm, and _____.

3. A meter that uses a moving pointer or needle to indicate a value on a scale is called a(n) _____ meter.
   a. LCD
   b. analog
   c. digital
   d. parallax

4. One megohm is equal to _____ ohms.

5. _____ are load types where the load impedance is not constant.
   a. Harmonic loads
   b. Linear loads
   c. Non-linear loads
   d. True RMS loads

6. _____ meters provide accurate measurements of AC values in environments where the basic AC waveform is distorted.
   a. Harmonic
   b. Linear
   c. Non-linear
   d. True RMS

7. Electrical power is measured in _____.

8. A voltmeter is used to measure _____.

9. A Wiggy® is an electrical trade name for a _____.
   a. solenoid type voltmeter
   b. solenoid type voltage tester
   c. solenoid type ammeter
   d. ohmmeter

10. A(n) _____ is used to determine the amount of resistance to the earth through a grounding electrode.
    a. multimeter
    b. ground resistance meter
    c. Megger®
    d. Wiggy®

11. A(n) _____ can indicate that voltage is present in a circuit without actually being connected to the circuit.
    a. Wiggy®
    b. ohmmeter
    c. ground resistance tester
    d. non-contact voltage tester

12. Never attach a continuity tester to a circuit that is energized. True or False?

13. Always read and follow any operating instructions that come with a meter. True or False?

14. Safety glasses are always required when using test and measurement instruments. True or False?

15. A _____ tester is a device used to indicate if there is a continuous path for current to flow in an electrical circuit or electrical device.

16. A voltmeter is connected in _____ with the circuit or component being tested.

17. A(n) _____ is designed to measure the amount of current flowing in a circuit.

18. An ohmmeter is used to measure the _____ of a circuit or circuit component.

19. The two basic types of ammeters are in-line and _____.

20. The positive or negative direction of DC voltage and current is referred to as _____.

Troubleshooting/Quality Exercises

DISCUSSION QUESTIONS

1. A common practice for electricians is to use a clamp-on ammeter to measure the current flow on each ungrounded conductor of a service entrance when all the house electrical loads are connected. Discuss what kind of information can be found by using a clamp-on ammeter in this way.

_____

_____

_____

_____

2. Discuss what can happen if a voltmeter or voltage tester is used on a circuit that exceeds the meter's maximum voltage rating.

_____

_____

_____

_____

HANDS-ON APPLICATIONS

1. Match the following test and measurement instruments to their common residential electrical use. Write the correct number for the meter in the blanks.

   1. Ammeter
   2. Voltmeter
   3. Megohmmeter
   4. Wiggy®
   5. Kilowatt-hour meter

   6. Continuity tester
   7. Non-contact voltage tester
   8. Multimeter
   9. Ohmmeter
   10. True RMS meter

_____ a. A trade name for a solenoid type of voltage tester

_____ b. A meter type that allows accurate measurement of AC values in harmonic environments

_____ c. An instrument that measures the amount of electrical energy supplied by an electric utility to a house

_____ d. A measuring instrument that measures values of resistance

_____ e. An instrument that is capable of measuring many different values

_____ f. A tester that indicates a voltage is present once it comes in close proximity to energized conductors or equipment

_____ g. A measuring instrument that measures current flow

_____ h. A measuring instrument that measures large amounts of resistance

_____ i. A measuring instrument that measures a precise amount of voltage

_____ j. A testing device that indicates a continuous path for current flow

REAL-WORLD SCENARIOS

1. Good quality test and measurement instruments are expensive. If you follow a few rules in the area of meter care and maintenance, your expensive meter will last in the harsh environment of residential electrical construction. List at least six rules to follow for the proper care and maintenance of test and measurement instruments.

    Rule 1:

    Rule 2:

    Rule 3:

    Rule 4:

    Rule 5:

    Rule 6:

2. An electrician is installing switches and receptacles in a house where the wall and ceiling coverings are installed. She knows that one end of a two-wire Type NM cable is in Box A but does not know whether the other end is in Box B or Box C. Using the procedure for using a continuity tester outlined at the end of Chapter 4 in the textbook as a guide, explain how she could use the continuity tester to determine which box contains the other end of the cable.

    _____

    _____

    _____

    _____

3. Electricians often use a clamp-on ammeter to determine the current flow in a conductor. List the four steps to follow when using a clamp-on ammeter. The procedure is outlined at the end of Chapter 4 in the textbook.

Step 1:

Step 2:

Step 3:

Step 4:

Chapter 5 Understanding Residential Building Plans

Outline

Objective: Demonstrate an understanding of residential building plans.

1) Understand residential building plans

Key terms: blueprints, plot plan, floor plan, elevation drawings, interior elevation drawings, sectional drawings, detail drawings, electrical drawings, schedules, scale

- A residential building plan consists of a set of drawings that craft people use as a guide to build the house.

- Building plans go by many names, such as prints, blueprints, drawings, construction drawings, and working drawings

- Plot plans show information such as the location of the house, walkway, or driveway on the building lot.

  o Information such as where a trench will be dug for an underground service entrance can be found on a plot plan.

- Floor plans are drawings that show building details from a "bird's eye" view, directly above the house.

  o A floor plan shows the length and width of the floor it is depicting.

  o Electricians can use the dimensions drawn on the floor plans to determine the exact size and location of various parts of the building, such as doors, windows, and walls.

- Elevation drawings show the side of the house facing a certain direction.

  o Use elevation drawings to determine the heights of windows, doors, porches, and other parts of the structure needed when installing electrical items such as outside lighting fixtures and outside receptacles.

- Interior elevation drawings are often included in a set of building plans.

  o The most common are kitchen and bathroom wall elevations.

  o These drawings show the design and size of cabinets and appliances located on or against the walls.

- Sectional drawings allow an electrician to see the inside of a building.

  o The section line is the point on the floor plan or elevation drawing depicted by the sectional drawing and is shown with a dashed line with arrows at each end.

  o A wall section drawing can allow the electrician to determine how he may run cable.

  o A sectional drawing of a floor may show how thick the wood will be for her to drill.

- Detail drawings are an enlarged view that makes detail much easier to see than in a sectional drawing.
  - Shows very specific details of a particular part of the building structure.
  - Used to determine exact locations for placement of electrical equipment.
- Electrical drawings show exactly what is required for the complete installation of the electrical system.
  - Electrical symbols used on the plans are a type of shorthand to show which electrical items are required and where they are located.
  - Used to estimate the amount of material and labor needed to install an electrical system.
  - Used in the bidding process to project the total cost for system installation.
- Schedules list and describe various items used in the construction of the building.
  - Provide specific information about electrical equipment to be installed in the building.
  - For example, a lighting fixture schedule lists and describes the various types of lighting fixtures used in the house and tells what type and how many lamps are used with each lighting fixture.

Objective: Identify common architectural symbols found on residential building plans.

2) Identify common architectural symbols

Key terms: symbols, legend

- Architectural symbols are used on building plans to depict the location of various items in a house.
  - Notes are used to provide information about a specific symbol.
  - Each trade has a specific set of architectural symbols.
- Building plans often have a legend that shows the types of architectural symbols used.
- Electrical symbols are used to show the location and type of electrical equipment required to be installed in a house.
  - The standard *Symbols for Electrical Construction Drawings* is published by ANSI (American National Standards Institute).
- Wiring symbols are used to depict the type of wiring method to be used or where the wiring will actually be run.
  - Curved dashed line that goes from a switch symbol to a lighting outlet symbol indicates that the outlet is controlled by that switch.
  - Slashes on a curved solid line are used in a cable diagram to indicate how many conductors are in the cable.

Objective: Determine specific dimensions on a building plan using an architect's scale.

3) Determine dimensions using an architect's scale

Key term: architect's scale

- An architect's scale is used to determine dimensions on a set of building plans.
- It is usually a three-sided device that has each side marked with specific calibrated scales.
- To use an architect's scale you must first determine the scale of the drawing.

Objective: Demonstrate an understanding of residential building plan specifications.

4) Understand building plan specifications

Key terms: specifications

- Specifications provide extra details about equipment and construction methods not in the regular building plans.

- They often include the specific manufacturer's catalog numbers and other information so the electrical items will be the right size and type as well as have the proper electrical rating.

Objective: Demonstrate an understanding of basic residential framing methods and components.

5) Understand framing methods and components

Key terms: platform frame, balloon frame, rafters, ceiling joists, draft-stops, top plate, wall studs, bottom plate, subfloor, floor joists, bridging, band joist, girders, sill, sheathing, foundation, footing

- Electricians will encounter two construction framing methods.

  - The platform frame (western frame) has the floor built first and then the walls are erected on top of it. If there is a second floor, it is simply built the same way as the first floor and placed on top of the completed first floor.

  - The balloon frame has the wall studs and the first floor joists both resting on the sill. If there is a second floor, the second floor joists rest on a one-by-four-inch ribbon cut flush with the inside edges of the studs.

- Rafters are used to form the roof structure of the building and are supported by the top plate.

- Ceiling joists are horizontal framing members that sit on top of the wall framing.

  - They form the structural framework for the ceiling.

  - If there is a floor above, these ceiling joists become floor joists and form the structural framework for the floor above.

- Draft-stops (fire-stops) are used to curtail the spread of fire in a house.

  - Usually pieces of lumber placed between studs or joists to block the path for fire through the cavities formed between studs and joists.

- The top plate is located at the top of the wall framework.

  - Usually a 2-by-4-inch or a 2-by-6-inch piece of lumber.

  - May use two pieces of lumber to form the top plate. When two pieces of lumber are used to form the top plate, it is called the double plate.

- The wall studs are used to form the vertical section of the wall framework.

  - Interior walls are usually 2-by-4-inch lumber, but they may be 2-by-6-inch.

- The bottom plate is the bottom of the wall framework. It rests on top of the subfloor.

  - Usually a 2-by-4-inch or a 2-by-6-inch piece of lumber.

- The subfloor is the first layer of flooring that covers the floor joists.

  - Usually made of 4-by-8-foot sheets of plywood or particleboard (some may still be using 1-by-6-inch boards).

- Floor joists are horizontal framing members that attach to the sill and form the structural support for the floor and walls.
    - o Usually 2-by-8-inch, 2-by-10-inch, or 2-by-12-inch pieces of lumber.
- The band joist, also called a rim joist, is used to stiffen the ends of the floor joists in platform framing.
- The diagonal braces or solid blocks of wood installed between floor joists are called bridging and are used to distribute the weight put on the floor.
- The girders in a house are the heavy beams (metal or wood) that support the inner ends of the floor joists.
    - o Usually made from several 2-by-10-inch or 2-by-12-inch lengths of lumber fastened together.
    - o There are two ways a girder can support the floor joists: the joists can rest on top of the girder or joist hangers can be nailed to the sides of the girder and the hangers support the joists.
- The sill is a piece of wood that lies on the top of the foundation and provides a place to attach the floor joists.
- The sheathing is the boards or sheet material attached to the outside of studs or rafters that add rigidity to the framed structure.
    - o The roof and wall outside finish material is attached to the sheathing.
- A foundation, usually made of poured concrete or concrete block, is the part of the house that supports the framework of the building.
- The footing is located at the bottom of the foundation and provides a good base for the foundation to be constructed.

Quizzes

CODE/REGULATIONS/STANDARDS

1. Common symbols used on blueprints are usually based on the symbols developed by the
   _____.
   - a. National Fire Protection Association
   - b. *National Electrical Code*®
   - c. American National Standards Institute

TOOLS

1. A(n) _____ is a standardized drawing on the building plan that shows the location and type of a particular material or component.
2. Because many plans may use symbols that are not standard, a(n) _____ is usually included in the plans to show the symbols used on the building plans and what they mean.
3. On residential blueprints, _____ are used to provide the sizes and other pertinent information about the various doors and windows used in the building.
4. A circle with a dot in the middle is the standard electrical symbol for a duplex receptacle. True or False?
5. The _____ is a drawing that shows building detail from a view directly above the house.
   - a. elevation drawing
   - b. sectional drawing
   - c. floor plan

6. Most residential plans are drawn on a scale of _____ = 1 foot.
   a. ⅛ inch
   b. ¼ inch
   c. ½ inch

7. A(n) _____ shows the location of items such as kitchen cabinets and appliances located on or against a wall in a kitchen.
   a. exterior elevation drawing
   b. interior elevation drawing
   c. floor plan

9. A(n) _____ shows information such as the location of a house, walkway, or driveway on the building lot.
   a. exterior elevation drawing
   b. floor plan
   c. plot plan

Chapter Quiz

1. The _____ is a table used on building plans to provide information about specific equipment or materials used in the construction of a house.

2. Which of the following are names for the building plans used by craftspeople to build houses?
   a. blueprints
   b. construction drawings
   c. working drawings
   d. all of the above

3. A(n) _____ is a cutaway view that allows you to see the inside of a building.

4. A(n) _____ is a solid dark line that is used to show the main outline of the building, including exterior walls, interior partitions, and interior walls.

5. A(n) _____ is a straight dashed line that is used to show the lines of an object that are not visible from the view shown in the plan.

6. Most residential plans are drawn on a scale of _____ = 1 foot.
   a. ⅛ inch
   b. ¼ inch
   c. ½ inch
   d. 1 inch

7. If a blueprint is drawn with a scale of ⅛ inch = 1 foot, a line 1 inch long will equal _____ on the blueprint.
   a. 8 inches
   b. 2 feet
   c. 8 feet
   d. 4 feet

8. Common symbols used on blueprints are usually based on the symbols developed by the
   _____.

   a. NFPA
   b. *NEC®*
   c. NAEM
   d. ANSI

9. The _____ is the bottom of the wall framework and rests on the top of the subfloor.

10. A(n) _____ is often shown as a broad line of long dashes followed by two short dashes. Arrows are located at each end of the line.

11. A(n) _____ shows the location of items such as kitchen cabinets and appliances located on or against a wall in a kitchen.

12. A(n) _____ shows information such as the location of a house, walkway, or driveway on the building lot.

13. A symbol is a standardized drawing on a building plan that shows the location and type of a particular material or component. True or False?

14. The diagonal braces or solid wood blocks installed between floor joists and used to distribute the weight put on the floor are called _____.

15. The scale to which a drawing has been done is usually found in the _____.

16. _____ are used to form the roof structure of the building and are supported by the top plate.

17. Draw the symbol for a duplex receptacle. _____

18. Draw the symbol for a three-way switch. _____

19. On residential blueprints, _____ are used to provide the sizes and other pertinent information about the various doors and windows used in the building.

20. In a set of blueprints, the drawing scale, the name of the building project, the name of the architectural firm, and the date of completion are found in the _____.

Troubleshooting/Quality Exercises

DISCUSSION QUESTION

1. Most residential building plans are drawn to a ¼ inch=1 foot scale. Knowing this, discuss what you could use out in the field to get measurements from a residential building plan drawn to this scale.

   _____

   _____

   _____

   _____

HANDS-ON APPLICATIONS

1. Look at the floor plan shown in Figure 5-2 of the textbook and answer the following:

   How many bedrooms are there in this house? _____

   How many sinks are there in the master bathroom? _____

   What area do you first walk into when entering the main house from the garage? _____

   Does the kitchen have a built-in dishwasher? _____

   How many skylights are there in the living room? _____

2. Look at the detail drawing shown in Figure 5-6 of the textbook and answer the following:

   What is the slope of the roof? _____

   What size are the floor joists? _____

   What size are the wall studs? _____

   How often are the wall studs placed along the wall? _____

   What is the height from the floor to the ceiling? _____

3. Look at the electrical floor plan shown in Figure 5-7 of the textbook and answer the following:

   The switch box installed in the master bathroom must be ganged how many times? _____

   How many duplex receptacles will be installed in bedroom #2? _____

   What does the "WP" indicate next to the symbol for a duplex receptacle located outside the front entrance door? _____

   From how many locations will the lighting fixture in the dining room be controlled?
   _____

   How many TV outlets will be installed in the master bedroom? _____

REAL-WORLD SCENARIOS

1. Electricians have to install electrical wiring in a variety of home designs. Older homes were built using balloon framing and newer homes are built using platform framing. Describe the basic difference between the two framing styles.

   _____

   _____

   _____

   _____

2. The building specifications provide detailed information for all of the construction trades involved with building a house. The electrical specifications are what electricians are most concerned with. What kinds of information can an electrical contractor get from the specifications that can help with the estimate of the total cost of installing the electrical system?

   _____

   _____

   _____

   _____

Chapter 6

Determining Branch Circuit, Feeder Circuit, and Service Entrance Requirements

Outline

Objective: Determine the minimum number and type of branch circuits required for a residential wiring system.

Objective: Demonstrate an understanding of the basic *NEC*® requirements for calculating branch circuit sizing and loading.

1) Determine the minimum number of branch circuits

2) Understand *NEC*® requirements for branch circuit sizing and loading

Key terms: general-purpose branch circuit, general lighting circuit, dwelling unit, volt-amperes, small-appliance branch circuit, laundry branch circuit, bathroom branch circuit, individual branch circuit, ampacity, outlet, ambient temperature, bundled

- The *NEC*® defines a branch circuit as the circuit conductors between the final overcurrent device (fuse or circuit breaker) and the power and/or lighting outlets.

- Types of residential branch circuits
  - Lighting branch circuits
  - Receptacle branch circuits
  - General-purpose branch circuits
  - General lighting branch circuits
  - Small-appliance branch circuits
  - Laundry branch circuits
  - Bathroom branch circuits
  - Individual branch circuits

- To determine the minimum number of general lighting circuits according to Section 220.12:
  - Calculate habitable floor area to determine minimum number of general lighting circuits required.
  - Multiply floor area by the unit load per square foot for general lighting to get the total general lighting load in volt-amperes.
  - Unit load for a dwelling unit, according to Table 220.12, is 3 volt-amperes per square foot.
  - Divide total general lighting load in volt-amperes by 120 volts for the voltage of residential general lighting circuits. This gives the total general lighting load in amps.
  - Divide total general lighting load in amperes by the size of the fuse or circuit breaker providing the overcurrent protection for the general lighting circuits. This will either be 15 or 20 amps.

- Small-appliance branch circuits
  - Section 210.11(C)(1) requires a minimum of two 20-amp-rated small-appliance branch circuits in each house.
  - Section 220.52(A) states that in each dwelling unit, the load must be computed at 1500 volt-amperes for each 2-wire small-appliance branch circuit.
  - This means that a minimum load for small-appliance branch circuits in a house would be 3000 volt-amperes.
- Laundry branch circuits
  - Supplies 120-volt electrical power to laundry areas.
  - Section 210.11(C)(2) requires a minimum of one 20-amp-rated laundry branch circuit in each house.
  - Section 220.52(B) states that in each dwelling unit, the load must be computed at 1500 volt-amperes for each 2-wire laundry branch circuit.
- Bathroom branch circuits
  - Provides 120-volt electrical power to the bathroom receptacle(s), but may feed lighting in the bathroom if nothing else outside the bathroom is fed from the circuit.
  - Section 210.11(C)(3) requires at least one 20-amp-rated bathroom branch circuit in each house.
  - There is no volt-ampere value for this circuit type included in your calculation for the total electrical load of a house.
- Individual branch circuits
  - Compute circuit rating for a specific appliance based on the ampere rating of the appliance (found on the appliance nameplate). For example, a garbage disposal being installed may have a nameplate rating of 6 amps at 120 volts. By multiplying the amperage of the appliance by the voltage, you get a total load for that appliance in volt-amperes.
  - Electric dryer branch circuits are calculated according to Section 220.54.
  - Electric cooking branch circuits are calculated according to Section 220.55.
- Determining the ampacity of a conductor
  - Section 210.19(A) requires branch circuit conductors to have an ampacity of at least equal to the maximum electrical load to be served.
  - Use Table 310.15(B)(16) to determine the ampacity of a conductor for use in residential wiring.
  - As the conditions of use change for a conductor, adjust the ampacity so that at no time could the temperature of the conductor exceed the temperature of the insulation on the conductor.
- Branch-circuit ratings
  - Based on the size of the fuse of circuit breaker protecting the circuit.
  - Standard branch-circuit ratings for receptacle and lighting circuits include:
    - 15-amp, 20-amp, 30-amp, 40-amp, and 50-amp
  - Circuit ratings for receptacles and lighting in a residential wiring system are usually 15 or 20 amperes.
  - Individual branch-circuit ratings can be any ampere rating, depending on the size of the load served.
- Sizing of the overcurrent protection devices
  - Table 210.24 summarizes requirements for the size of conductors and the size of the overcurrent protection for branch circuits where two or more outlets are required.

Objective: Calculate the minimum conductor size for a residential service entrance.

Objective: Demonstrate an understanding of the steps required to calculate a residential service entrance using the standard or optional method as outlined in Article 220 of the *NEC®*.

Objective: Determine the proper size of the service entrance main disconnecting means.

3) Calculate the minimum conductor size for a service entrance

4) Understand the steps used to calculate a residential service entrance

5) Determine the size of the service entrance main disconnect

Key terms: feeder

- Standard method service entrance calculation
  - ○ Calculate the square foot area by using the outside dimensions of the dwelling.
  - ○ Multiply the square foot area by the unit load per square foot as found in Table 220.12. Use 3 VA per square foot for dwelling units.
  - ○ Add load allowance for the required two small-appliance branch circuits and the one laundry circuit (1500 VA each or a minimum total of 4500 VA). Be sure to add 1500 VA each for any additional small-appliance or laundry circuits.
  - ○ Apply the demand to the sum of the loads up to this point. According to Table 220.42, the first 3000 VA is taken at 100%; 3001 to 120,000 VA at 35%; and anything over 120,000 VA at 25%.
  - ○ Add in the loading for electric ranges, counter-mounted cooking units, or wall-mounted ovens. Use Table 220.55 and the Notes for cooking equipment.
  - ○ Add in the electric clothes dryer load, using Table 220.54. Remember to use 5000 VA or the nameplate rating of the dryer if it is greater than 5 kW.
  - ○ Add the loading for heating (H) or the air-conditioning (AC), whichever is greater. Remember, both of these loads will not be operating at the same time.
  - ○ Add the loading for all other fixed appliances, such as dishwashers, food waste disposals, trash compactors, water heaters, etc.
    - ■ Section 220.53 allows you to apply an additional demand of 75% to the fixed appliance load total if there are four or more appliances.
    - ■ If there are three or less, take their total at 100%.
  - ○ According to Section 220.50, you must add an additional 25% of the largest motor load.
  - ○ The sum of the loads is called the total computed load. Divide this figure by 240 volts to find the minimum ampacity required for the ungrounded service entrance conductors.
  - ○ Calculate the minimum neutral ampacity by referring to Section 220.61.
    - ■ Add volt-amp load for the 120-volt general lighting, small-appliance, and laundry circuits.
    - ■ Add the cooking and drying loads and multiply the total by 70%.
    - ■ Add in loads of all appliances operating on 120 volts.
    - ■ Remember to apply the additional demand of 75% to the total neutral loading of the appliances if there are four or more of them.

- Divide the total neutral load by 240 volts to find the minimum ampacity of the service neutral.

    - Section 310.15(B)(7) allows the neutral to be smaller than the ungrounded service entrance conductors, but in no case can it be smaller than the grounding electrode conductor.

  - Calculate the minimum size copper grounding electrode conductor by using the largest size ungrounded service entrance conductor and Table 250.66.

- Optional method service entrance calculation

- Because sizing the service entrance neutral using the optional method is done exactly the same way as in the standard method, only the steps used to get the minimum ungrounded service conductor size, the minimum service rating, and the grounding electrode conductor will be listed.

  - Calculate the square foot area by using the outside dimensions of the dwelling.

  - Multiply square foot area by the unit load per square foot as found in Table 220.12. Use 3 VA per square foot for dwelling units.

  - Add load allowance for the required two small-appliance branch circuits and one laundry circuit (1500 VA each or a minimum total of 4500 VA). Be sure to add 1500 VA each for any additional small-appliance or laundry circuits.

  - Add nameplate ratings of all appliances fastened in place, permanently connected, or located to be on a specific circuit, such as electric ranges, wall-mounted ovens, counter-mounted cooktops, electric clothes dryers, trash compactors, dishwashers, and electric water heaters.

  - Take 100% of the first 10,000 VA and 40% of the remaining load. Refer to Section 220.82(B).

  - Add the value of the largest heating and air-conditioning load from the list included in Section 220.82(C).

  - The sum of these loads is called the total computed load. Divide this figure by 240 volts to find the minimum ampacity required for the ungrounded service entrance conductors.

  - The neutral size is calculated exactly the same as for the standard method. There is no optional method for calculating the neutral size.

  - Calculate the minimum size copper grounding electrode conductor by using the largest size ungrounded service entrance conductor and Table 250.66.

- For dwelling units, wire sizes as listed in Table 310.15(B)(7) are permitted as:

  - 120/240-volt, 3-wire, single-phase service entrance conductors

  - Service lateral conductors

  - Feeder conductors that serve as the main power feeder to a dwelling unit

Objective: Determine the proper size of a loadcenter used to distribute the power in a residential wiring system.

6) Determine the size of a loadcenter

Key terms: loadcenter

- Sizing the loadcenter

  - Once the service entrance calculation has been done and the minimum size service entrance has been calculated, a loadcenter size can be determined. (Loadcenters are also called panels.)

  - Section 408.30 of the *NEC*® says that all panelboards must have a rating not less than the minimum feeder capacity required for the load as computed in accordance with Article 220.

- The rating of a panelboard is based on the ampacity of the bus bar(s).
- One company offers a loadcenter suitable for residential use that can accommodate up to 60 circuits.
- A single-pole circuit breaker provides overcurrent protection to circuits that operate on 120 volts.
- A two-pole circuit breaker provides overcurrent protection to circuits that operate on 240 volts.
- Once you have determined the number of circuits that will need to be installed, you can choose a panel that is designed to accommodate the number of circuit breakers you will install.

Objective: Calculate the minimum size feeder conductors delivering power to a subpanel.

7) Calculate the size feeder supplying a subpanel

Key terms: subpanel

- Sizing feeders and subpanels

   o Subpanels are usually installed to locate a loadcenter closer to an area of the house where several circuits are required.

   o The wiring from the main panel to the subpanel is called a feeder.

   o Sizing the feeder to a subpanel is done exactly the same way as sizing service entrance conductors feeding a main service panel.

   o Use either the standard or optional method, including only those electrical loads fed from the subpanel.

Quizzes

CODE/REGULATIONS/STANDARDS

1. The center of the hole changes each time a ring is removed from a concentric knockout in an electrical enclosure. True or False?

2. _____ is the current in amperes that a conductor can carry continuously under certain conditions of use without exceeding its temperature rating.

3. The *NEC*® requires at least _____ small-appliance circuits be installed in each dwelling unit.

4. The *NEC*® states that the load for a household dryer must be computed at _____ watts or the nameplate rating, whichever is greater.

5. When cables or conductors are physically tied, wrapped, or taped together, they are considered to be _____.

   a. bundled

   b. messy

   c. supported

6. The minimum service entrance rating for a single-family home permitted by the *NEC*® is _____ amperes.

   a. 60

   b. 75

   c. 100

7. If cables are bundled together for more than 24 inches and the total number of current-carrying conductors exceeds _____, the ampacity of the conductors must be derated.

   a. 2

   b. 3

   c. 4

8. In residential wiring, which of the following house areas does not have its power supplied by a small-appliance circuit?

   a. bathroom

   b. dining room

   c. kitchen

9. According to the *NEC*®, the unit load used to calculate the minimum number of general lighting circuits in a dwelling is _____ volt-amperes per square foot.

   a. 2

   b. 3

   c. 4

10. The calculated load for a small-appliance branch circuit is computed at _____.

   a. 1200 volt-amperes

   b. 1500 volt-amperes

   c. 3000 volt-amperes

Chapter Quiz

1. A(n) _____ is the circuit conductors between the final overcurrent device and the outlets.

   a. open circuit

   b. branch circuit

   c. feeder circuit

   d. service conductor

2. The circuit conductors between the service equipment and the final branch circuit overcurrent protection device is a(n) _____.

   a. open circuit

   b. branch circuit

   c. feeder circuit

   d. service conductor

3. Which one of the following areas is included when computing the general lighting load of a house?

   a. open porches

   b. garages

   c. crawl spaces

   d. finished basements

4. A house with 2000 square feet of habitable living space will require at least _____ general lighting circuits if 15-ampere circuit breakers are used to provide circuit protection.

5. Which one of the following is not permitted to be connected to a small-appliance branch circuit?
   a. refrigerator
   b. electric clock
   c. gas stove
   d. kitchen lighting fixture

6. If cables are bundled together for more than 24 inches and the total number of current-carrying conductors exceeds _____, the ampacity of the conductors must be derated.

7. When you work on a service entrance calculation, if a residence has four fixed appliances, a demand factor of _____ % is applied to the total load of these appliances.

8. The rating of a panelboard is based on the ampacity of the _____.
   a. main breaker
   b. bus bar
   c. service conductor
   d. total amperage of each breaker in the panelboard

9. A 200-ampere-rated panel can accommodate a maximum of _____ circuits.
   a. 30
   b. 40
   c. 42
   d. There is no maximum number

10. The minimum service entrance rating for a single-family home permitted by the *NEC®* is _____ amperes.

11. When cables or conductors are physically tied, wrapped, or taped together, they are considered to be _____.
    a. bundled
    b. messy
    c. supported
    d. secured

12. According to Table 310.15(B)(16) in the *NEC®*, the ampacity for a 12-AWG copper conductor using the 60°C column is _____.

13. According to Table 310.15(B)(16) in the *NEC®*, the ampacity for a 3-AWG copper conductor using the 75°C column is _____.

14. According to *NEC®* Table 310.15(B)(7), the minimum conductor size for a 200-ampere residential service entrance is _____ CU or _____ AL.

15. According to *NEC®* Table 310.15(B)(7), the minimum conductor size for a 300-ampere residential service entrance is _____ CU or _____ AL.

16. _____ temperature is the temperature of the air that surrounds an object on all sides.

17. The *NEC®* states that each two-wire small-appliance circuit shall be computed at _____ volt-amperes.

18. The *NEC®* states that the load for a household electric clothes dryer must be computed at _____ watts or the nameplate rating, whichever is greater.

19. If an electric clothes dryer has a nameplate rating of 4500 watts, _____ watts will be used in the standard method service calculation.

20. If the total computed load of a residence is 45,840 volt-amperes, the computed amperage of the residence is _____ amps.

Troubleshooting/Quality Exercises

DISCUSSION QUESTIONS

1. Considering the termination temperature ratings of the equipment used in residential wiring, discuss why electricians usually determine conductor ampacity based on the 60°C column of Table 310.15(B) (16).

   _____

   _____

   _____

2. Some electricians do not bother to calculate the minimum size of neutral service conductor. They size the minimum ungrounded conductor size and then use this information to determine the minimum neutral conductor size. Explain how they determine the minimum neutral size using the minimum ungrounded conductor size.

   _____

   _____

   _____

3. An electrician does a standard service entrance calculation and an optional service entrance calculation for the same house. The optional service calculation results in a smaller size service entrance. Is this normal? Explain.

   _____

   _____

   _____

   _____

HANDS-ON APPLICATIONS

1. Match the following terms to their definitions. Write the correct number for the term in the blanks.

   1. Small-appliance branch circuit
   2. Ambient temperature
   3. Outlet
   4. Ampacity
   5. Nipple

   6. Branch circuit
   7. Laundry branch circuit
   8. Bundled
   9. Individual branch circuit
   10. Feeder

   _____ a. The temperature of the air that surrounds an object

   _____ b. The circuit conductors between the final overcurrent device and the outlets

   _____ c. Cables or conductors that are physically tied, wrapped, taped, or otherwise periodically bound together

   _____ d. The circuit conductors between the service equipment and the final branch circuit overcurrent device

   _____ e. A circuit that supplies only one piece of equipment

   _____ f. A branch circuit type that supplies electrical power to a clothes washer

   _____ g. An electrical conduit of 2 feet or less in length used to connect two electrical enclosures

_____ h. A point on the wiring system where current is taken to supply electrical equipment

_____ i. A branch circuit type that supplies electrical power to receptacles located in a kitchen

_____ j. The current that a conductor can carry continuously without exceeding its temperature rating

REAL-WORLD SCENARIOS

1. Most residential general lighting circuits are wired with 14-AWG copper conductors and are protected by 15-amp circuit breakers. However, there are many electricians who install 12-AWG copper conductors and protect them with 20-amp circuit breakers. How will you know which conductor size and which circuit breaker size will be used for the general lighting branch circuits in the house you are wiring?

_____

_____

_____

_____

2. One way to determine the minimum number of general lighting circuits in a house is to use the following information: each 600 square feet requires one 15-amp general lighting circuit and each 800 square feet requires one 20-amp general lighting circuit. If a house has 2400 square feet of living space, calculate the minimum number of 15-amp and 20-amp general lighting circuits required.

_____

_____

_____

_____

3. An electrician has installed an 18-inch length of conduit, called a nipple, between two electrical enclosures. There will be more than 10 current-carrying conductors installed in the nipple. Will the electrician have to derate the ampacity of the conductors because there are more than three? List the section in the *NEC®* that addresses this application.

_____

_____

_____

_____

4. It is a good idea to oversize a panel when it comes to the number of circuits it is rated for. This practice will help ensure that there is room for expansion of the electrical system in the future. A good rule of thumb is to never fill a panel to more than _____ percent of its capacity.

_____

_____

_____

_____

Chapter 7 — Introduction to Residential Service Entrances

Outline

Objective: Demonstrate an understanding of an overhead and an underground residential service entrance.

1) Understand overhead and underground residential service entrances

Key terms: overhead service, underground service, equipment

- The *NEC*® defines a service as the conductors and equipment for delivering electric energy from the serving electric utility to the wiring system of the premises served.
- There are two types of service entrances used to deliver electrical energy to a residential wiring system:
  - An overhead service
  - An underground service
- Overhead service
  - Used most often.
  - Less expensive and takes less time to install than underground service.
  - Includes service conductors between the terminals of the service equipment main disconnect and a point outside the home where they are connected to overhead wiring.
  - Overhead wiring connected to the electric utility's electrical system.
- Underground service
  - Service conductors between terminals of the service main disconnect and point of connection to utility wiring buried in the ground to protect the conductors from physical damage.
  - Also prevents accidental contact with the conductors by people.
  - More costly and time-consuming procedure for repair than overhead service.

Objective: Define common residential service entrance terms.

2) Define common residential service entrance terms

Key terms: service head, service point, drip loop, service mast, service drop, service entrance cable, overhead system service entrance conductors, underground system service entrance conductors, service raceway, meter enclosure, service equipment, grounding electrode, supplemental grounding electrode, transformer, pad-mounted transformer, utility pole, riser, lateral

- Service head (weatherhead)
  - Fitting placed on service drop end of service entrance cable or service entrance raceway designed to minimize the amount of moisture that can enter the cable or raceway.

- Service point
  - Point of connection between the wiring of the electric utility and the house wiring.
- Drip loop
  - Intentional loop put in service entrance conductors that extend from a weatherhead at the point where they connect to the service drop conductors.
  - Conducts rainwater to a lower point than the weatherhead so that no water will drip down the service entrance conductors and into the meter enclosure.
- Service mast
  - Piece of rigid metal conduit (RMC) or intermediate metal conduit (IMC), usually 2 or 2½ inches in diameter, which provides protection for service conductors and the proper height requirements for service drops.
  - Mast usually extends from meter enclosure, allowing attached service drop to have the required distance above grade.
- Service drop
  - Overhead service conductors from utility pole to the point where the connection is made to the service entrance conductors at the house.
  - Usually owned by the utility company.
- Service entrance cable
  - Type SEU cable: service conductors designed to be used outdoors on the side of house.
  - Type USE cable: service conductors designed to be buried in a trench for an underground service.
- Service entrance conductors
  - Conductors from the service point to the service disconnecting means.
  - Can be enclosed in a raceway or be part of a service entrance cable assembly.
- Overhead system service entrance conductors
  - Service conductors between the terminals of the service equipment and the point where they are joined by tap or splice to the service drop.
- Underground system service entrance conductors
  - Service conductors between the terminals of the service equipment and the point of connection to the service lateral.
- Service raceway
  - Rigid metal conduit, intermediate metal conduit, electrical metallic tubing, rigid PVC conduit, or any other approved raceway that encloses the service entrance conductors.
- Meter enclosure
  - Weatherproof electrical enclosure that houses the kilowatt-hour meter, also known as the "meter socket," "meter base," or "meter trim."
- Service equipment
  - Necessary equipment connected to the load end of the service conductors supplying a building.
  - Intended to be main control and cutoff of the supply.
  - Equipment can consist of fusible disconnect switch or main breaker panel accommodating branch-circuit overcurrent protection devices (fuses or circuit breakers).

- Grounding electrode

  ○ Part of service entrance that provides a direct connection to the earth.

  ○ Usually the metal water pipe that brings water to the home. If there is no metal water pipe, a ground rod is usually used.

  ○ Limits the voltage imposed by lightning, line surges, or unintentional contact with higher voltage lines and stabilizes the voltage to earth during normal operation.

- Supplemental grounding electrode

  ○ *NEC*® requires that a metal water pipe electrode be supplemented by another electrode, usually an 8-foot rod driven into the ground.

- Transformer

  ○ Electrical equipment used by the electric utility to step-down the high voltage of the utility system to the 120/240 volts required for a residential electrical system.

  ○ When installed on a pad at ground level the transformer is called a pad-mounted transformer.

- Utility pole

  ○ Circular column usually made of treated wood and set in the ground for the purpose of supporting utility equipment and wiring.

  ○ Typically support transformers and electrical system wiring for communication utilities and fiber-optic cable and coaxial cable for cable television providers.

- Riser

  ○ Length of raceway that extends up a utility pole and encloses the service entrance conductors in an underground service entrance.

  ○ "Standoff" supports multiple risers for such things as telephone lines and cable television lines.

- Lateral

  ○ Underground service conductors between the electric utility transformer, including any risers at a pole or other structure, and the first point of connection to the service entrance conductors in a meter enclosure.

Objective: Demonstrate an understanding of *NEC*® requirements for residential service entrances.

Objective: List several *NEC*® requirements that pertain to residential service entrances.

3) Understand and list several NEC® requirements for residential service entrances

- Article 230 of the *NEC*® covers many of the requirements for the installation of service entrances.

- Section 230.7

  ○ Wiring other than service conductors must not be installed in the same service raceway or service cable.

  ○ Other residential wiring system conductors must be separated from the service.

- Section 230.8

  ○ Must install a raceway seal where a service raceway enters a residential building from an underground distribution system.

  ○ Also see Section 300.5(G) and Section 300.7(A)

- Section 230.9

- Service conductors must have a minimum clearance of 3 feet (900 millimeters) from windows that can be opened, doors, porches, balconies, ladders, stairs, fire escapes, or similar locations in a residential building.

- Section 230.10
  - Vegetation such as trees shall not be used to support overhead service conductors.

- Section 230.22
  - Individual service entrance conductors must be insulated.
  - Grounded (neutral) conductor of a multi-conductor service entrance cable can be bare.

- Section 230.23
  - Determines the requirements for the minimum size of the service drop conductors.
  - Service drop conductors must have sufficient ampacity to carry the current for the computed residential electrical load and must have adequate mechanical strength.

- Section 230.24
  - Provides dimensions for the minimum service drop clearance over roofs.
  - Provides dimensions for the minimum amount of vertical clearance for service drop conductors.

- Section 230.28
  - When a service mast is used for support of the service drop conductors, it must be strong enough to withstand strain of the service drop.
  - Only electrical power service drops are allowed to be attached to the mast.

- Section 300.5 and Table 300.5
  - Gives minimum cover and protection requirements for underground service conductors.

- Section 230.43
  - Lists wiring methods that could be used to install a residential service entrance.

- Section 230.51
  - Gives requirements for supporting service entrance cable.

- Section 230.54
  - Lists the rules that apply to overhead service locations.

- Section 230.70
  - Means must be provided to disconnect all conductors in a building or other structure from the service entrance conductors.
  - Applies to both overhead and underground service entrance type.
  - Covers maximum number of disconnects permitted as the disconnecting means for the service conductors that supply a building.

- Section 230.79
  - Service disconnecting means shall have a rating not less than the load to be carried as determined in accordance with Article 220.
  - For a one-family house, the minimum size service is 100 amperes.

- Section 230.90
  - Requires each ungrounded service conductor to have overload protection.

Objective: Demonstrate an understanding of grounding and bonding requirements for residential service entrances.

4) Understand grounding and bonding for residential service entrances

Key terms: ground, grounded, neutral, bonding

- Section 250.21

  - States that you must ground alternating current systems of 50 to 1000 volts supplying premises wiring systems so the maximum voltage to ground on ungrounded conductors does not exceed 150 volts.

- Section 250.24(A)

  - Residential wiring system supplied by a grounded alternating current service must have a grounding electrode conductor connected to the grounded service conductor.

  - The grounded service conductor is called the neutral.

- Section 250.24(C)

  - Grounded (neutral) conductor of a residential service must run to the service disconnecting means and be bonded (attached) to the disconnecting means enclosure.

- Section 250.28

  - Covers requirements for the main bonding jumper.

  - For a grounded system, an unspliced main bonding jumper must be used to connect the equipment grounding conductor(s) and the service disconnect enclosure to the grounded conductor of the system within the enclosure for each service disconnect.

- Section 250.50

  - Covers requirements for grounding electrode system.

  - If present, each item in 250.52(A)(1) through (A)(7) must be bonded together to form the grounding electrode system.

  - Where none of these electrodes are present, one or more electrodes specified in 250.52(A)(4) through (A)(8) must be installed and used.

- Section 250.52

  - Lists electrodes permitted for grounding:

    - Metal underground water pipe in direct contact with the earth for 10 feet or more (3.0 meters or more).

    - The metal frame of the building or structure, where effectively grounded, can be used as a grounding electrode.

    - A concrete-encased electrode can be used and is an excellent choice.

    - A ground ring encircling the house, in direct contact with the earth, consisting of at least 20 feet (6.0 meters) of bare copper conductor (at least 2 AWG) must be used.

    - Rod and pipe electrodes, commonly called "ground rods," can be used.

    - Other listed electrodes are permitted.

    - Plate electrodes can be used, but rarely are used in residential work.

- Section 250.53

  - Covers some installation rules for grounding electrode system:

  - Unless a single rod, pipe or plate electrode has a resistance of 25 ohms or less to earth, it must be supplemented by an additional electrode of a type specified in Section 250.52(A)(2) through (A)(8).

- Where practicable, embed rod, pipe, and plate electrodes below permanent moisture level.

- Where more than one ground rod, pipe, or plate is used, place each electrode type at least 6 feet (1.83 meters) from any other electrode.

- When used as grounding electrode, metal underground water pipe must meet the following requirements:

  - Bonding around equipment such as water meters and filtering equipment is required.

  - Supplement metal underground water pipe.

  - Where the supplemental electrode is a rod, pipe, or plate electrode, portion of bonding jumper that is the sole connection to supplemental grounding electrode not required to be larger than 6-AWG copper wire or 4-AWG aluminum wire.

- If a ground ring is installed as the grounding electrode, bury it at least 30 inches (750 millimeters) deep.

- Install rod and pipe electrodes so that at least 8 feet (2.44 meters) of length is in contact with the soil.

- Bury plate electrodes at least 30 inches (750 millimeters) below the surface of the earth.

- Section 250.64

  - Covers the installation of grounding electrode conductor:

    - Do not use bare aluminum or copper-clad aluminum grounding electrode conductors in direct contact with masonry or the earth or where subject to corrosive conditions.

    - Securely fasten grounding electrode conductor (GEC) or its enclosure to the surface on which it is carried.

      - Protect 4 AWG or larger GEC if subject to physical damage.

      - Protect 6 AWG GEC if subject to physical damage.

      - Always protect 8 AWG GEC.

    - Install GEC in one continuous length without a splice.

    - Splicing is allowed only by irreversible compression-type connectors listed for the purpose or by the exothermic welding process.

    - Metal raceways for GECs must be electrically continuous from the point of attachment to cabinets or equipment to the grounding electrode.

    - Securely fasten them to the ground clamp or fitting.

    - Bond each end of a metal raceway that is not continuous.

- Section 250.66

  - Specifies how to determine the size of the GEC.

  - Table 250.66 is used to size the GEC of a grounded AC system and is based on the largest size service entrance conductor.

- Section 250.68

  - Covers the GEC connection to the grounding electrode.

  - Interior metal water piping located not more than 5 feet (1.52 m) from the point of entrance to the building is permitted to be used as a conductor to interconnect electrodes that are part of the grounding electrode system.

- Section 250.70v
  - ○ Lists methods for connecting grounding conductor to an electrode.
- Section 250.80
  - ○ Must ground metal enclosures and raceways for service conductors and equipment.
- Section 250.92(A)
  - ○ Bond noncurrent-carrying metal parts of the service equipment.
- Section 250.92(B)
  - ○ Allowed methods of bonding at the service.
- Section 250.94
  - ○ Must provide an intersystem bonding termination for telephone and television system wiring.

Objective: Demonstrate an understanding of common electric utility company requirements.

Objective: Demonstrate an understanding of how to establish temporary and permanent power with an electric utility company.

5) Understand common utility company requirements and how to establish temporary and permanent power

- Once the type of service entrance being installed is determined, contact the local electric utility.
- Coordinate the service type, location of the service, and service installation.
- Ask the utility company for their publication that outlines the rules to follow when installing services.

Quizzes

SAFETY

1. Explain why the *NEC*® and local utility companies require the service entrance disconnecting means, if located inside a house, to be installed as close as possible to where the service conductors enter the house.

   _____

   _____

   _____

   _____

2. Always be sure to check with the electric utility company and the authority having jurisdiction in your area before starting to install an overhead or underground service entrance. True or False?

CODE/REGULATIONS/STANDARDS

1. Overhead service conductors are permitted to be attached to trees for support. True or False?

2. The overhead service conductors from the last pole to the point connecting them to the service entrance conductors at the building are called _____ conductors.

    a. service drop

    b. riser

    c. service lateral

3. The *NEC*® requires a minimum clearance of _____ feet above residential property and driveways where the voltage to ground does not exceed 300 volts.

    a. 18

    b. 12

    c. 15

4. When a metal underground water pipe is used as a grounding electrode, it must be in contact with the earth for at least _____ feet.

    a. 5

    b. 8

    c. 10

5. Ground rods must be at least _____ feet long.

    a. 6

    b. 8

    c. 10

6. A(n) _____ is the conductor connected to the neutral point of a system and is intended to carry current under normal conditions.

    a. grounding conductor

    b. ungrounded conductor

    c. neutral conductor

7. The type of cable normally used as the service drop for a residential service entrance is called _____.

    a. quadplex cable

    b. Romex™

    c. triplex cable

8. A piece of electrical equipment used by the utility company to step down the high voltage of the utility system to the 120/240 volts required for a residential electrical system is called a(n) _____.

    a. transformer

    b. utility pole

    c. step-down device

Chapter Quiz

1. Overhead service conductors are permitted to be attached to trees for support. True or False?

2. The enclosure housing a temporary service entrance disconnecting means must be dust-tight. True or False?

3. If a single grounding electrode that is a rod, pipe, or plate does not have resistance to earth of 25 ohms or less, an additional electrode must be installed. True or False?

4. _____ is a general term used for material, fittings, devices, appliances, lighting fixtures, apparatus, machinery, and other parts used in connection with an electrical installation.
   a. Circuit
   b. Equipment
   c. Enclosure
   d. Electrode

5. When installing service drop conductors above a residential roof, conductors must have a vertical clearance of at least _____ feet (general rule).

6. The *NEC*® requires a clearance of at least _____ feet above final grade for service conductors where they are connected to the building, above sidewalks, and other areas accessible to pedestrians.

7. The *NEC*® requires a minimum clearance of _____ feet above residential property and driveways where the voltage to ground does not exceed 300 volts.

8. The *NEC*® requires a vertical clearance of at least _____ feet above public streets, alleys, roads, and parking areas subject to truck traffic.

9. Underground service conductors are required to be buried at least _____ feet by the *NEC*®.

10. Service entrance cables must be supported by straps or staples within _____ inches of any electrical enclosure and at intervals not exceeding _____ inches.
    a. 8, 24
    b. 12, 24
    c. 8, 30
    d. 12, 30

11. When a metal underground water pipe is used as a grounding electrode, it must be in contact with the earth for at least _____ feet.

12. When a ground ring is used as a grounding electrode, it must be at least a 2-AWG copper conductor at least _____ feet long.

13. A(n) _____ is a conducting object through which a direct connection to the earth is established.
    a. bond wire
    b. grounding electrode
    c. main bonding jumper
    d. grounded conductor

14. A(n) _____ is the conductor connected to the neutral point of a system and is intended to carry current under normal conditions.

   a. grounding conductor

   b. ungrounded conductor

   c. service conductor

   d. neutral conductor

15. The type of cable normally used as the service drop for a residential service entrance is called _____.

   a. quadplex cable

   b. Romex™

   c. triplex cable

   d. UF cable

16. When bonding together the noncurrent-carrying metal parts of a service entrance, which of the following methods may be used?

   a. exothermic welding

   b. listed pressure connectors

   c. listed clamps

   d. all of the above

17. The _____ is the conductor used to connect the grounding electrode to a system conductor or equipment.

   a. grounding electrode conductor

   b. grounding conductor

   c. neutral conductor

   d. service conductor

18. According to *NEC*® Table 250.66, the minimum size copper grounding electrode conductor for a service entrance installed with 2/0 copper conductors is _____.

19. The _____ conductor of an AC service is connected to a grounding electrode system to limit the voltage to ground from lightning, line surges, and unintentional high-voltage crossovers, and to stabilize the voltage to ground during normal operation.

20. A length of raceway that extends up a utility pole and encloses the service entrance conductors in an underground service is called a(n) _____.

Troubleshooting/Quality Exercises

DISCUSSION QUESTIONS

1. Most residential electricians do not own the necessary ground resistance measuring instrument to determine the resistance to the earth through a ground rod. Because of this, some electric utilities have a requirement that must be followed when installing ground rod grounding electrodes. Discuss this electric utility requirement.

   _____

   _____

   _____

   _____

2. There is no *NEC®* rule for how often an electrician must secure and support the grounding electrode conductor. Discuss what is considered to be a good rule to follow that will provide adequate securing and supporting for the grounding electrode conductor.

_____

_____

_____

_____

HANDS-ON APPLICATIONS

1. Match the following service entrance terms to their definitions. Write the correct number for the term in the blanks.

1. Bonding	10. Service mast
2. Utility pole	11. Grounding electrode conductor
3. Drip loop	12. Service lateral
4. Transformer	13. Meter enclosure
5. Ground	14. Service head
6. Service raceway	15. Neutral conductor
7. Grounded	16. Service drop
8. Triplex	17. Intersystem Bonding Termination
9. Grounding electrode	

_____ a. Connected to establish electrical continuity and conductivity

_____ b. An intentional loop put in service conductors at the point where they exit from the weatherhead

_____ c. The earth

_____ d. Connected to earth

_____ e. A conducting object through which a direct connection to earth is established

_____ f. The conductor used to connect the grounding electrode to a system conductor or to equipment

_____ g. The weatherproof electrical enclosure that houses the kilowatt-hour meter

_____ h. The conductor connected to the neutral point of a system that is intended to carry current under normal conditions

_____ i. The overhead service conductors from the pole to the building

_____ j. The fitting that is placed on the service drop end of service entrance cable or service entrance raceway

_____ k. The underground service conductors between the transformer and the first point of connection to the service entrance conductors

_____ l. A piece of rigid metal or intermediate metal conduit that provides service conductor protection and the proper height for a service drop

_____ m. The conduit or other approved raceway that encloses the service entrance conductors

_____ n. A piece of electrical equipment that steps down the high utility voltage to the 120/240 volts required for a residential electrical system

_____ o. A device that provides a means for connecting bonding conductors for communications systems to the grounding electrode system

_____ p. A cable type used as the service drop

_____ q. A wooden circular column used to support electrical, video, and telecommunications utility wiring

REAL-WORLD SCENARIOS

1. Setting up a temporary service entrance is not necessary on many new residential construction sites. Explain why many residential construction sites do not need a temporary service.

_____

_____

_____

_____

2. An electrical plan may show a particular type of service entrance, but the type of service may not be able to be installed because of local electric utility restrictions. Give an example of when this might be the case.

_____

_____

_____

_____

Chapter 8　Service Entrance Equipment and Installation

Outline

Objective: Identify common overhead service entrance equipment and materials.

1) Identify overhead service equipment and materials

Key terms: cable hook, threaded hub, raintight, enclosure, sill plate, conduit "LB," mast kit, porcelain standoff, roof flashing

- Overhead service equipment and materials
  - There are three different ways to install an overhead service entrance:
    - Using service entrance cable installed on the side of the house
    - Using electrical conduit installed on the side of the house
    - Using a mast-style service installation
- Equipment when using service entrance cable
  - Service drop conductors
    - Bring electrical energy from local electric utility system to the house.
    - The overhead cable, typically called triplex cable, is usually installed and owned by the electric utility company.
  - Weatherhead
    - Used to stop entrance of water into the end of a service entrance cable.
    - Each service entrance conductor must exit from a separate hole in the weatherhead.
  - Cable hook
    - Attaches service drop cable to the side of the house.
    - Install hook below the weatherhead.
  - Service entrance cable
    - Overhead services will usually use an SEU type service entrance cable to bring the electricity down the side of the house from the point of attachment to the meter socket, and then on into the main service equipment.
  - Clips for the service entrance cable
    - Used to provide support for the cable.
    - Clip types that are often used include fold-over clips and one- or two-hole straps.
  - Meter socket and threaded hub
    - Used to provide a location in the service entrance for the local electric utility to install a kWh meter that will measure the amount of electrical energy used by the house electrical system.

- Raintight service entrance cable connector
  - Used to secure the service entrance cable to the top of the meter socket.
  - Special design that allows a raintight connection so no water enters the meter socket around the service entrance cable.
- Regular service entrance cable connectors
  - Used to connect service entrance cable to the bottom of the meter socket.
  - Used to connect service cable to main service equipment enclosure when the enclosure is located inside the house.
- Sill plate
  - Used to protect service entrance cable at the point where it enters a house through an outside wall.
- Service equipment
  - Contains the main service disconnecting means, usually a main circuit breaker, used to disconnect the supply electricity from the house electrical system.
  - There are two types of service equipment commonly used in residential wiring systems:
    - Main breaker panel installed inside the house.
    - Combination meter socket/main breaker disconnect installed outside the house.
- Equipment when using electrical conduit
  - Service raceway
    - Provides protection for the service entrance conductors.
  - Service entrance conductors
    - Installed in service raceway as individual conductors.
  - Used to bring electricity down the side of the house from the point of attachment to the meter socket and then on into the main service panel.
  - Service raceway weatherhead
    - Designed to fit the size and type of electrical conduit being used for service raceway.
    - Used to stop the entrance of water into the end of the service entrance raceway.
    - Conduit clamps
    - Used to support the service raceway.
    - Types often used include one-hole straps, two-hole straps, and conduit hangers.
  - Conduit connector fittings
    - Used to connect service raceway to the various enclosures and conduit fittings used in service entrance installation.
  - Meter socket with a threaded hub
    - Used to provide a location in the service entrance for the local electric utility to install a kWh meter.
  - Conduit LB
    - Allows for a 90° change of direction in a raceway path.
    - Allows the raceway coming from the bottom of the meter socket to go through the side of the house and continue to the service equipment enclosure.

- ○ Insulated bushing

  - Fiber or plastic fitting designed to screw onto the ends of conduit or cable connector to provide protection to the conductors.

- Equipment when using a service entrance mast

  - ○ Mast kit

    - Often purchased through an electrical distributor.

    - Includes required additional pieces of equipment for a mast-type service.

  - ○ The mast

    - Extends above the roof line so that the minimum required service drop clearance can be achieved when bringing a service drop to a low building like a ranch-style house.

    - Weatherhead

    - Must be designed to fit the size of electrical conduit being used for the service mast.

    - Used to stop entrance of water into the end of the service entrance mast.

  - ○ Porcelain standoff fitting

    - As the mast extends above the roof in this service entrance style, there is no convenient place to install the cable hook for the attachment of the service drop.

    - A porcelain standoff fitting is placed on the mast and provides a spot for the service drop to be attached.

  - ○ Weather collar and roof flashing

    - The roof flashing is placed around the hole in the roof that is made for the service mast to extend through.

    - Once the mast is installed through the hole, a rubber weather collar is installed on the mast.

    - Together, the roof flashing and weather collar provide a seal against rain or melting snow leaking around the mast and into the building.

- Section 250.94 requires that an intersystem bonding termination be installed.

  - This device is used to provide a connection point for cable television, satellite TV, telephone, and any other system that requires being bonded to the house electrical power system grounding electrode.

Objective: Demonstrate an understanding of common installation techniques for overhead services.

2) Understand installation techniques for overhead services

Key terms: line side, load side

- Installing an overhead service using service entrance cable

  - ○ Install threaded hub and raintight connector to the meter socket.

  - ○ Locate and install the meter socket.

  - ○ Measure and cut the service entrance cable to the length required to go from the meter socket up to a height on the side of the house to keep the service drop conductors at the minimum required clearance from grade.

  - ○ Strip off approximately 3 feet of outside sheathing from one end of the SEU cable and 1 foot of outside sheathing from the other end.

○ Install the cable weatherhead on the end that has 3 feet of free conductor.

○ Mount assembled service entrance cable with attached weatherhead on the side of the house directly above and vertically aligned with the raintight connector previously installed in the meter socket hub.

○ Insert the other stripped end into the meter socket through the raintight connector and tighten the connector around the cable.

○ Secure the cable to the side of the building with clips or straps.

○ Install the cable hook at a proper location in relation to the weatherhead.

○ Locate and drill a hole for service entrance cable to enter the house and be attached to the service entrance main panel enclosure.

○ Remove knockout from the bottom of the meter socket to attach the service cable to the bottom of the meter socket.

○ Measure and cut service entrance cable to go from the bottom of the meter socket to the service panel enclosure.

○ Insert the 1 foot stripped end of service entrance cable through the cable connector on the bottom of the meter socket and tighten the connector onto the cable.

○ Insert the other end of the cable through the hole you previously drilled in the wall.

○ Install a sill plate over the cable at the point where the cable goes through the hole in the side of the house.

○ Make the meter socket connections. Connect the conductors from the weatherhead end to the line-side lugs. Connect the conductors from the service panel end to the load-side lugs.

○ Secure service entrance cable to service entrance panel and make the proper connections in the service entrance panel.

- Installing an overhead service using service entrance raceway

  ○ Install a threaded hub to the meter socket.

  ○ Locate and install the meter socket.

  ○ Measure and cut the electrical conduit to go from the meter socket up to a height on the side of the house that will keep the service drop conductors at the minimum required clearance from grade.

  ○ Measure and cut the length of service entrance conductor wire needed.

  ○ Insert one end of the service raceway into the meter socket hub and tighten it.

  ○ Mount the service entrance raceway on the side of the house directly above and vertically aligned with the meter socket hub.

  ○ Install the raceway weatherhead on the top of the raceway and then remove the top cover of the weatherhead.

  ○ At the meter socket end, insert all three service entrance conductors and push them up through the raceway until approximately 3 feet of the conductors come out of the top.

  ○ At the weatherhead end, insert the two black ungrounded conductors and the white tape identified neutral conductor through separate holes in the weatherhead.

  ○ Install the cable hook at a proper location in relation to the weatherhead of the service raceway.

  ○ Drill a hole for the service entrance raceway to enter the house and be attached to the service entrance main panel enclosure.

  ○ Remove the knockout from the bottom of the meter socket that will be used to attach the service raceway to the bottom of the meter socket.

- ○ Install the service raceway between the meter socket and the service panel.
  - ■ Use LB fittings to make the 90° turn through the drilled hole and into the house.
- ○ Measure and cut the service entrance conductors to go from the bottom of the meter socket to the service panel enclosure.
- ○ Install the conductors between the meter socket and the service panel.
- ○ Make the meter socket connections. Connect the conductors from the weatherhead end to the line-side lugs. Connect the conductors from the service panel end to the load-side lugs.
- ○ Make the proper connections in the service entrance panel.
- Installing an overhead service using a mast
  - ○ Install a threaded hub to the meter socket.
  - ○ Locate and install the meter socket.
  - ○ Cut a hole in the roof aligned with the threaded hub of the meter socket.
  - ○ Measure and cut the electrical conduit to go from the meter socket up through the hole in the roof to a height that will keep the service drop conductors at the minimum required clearance from grade.
  - ○ Insert the mast pipe through the hole you cut in the roof and raise the mast into position.
  - ○ Install the roof flashing and weather collar (rubber boot).
  - ○ Measure and cut the length of service entrance conductor wire needed.
  - ○ At the meter socket end, insert all three service conductors and push them up through the raceway until approximately 3 feet of the conductors come out of the top of the mast.
  - ○ Remove the top of the weatherhead and install the weatherhead body on the top of the mast.
  - ○ Install the porcelain standoff onto the mast and align it in the direction from which the service drop will be coming.
  - ○ The installation of a mast style service entrance from the meter socket to the service entrance panel is the same as the installation steps already outlined.

Objective: Identify common underground service entrance equipment and materials.

Objective: Demonstrate an understanding of common installation techniques for underground services.

3) Identify underground service equipment and materials

4) Understand the installation techniques for underground services

- Underground service equipment
  - ○ An underground service entrance will consist of many of the same kinds of equipment and materials used in an overhead service installation.
  - ○ There are some pieces of equipment needed for underground service that are not needed for an overhead service.
    - ■ Meter socket for an underground service
      - • Must be large enough to allow a minimum of two service entrance raceways to enter the bottom of the socket: one raceway for the incoming electrical power, and one raceway to come back out of the meter socket to the service panel.

○ Service raceways used in an underground service.

■ Where the underground service conductors exit the trench at the meter socket side or at the transformer side, Section 300.5(D)(4) requires the conductors to be protected from physical damage by using:

- Rigid metal conduit (RMC)

- Intermediate metal conduit (IMC)

- Schedule 80 rigid PVC conduit (PVC)

○ Underground service conductors

■ Two ways to install underground service entrance conductors:

- Using Type USE service entrance cable, which can be buried at a depth required by the local electric utility and the *NEC*®.

- Running a raceway for the entire length of the service lateral and pulling in suitable conductors.

- Because a conduit located underground is considered a wet location by the *NEC*®, the insulation on conductors pulled into such a conduit must have a "W" as part of their insulation letter designation.

○ Underground raceway bushings

■ Section 300.5(H) requires a bushing to be placed on the end of a raceway that ends underground and provides protection for the underground service entrance conductors emerging from the ground.

○ Marking tape

■ Section 300.5(D)(3) requires a warning ribbon to be used when installing an underground service.

Objective: Demonstrate an understanding of voltage drop in underground service laterals.

5) Understand voltage drop in underground service laterals

Key terms: voltage drop

- The length of underground service conductors can be very long.

- This can result in the voltage at the house being much lower than what it is at the utility transformer.

- Voltage drop is the amount of voltage needed to "push" the house electrical load current through the service conductors from the transformer, to the service panel, and back to the transformer.

- The *NEC*® recommends a voltage drop of no more than 3%.

- To compensate for voltage drop, larger underground service conductors are installed.

Objective: Demonstrate an understanding of service panel installation techniques.

6) Understand service panel installation techniques

- Article 110 of the *NEC*® provides specifications for working space around electrical equipment

- Section 110.26(A)(1) tells us that the depth of the working space in front of residential electrical equipment must not be less than that specified in Table 110.26(A)(1)

- Service panel installation
  - Since the most common location for the service entrance panel is inside the house, discussion on the installation of the service entrance panel will be limited to an inside location.
  - Fasten the service panel in place.
  - Attach the service conductors from the meter socket to the service panel.
  - Connect both the ungrounded (hot) service entrance conductors to the main circuit breaker terminals.
  - Connect the grounded neutral service entrance conductor to the proper terminal on the neutral bus bar.
  - Bond the enclosure to the equipment grounding bus bar and the neutral bus bar using the main bonding jumper.
  - Connect the grounding electrode to the proper terminal on the neutral bus. Route this conductor to the grounding electrode.

Objective: Demonstrate an understanding of subpanel installation techniques.

7) Understand subpanel installation techniques

- Subpanels
  - Subpanels have some specific design differences as compared to a panel used as the main service panel.
    - Separate grounding bar and separate grounded (neutral) bar.
    - Four-conductor feeder cable used to provide electrical power to a subpanel.
    - They do not normally have a main circuit breaker.
    - Ungrounded feeder conductors are terminated at the panel's main lugs.
- Subpanel installation
  - The installation procedures for a subpanel are similar to those of a main service panel but there are a few differences.
  - Fasten the subpanel in place.
  - Run the proper size feeder cable from the main service panel to the subpanel location.
  - In the main service panel:
    - Connect both ungrounded (hot) feeder conductors to the two-pole circuit breaker terminals.
    - Connect both the feeder grounded neutral conductor and the feeder grounding (bare) conductor to the grounded neutral bus bar.
  - In the subpanel:
    - Connect both ungrounded (hot) feeder conductors to the main lug terminals.
    - Connect the grounded neutral feeder conductor to the neutral bus bar.
    - Connect the grounding (bare) conductor of the feeder to the equipment grounding bar.

Objective: Demonstrate an understanding of service entrance upgrade techniques.

8) Understand service upgrade techniques

Key terms: backfeeding

- Service entrance upgrading

  ○ Upgrading an existing service entrance requires working closely with the local electric utility.

  ○ The electrician will typically need to continue electrical service to the house until the switchover from old service to new service is accomplished.

  ○ Do this by running a jumper cable from a load-side circuit breaker in the old service entrance main panel to a load-side circuit breaker in the new service entrance panel.

  ○ Called "backfeeding," this procedure should be done only by an experienced electrician.

Quizzes

SAFETY

1. When a residential service entrance is upgraded, backfeeding the existing panel should be done only by an experienced electrician. True or False?

2. If aluminum wire is being used as service conductors or feeder conductors, be sure to use an antioxidant where required and follow the instructions that come with the antioxidant. True or False?

3. Proper personal protective equipment for an electrician when installing a service entrance includes only safety glasses and a hard hat. True or False?

CODE/REGULATIONS/STANDARDS

1. When mounting the service entrance panel, make sure that there will be a working space in front of the panel of at least _____ inches wide, _____ feet high, and 3 feet deep measured from the panel front.
   a. 24, 6
   b. 30, 6½
   c. 36, 7

2. If a metal water pipe is used as a grounding electrode, a supplemental grounding electrode must be used only if the resistance to ground exceeds 25 ohms. True or False?

3. When using service entrance cable for the run between the weatherhead and the meter socket, the cable must be supported within _____ of both the weatherhead and the meter socket.
   a. 12 inches
   b. 18 inches
   c. 24 inches

4. When installing an underground service, a warning tape must be installed at least _____ above the buried service entrance conductors.
   a. 8 inches
   b. 12 inches
   c. 18 inches

5. When Type NM or Type SER is used as a feeder cable supplying power to a subpanel, the cables must be supported within _____ inches of both the main panel and the subpanel.

    a. 8

    b. 12

    c. 24

TOOLS

1. When mounting a meter socket on the side of a house, a _____ is typically used to make sure it is level and plumb on the top and bottom.

    a. plumb bob

    b. torpedo level

    c. tape measure

2. When installing a raintight cable connector into a threaded hub on the top of a meter socket, tighten it securely using an adjustable wrench or _____.

    a. pump pliers

    b. lineman pliers

    c. long-nose pliers

Chapter Quiz

1. Electrical metallic tubing is permitted to be used as a service raceway. True or False?

2. If a metal water pipe is used as a grounding electrode, a supplemental grounding electrode must be used only if the resistance to ground exceeds 25 ohms. True or False?

3. In a subpanel, the grounding bar and the neutral bar are electrically connected together so that the white grounded conductors and the bare or green grounding conductors can all terminate at the same location. True or False?

4. When connecting subpanel feeder conductors in the main panel, both the grounded neutral conductor and the grounding conductor are connected to the main panel grounded neutral bar. True or False?

5. A porcelain standoff is used to attach the service drop cable to the side of a house. True or False?

6. When installing a service entrance cable to the bottom of a meter socket enclosure located on the side of a house, you must use a raintight connector like the one used to connect the cable to the threaded hub. True or False?

7. The conductors that run overhead from the utility transformer to the residence are called the _____.

    a. service entrance cables

    b. service drop conductors

    c. service laterals

    d. SEU cables

8. _____ is a common wiring method used to bring the electrical power from the point of attachment to the meter socket and then on to the main service equipment.

    a. Triplex cable

    b. Service drop conductors

    c. Service laterals

    d. Service entrance cable

9. When installing an overhead service, _____ feet of conductor should stick out of the weatherhead for the formation of a drip loop.

10. Where underground service conductors exit the trench to run up to a meter socket on the side of a house, _____ is not permitted to be used to provide the conductors with protection from physical damage.
    a. rigid metal conduit
    b. intermediate metal conduit
    c. electrical metallic tubing
    d. schedule 80 rigid PVC conduit

11. The *NEC*® provides the requirements for working clearances around electrical equipment in Article _____.

12. The grounding electrode conductor must be attached to a metal water pipe grounding electrode no more than _____ feet from the point where the metal water pipe enters the house.

13. When Type NM or Type SER is used as a feeder cable supplying power to a subpanel, the cables must be supported within _____ inches of both the main panel and the subpanel.

14. The working space in front of residential service equipment as stated in *NEC*® Section 110.26 is _____ feet.

15. Outgoing electrical power is connected to the _____ side of electrical equipment.

16. If aluminum service conductors are being used, make sure a(n) _____ is used at each termination point to prevent oxidation of the conductors.

17. A(n) _____ is a surface on which a service panel or subpanel is mounted. It is usually made of plywood and is often painted a flat black color.

18. Explain the purpose of a weatherhead when used on either a service raceway or service entrance cable.

    _____

    _____

    _____

    _____

19. Explain what an M.L.O. panel is.

    _____

    _____

    _____

    _____

20. Explain the purpose of a porcelain standoff fitting.

    _____

    _____

    _____

    _____

Troubleshooting/Quality Exercises

DISCUSSION QUESTIONS

1. Some electric utilities supply the meter enclosure to the electrician. If you are working in an area and are not sure whether you supply the meter enclosure or whether the utility company supplies the meter enclosure, how would you find out?

_____

_____

_____

_____

2. Some homes have an attached garage and the service entrance is installed on the side of the garage. Discuss the different ways that the main service disconnect can be installed.

_____

_____

_____

_____

3. Discuss what common wiring practice is followed in areas of the country where the house building style does not provide a good location for the service equipment inside the house.

_____

_____

_____

_____

HANDS-ON APPLICATIONS

1. Match the following terms to their definitions. Write the correct number for the term in the blanks.

   1. Voltage drop
   2. Backboard
   3. Threaded hub
   4. Sill plate

   5. Mast kit
   6. Cable hook
   7. Load side
   8. Line side

   _____ a. The surface on which a service panel or subpanel is mounted

   _____ b. A piece of equipment that will help keep water from entering the hole in the side of a house that the service entrance cable from the meter socket to the service panel goes through

   _____ c. The part used to attach the service drop cable to the side of a house

   _____ d. The location in electrical equipment where the incoming electrical power is connected

   _____ e. The location in electrical equipment where the outgoing electrical power is connected

   _____ f. A package of additional equipment that is needed for the installation of a mast-type service entrance

   _____ g. The piece of equipment that is attached to the top of a meter socket so that a raceway or cable connector can be attached

   _____ h. The amount of voltage that is needed to "push" the house electrical load current through the wires from the transformer, to the service panel, and back to the transformer

REAL-WORLD SCENARIOS

1. Sometimes it is necessary to mount the meter socket on the side of a house before any outside siding has been installed. What can you use to make sure the meter socket sets out from the wall surface so that it is easier for the siding to be installed around the meter socket?

_____

_____

_____

_____

2. As long as the AHJ does not object, electricians sometimes use URD cable as the service entrance conductors in an overhead service entrance installed with conduit. Why?

_____

_____

_____

_____

3. Even though the *NEC*® recommends a voltage drop of no more than 3% for a branch circuit or feeder, electricians usually do not worry too much about this in residential wiring. Why not?

_____

_____

_____

_____

4. Main-lug-only panels used as subpanels in residential wiring do not come with an equipment grounding terminal bar already installed. What must an electrician do if this is the case and a separate equipment grounding bar is required?

_____

_____

_____

_____

Chapter 9

General Requirements for Rough-In Wiring

Outline

Objective: Discuss the selection of appropriate wiring methods, conductor types, and electrical boxes for a residential electrical system rough-in.

Objective: List several general requirements that apply to wiring methods, conductors, and electrical boxes installed during the rough-in stage of a residential wiring system.

1) Discuss the selection of wiring methods, conductor types, and electrical boxes

2) List several general requirements for wiring methods, conductors, and electrical box installation

Key terms: deteriorating agents, knockout plug, rough-in wiring

- General wiring requirements:
  - Chapter 2 covered many types of equipment and material used to install a residential electrical system.
  - The majority of residential electrical system wiring is done using nonmetallic sheathed cable (Type NM).
  - Type NM cable may not be allowed in some areas of the country and the rough-in wiring is usually installed using Type AC cable, Type MC cable, or EMT.
- Article 110 of the *NEC*® covers many of the requirements for the installation of the rough-in wiring.
  - Section 110.3(B) states that listed or labeled equipment must be installed and used in accordance with any instructions included in the listing or labeling.
  - Section 110.7 addresses wiring integrity.
    - Completed wiring installations must be free from short circuits and from ground faults.
  - Section 110.11 covers deteriorating agents and states that unless identified for use in the operating environment, no conductors or equipment can:
    - Be located in damp or wet locations.
    - Be exposed to gases, fumes, vapors, liquids, or other agents that have a deteriorating effect on the conductors or equipment.
    - Be exposed to excessive temperatures.
  - Section 110.12 states that all electrical equipment must be installed in a neat and workmanlike manner.
  - Section 110.12(A) covers unused openings and states that any unused cable or raceway openings in boxes, cabinets, or other electrical equipment must be effectively closed. A common piece of equipment used to meet this requirement is a knockout plug.
  - Section 110.12(B) requires that any internal parts of electrical equipment must not be damaged or contaminated by foreign materials such as paint, plaster, cleaners, abrasives, or corrosive residues.

Objective: Demonstrate an understanding of general requirements for wiring as they apply to residential rough-in wiring.

3) Understand general requirements for rough-in wiring

Key terms: inductive heating, Sheetrock®

- Wiring methods – Article 300

  ○ Section 300.3(A) states that single conductors with an insulation type listed in *NEC®* Table 310.104(A) can be installed only as part of an *NEC®*-recognized wiring method.

  ○ Section 300.3(B) states all conductors of the same circuit and, where used, the grounded conductor and all equipment grounding conductors and bonding conductors must be contained within the same raceway, trench, cable, or cord. This is designed to eliminate inductive heating.

  ○ Section 300.4 states that where a wiring method is subject to physical damage, conductors must be adequately protected. If a cable or raceway is going to be installed through a wood framing member, follow these rules:

    ▪ In both exposed and concealed locations, where a cable or raceway-type wiring method is installed through bored holes in joists, rafters, or wood members, holes must be bored so that the edge of the hole is not less than 1¼ inches (32 millimeters) from the nearest edge of the wood member.

    ▪ Where there is no objection because of weakening the building structure, in both exposed and concealed locations, cables or raceways are permitted to be laid in notches in wood studs, joists, rafters, or other wood members where the cable or raceway at those points is protected against nails or screws by a steel plate at least ¹⁄₁₆ inch (1.6 millimeters) thick installed before the building finish is applied. A listed and marked steel plate can be less than ¹⁄₁₆ inch (1.6 millimeters) if it still provides equal or greater protection against nail or screw penetration.

  ○ Section 300.4(B) covers the use of nonmetallic sheathed cables and electrical nonmetallic tubing (ENT) through metal framing members. If a Type NM cable or ENT raceway is going to be installed through a metal framing member, follow these rules:

    ▪ In both exposed and concealed locations where nonmetallic sheathed cables pass through either factory or field punched, cut, or drilled slots or holes in metal members, protect the cable with listed bushings or listed grommets covering all metal edges securely fastened in the opening prior to installation of the cable.

    ▪ Where nails or screws are likely to penetrate nonmetallic sheathed cable or electrical nonmetallic tubing, use a steel sleeve, steel plate, or steel clip not less than ¹⁄₁₆ inch (1.6 millimeters). A listed and marked steel plate can be less than ¹⁄₁₆ inch (1.6 millimeters) if it still provides equal or greater protection against nail or screw penetration.

  ○ Section 300.4(C) addresses the installation of cables or raceways in spaces behind panels.

    ▪ Support cables or raceway-type wiring methods, installed behind panels designed to allow access, according to their applicable articles.

  ○ Section 300.4(D) covers the requirements for installing cables and raceways parallel to framing members and furring strips.

    ▪ Maintain the 1¼ inches (32 millimeters) distance along the framing member.

  ○ Section 300.4(F) covers cables and raceways installed in shallow grooves.

    ▪ Protect cable or raceway-type wiring methods installed in a groove, and to be covered by wallboard, siding, paneling, carpeting, or similar finish, with a ¹⁄₁₆ inch (1.6 millimeters) thick steel plate or sleeve; or with not less than 1¼ inches (32 millimeters) free space for the full length of the groove in which the cable or raceway is installed. A listed and marked steel plate can be less than ¹⁄₁₆ inch (1.6 millimeters) if it still provides equal or greater protection against nail or screw penetration.

- ○ Section 300.10 covers the electrical continuity of metal raceways and enclosures.

  - ▪ Metallically join together into continuous electric conductor metal raceways, cable armor, and other metal enclosures for conductors.

  - ▪ Connect to all boxes, fittings, and cabinets to provide effective electrical continuity.

- ○ Section 300.11(A) states that raceways, cables, boxes, cabinets, and fittings must be securely fastened in place.

- ○ Section 300.11(B) states raceways must be used only as a means of support for other raceways, cables, or nonelectric equipment under certain conditions.

  - ▪ Where the raceway or means of support is identified for the purpose.

  - ▪ Where the raceway contains power supply conductors for electrically controlled equipment and is used to support Class 2 circuit conductors or cables solely to connect to the equipment control circuits.

  - ▪ Where the raceway is used to support boxes or conduit bodies in accordance with 314.23 or to support luminaires in accordance with Article 410.

- ○ Section 300.11(C) states cable wiring methods must not be used as a means of support for other cables, raceways, or nonelectric equipment.

- ○ Section 300.12 requires all metal or nonmetallic raceways, cable armors, and cable sheaths to be continuous between cabinets, boxes, fittings, or other enclosures or outlets.

- ○ Section 300.14 covers the length of free conductors at outlets, junctions, and switch points.

  - ▪ At least 6 inches (150 millimeters) of free conductor, measured from the point in the box where it emerges from its raceway or cable sheath, must be left at each outlet, junction, and switch point for splices or the connection of luminaires or devices.

- ○ Section 300.15 states that where the wiring method is conduit, tubing, Type AC cable, Type MC cable, or nonmetallic sheathed cable, install a box or conduit body at each conductor splice point, outlet point, switch point, junction point, termination point, or pull point.

- ○ Section 300.21 covers the spread of fire or products of combustion.

  - ▪ Make electrical installations in hollow spaces, vertical shafts, and ventilation or air-handling ducts so that the possible spread of fire or products of combustion (i.e., smoke) will not be substantially increased.

- ○ Section 300.22(C) addresses wiring in spaces used for environmental air handling, such as those found in forced hot-air furnace systems in dwelling units.

Objective: Demonstrate an understanding of general requirements for conductors as they apply to residential rough-in wiring.

4) Understand general requirements for conductors

- • General Requirements for Conductors – Article 310

  - ○ Section 310.106(A) addresses the minimum size of conductors.

    - ▪ Minimum size of conductors is shown in *NEC*® Table 310.106(A).

  - ○ Section 310.10(F) states conductors used for direct burial applications must be of a type identified for such use.

- Section 310.10(D) addresses those locations exposed to direct sunlight.

  - Insulated conductors and cables used where exposed to the direct rays of the sun must be of a type listed for sunlight resistance or listed and marked "sunlight resistant." Tapes or sleeving, listed or marked as sunlight resistant, may also be used.

- Section 310.15(A)(3) covers the temperature limitation of conductors.

  - No conductor can be used in such a manner that its operating temperature exceeds that designated for the type of insulated conductor involved.

- Section 310.120(A) states that all conductors and cables must be marked to indicate certain information.

  - Maximum rated voltage.

  - Proper type letter or letters for the type of wire or cable.

  - Manufacturer's name, trademark, or other distinctive marking by which the organization responsible for the product can be readily identified.

  - AWG size or circular mil area.

  - Cable assemblies where the neutral conductor is smaller than the ungrounded conductors.

- Section 310.110(A) states that insulated grounded conductors must be identified in accordance with Section 200.6.

  - A grounded conductor of 6 AWG or smaller is to be identified by a white or gray color along its entire length.

  - An alternative method of identification is described as "three continuous white stripes on other than green insulation along the conductor's entire length."

  - An insulated grounded conductor larger than 6 AWG must be identified either by a continuous white or gray outer finish, three continuous white stripes on other than green insulation along its entire length, or by a distinctive white marking at its terminations.

- Section 310.110(B) states that equipment grounding conductor identification must be in accordance with 250.119.

  - Equipment grounding conductors are permitted to be bare or insulated.

  - Individually insulated equipment grounding conductors of 6 AWG or smaller must have a continuous outer finish that is either green or green with one or more yellow stripes.

  - For equipment grounding conductors larger than 6 AWG, it is permitted, at the time of installation, to be permanently identified as an equipment grounding conductor at each end and at every point where the conductor is accessible.

- Table 310.104(A) (Conductor Applications and Insulations) lists and describes insulation types recognized by the *NEC*®.

- Section 310.15 covers the ampacities for conductors rated 0–2000 volts.

- Table 310.15(B)(16) is used to determine the ampacity of a conductor used in residential wiring.

Objective: Demonstrate an understanding of general requirements for electrical box installation as they apply to residential rough-in wiring.

5) Understand general requirements for electrical box installation

- Installation and Use of Boxes as Outlet, Device, or Junction Boxes – Article 314

  o Section 314.16 provides guidelines for calculating the maximum number of conductors in outlet, device, and junction boxes.

    ▪ Boxes must be of sufficient size to provide free space for all enclosed conductors.

  o Section 314.17 states any conductors entering electrical boxes must be protected from abrasion.

  o Section 314.17(B) applies to metal boxes and states that where a raceway or cable is installed with metal boxes, raceway or cable must be secured to boxes.

  o Section 314.17(C) applies to nonmetallic boxes and states that nonmetallic boxes must be suitable for the lowest temperature-rated conductor entering the box.

  o Section 314.20 requires boxes installed in walls or ceilings with a surface of concrete, tile, gypsum (Sheetrock®), plaster, or other noncombustible material, be installed so that the front edge of the box will not be set back from the finished surface more than ¼ inch (6 millimeters).

  o Section 314.23(A) states that an electrical box mounted on a building or other surface must be rigidly and securely fastened in place.

  o Section 314.23(B) says that a box supported from a structural member of a building must be rigidly supported either directly or by using a metal, polymeric, or wood brace.

  o 314.23(C) allows mounting an electrical box in a finished surface as long as it is rigidly secured by clamps, anchors, or fittings identified for the application.

  o Section 314.27(A) states boxes used at luminaire outlets must be designed for the purpose.

  o Section 314.27(B) allows outlet boxes to support luminaires weighing no more than 50 pounds (23 kilograms).

  o 314.27(B) states boxes listed specifically for floor installation must be used for receptacles located in the floor.

  o 314.27(C) states where a box is used as the sole support for a ceiling-suspended paddle fan, the box must be listed for the application and for the weight of the fan to be supported.

  o Section 314.29 states boxes must be installed so that the wiring contained in them can be rendered accessible without removing any part of the building.

- Required Receptacle Outlets – Section 210.52

  o Section 210.52(A) states that in every kitchen, family room, dining room, living room, parlor, library, den, sunroom, bedroom, recreation room, or similar room or area of dwelling units, receptacle outlets rated 125-volt, 15- or 20-amp, must be installed in accordance with the general provisions specified.

    ▪ Receptacles must be installed so that no point measured horizontally along the floor line in any wall space is more than 6 feet (1.8 meters) from a receptacle outlet.

    ▪ Any wall space 2 feet (600 millimeters) or more in width requires a receptacle outlet.

- ○ Section 210.52(C) covers the required location and requirements of receptacles at countertop locations in a dwelling room, kitchen, pantry, breakfast room, or dining room.

  - A receptacle outlet must be installed at each wall counter space that is 12 inches (300 millimeters) or wider.

  - Receptacle outlets must be installed so that no point along the wall line is more than 24 inches (600 millimeters) measured horizontally from a receptacle outlet in that space.

  - At least one receptacle outlet must be installed at each island countertop space with a long dimension of 24 inches (600 millimeters) or greater and a short dimension of 12 inches (300 millimeters) or greater.

  - At least one receptacle outlet must be installed at each peninsular countertop space with a long dimension of 24 inches (600 millimeters) or greater and a short dimension of 12 inches (300 millimeters) or greater.

  - Countertop spaces separated by range tops, refrigerators, or sinks are considered as separate countertop spaces.

  - Receptacle outlets must be located above, but not more than 20 inches (500 millimeters) above the countertop.

- ○ Section 210.52(D) requires one wall receptacle in each bathroom of a dwelling unit to be installed adjacent to and within 36 inches (900 millimeters) of the sink or installed on the side or face of the bathroom cabinet not more than 12 inches (300 millimeters) below the countertop.

- ○ Section 210.52(E) states that for a one-family dwelling and each unit of a two-family dwelling that is at grade level, at least one receptacle outlet accessible while standing at grade level and not more than 6.5 feet (2 meters) above grade must be installed at the front and back of the dwelling. Balconies, decks, and porches that are accessible from inside the house must have at least one receptacle.

- ○ Section 210.52(F) requires at least one receptacle outlet to be installed for the laundry.

  - A 20-ampere branch circuit, which can have no other outlets on the circuit, supplies the laundry receptacle outlet(s).

- ○ Section 210.52(G) states that for a one-family dwelling, at least one receptacle outlet, in addition to any provided for laundry equipment, must be installed in each basement and in each attached garage, and in each detached garage with electric power.

- ○ Section 210.52(H) requires that in dwelling units, hallways of 10 feet (3 meters) or more in length must have at least one receptacle outlet.

- ○ Section 210.52(I) requires a receptacle be installed in each wall space 3 feet (900 mm) or more in width in a foyer that has an area of more than 60 square feet.

- Required Lighting Outlets – Section 210.70

  - ○ Section 210.70(A)(1) requires at least one wall switch-controlled lighting outlet be installed in every habitable room and bathroom.

    - *Exception No. 1* to the general rule allows one or more receptacles controlled by a wall switch to be permitted in lieu of lighting outlets, but only in areas other than kitchens and bathrooms.

    - *Exception No. 2* allows lighting outlets to be controlled by occupancy sensors under some circumstances:

      - When they are in addition to wall switches.

      - When located at a customary wall switch location and equipped with a manual override allowing the sensor to function as a wall switch.

⚬ Section 210.70(A)(2) lists three additional locations where lighting outlets need to be installed.

■ In hallways, stairways, attached garages, and detached garages with electric power.

■ In attached garages and detached garages with electric power, to provide illumination on the exterior side of outdoor entrances or exits with grade-level access.

■ Where one or more lighting outlet(s) are installed for interior stairways, there must be a wall switch at each floor level, and landing level that includes an entryway, to control the lighting outlet(s) where the stairway between floor levels has six risers or more.

● An exception states that in hallways, stairways, and at outdoor entrances, remote, central, or automatic control of lighting is permitted.

⚬ Section 210.70(A)(3) addresses storage or equipment spaces.

■ In attics, crawl spaces, utility rooms, and basements, at least one lighting outlet containing a switch or controlled by a wall switch must be installed where these spaces are used for storage or contain equipment requiring servicing.

● Required communications outlet – Section 800.156

⚬ A minimum of one communications outlet must be installed in all newly constructed dwelling units.

Quizzes

CODE/REGULATIONS/STANDARDS

1. Boxes for floor receptacles must be specifically listed for that purpose and no other box type can be used in a floor installation. True or False?

2. A receptacle controlled by a wall switch is permitted to serve as a lighting outlet in kitchens. True or False?

3. When holes are bored through wood framing members for the purpose of installing an electrical cable, the holes must be bored so the edge of the hole is not less than _____ from the nearest edge of the wood framing member.
   a. 1 inch
   b. 1¼ inches
   c. 1½ inches

4. Where the opening to an outlet, junction, or switch point is less than 8 inches in any dimension, each conductor must be long enough to extend at least _____ outside the opening.
   a. 3 inches
   b. 6 inches
   c. 8 inches

5. Lighting fixtures weighing no more than _____ are permitted to be supported by outlet boxes.
   a. 35 pounds
   b. 45 pounds
   c. 50 pounds

6. Receptacles must be installed so that no point measured along the floor line in any wall space is more than _____ from a receptacle.

   a. 3 feet

   b. 6 feet

   c. 10 feet

7. Hallways of _____ or more in length must have at least one receptacle outlet.

   a. 6 feet

   b. 10 feet

   c. 12 feet

8. The _____ stage is when the raceways, cable, boxes, and other electrical equipment are installed.

   a. planning and preparation

   b. trim-out

   c. rough-in

9. The term used by electricians to describe the process of installing a residential electrical system is _____.

10. Where cables and raceways are installed less than 1¼ inches from the edge of framing members, a metal plate at least _____ thick must be installed to protect the cables and raceways. (general rule)

Chapter Quiz

1. An electrician discovers that a two-wire cable has been run instead of a three-wire cable. To remedy the problem, the electrician runs a single insulated conductor along with the two-wire cable. This method is acceptable according to the *NEC®*. True or False?

2. An electrician must cut a groove into a wood beam to run Type NM cable to a certain location. If the groove is cut into the beam 1⅛ inches, a steel plate at least 1¹⁄₁₆ inch thick is required to protect the cable. True or False?

3. Boxes for floor receptacles must be specifically listed for that purpose and no other box type can be used in a floor installation. True or False?

4. A receptacle controlled by a wall switch is permitted to serve as a lighting outlet in kitchens. True or False?

5. _____ heating will occur when current-carrying conductors of a circuit are brought through separate holes in a metal box or enclosure.

   a. Invective

   b. Inductive

   c. Convective

   d. Conductive

6. When holes are bored through wood framing members for the purpose of installing an electrical cable, the holes must be bored so the edge of the hole is not less than _____ inches from the nearest edge of the wood framing member.

7. Type _____ cable should be used when running a cable underground to an outside pole light.

8. Which of the following colors is *not* acceptable for an ungrounded conductor?

   a. brown

   b. yellow

   c. orange

   d. gray

9. Where nonmetallic sheathed cable is used with a single-gang nonmetallic device box, the cable sheath must extend inside the box not less than _____ inch(es).

10. Where device boxes are installed in walls or ceilings that have a *noncombustible* surface, the edge of the box must _____ the wall or ceiling surface.
    a. not be set back farther than ⅛ inch from
    b. not be set back farther than ¼ inch from
    c. not be set back farther than ½ inch from
    d. be flush with

11. When device boxes are installed in walls or ceilings with a *combustible* surface such as wood, the edge of the boxes must _____ the finished surface.
    a. not be set back farther than ⅛ inch from
    b. not be set back farther than ¼ inch from
    c. not be set back farther than ½ inch from
    d. be flush with

12. Receptacles that are installed to meet *NEC*® Section 210.52 placement requirements are not permitted to be more than _____ feet above the floor.
    a. 4.5
    b. 5
    c. 5.5
    d. 6

13. Receptacles must be installed so that no point measured along the floor line in any wall space is more than _____ from a receptacle.

14. Individual wall spaces of _____ feet or more in width are considered usable for the location of a lamp or appliance, and a receptacle outlet is required.

15. At least one receptacle outlet must be installed at each peninsular countertop space with a long dimension of _____ inches or greater and a short dimension of _____ inches or greater.

16. Receptacle outlets must be installed above, but not more than _____ above the kitchen countertop.

17. The required wall receptacle in each bathroom of a dwelling unit must be installed adjacent to, and within _____ inches of the sink. (general rule)

18. Hallways of _____ feet or more in length must have at least one receptacle outlet.

19. Where one or more lighting outlet(s) are installed for interior stairways, there must be a wall switch to control the lighting outlet(s) at each floor level and landing, where the stairway between floor levels has _____ risers or more.

20. Receptacles must be placed above a kitchen countertop so that no more than a measurement of _____ feet is between each receptacle.

Troubleshooting/Quality Exercises

DISCUSSION QUESTIONS

1. From time to time, depending on the conductor size, an electrician has to re-identify a conductor as a grounded conductor or a grounding conductor. Wrapping the conductor with green electrical tape is the most common way to re-identify a grounding conductor. White tape is used to re-identify a grounded conductor. Discuss the difference in the way a grounding conductor is re-identified with green tape and how a grounded conductor is re-identified with white tape.

_____

_____

_____

_____

HANDS-ON APPLICATIONS

1. Match the following 2011 *NEC*® sections to the topic they cover. Write the correct number for the section in the blanks.

   1. 110.3(B)
   2. 110.12(A)
   3. 300.3(B)
   4. 300.4(A)(1)
   5. 300.4(D)
   6. 300.11(A)
   7. 300.14
   8. 200.6(A)
   9. 314.16
   10. 314.20
   11. 314.27(A)
   12. 314.27(B)
   13. 210.52
   14. 210.70(A)(1)
   15. 800.156

   _____ a. States that all conductors of the same circuit must be contained in the same raceway, trench, cable, or cord

   _____ b. States that when cables are run parallel to framing members, there must be 1¼ inches minimum from the edge of the cable to the edge of the framing member

   _____ c. Requires at least one wall switch-controlled lighting outlet in every habitable room and bathroom

   _____ d. States that the minimum length of free conductor in a box is 6 inches

   _____ e. States that listed or labeled equipment must be installed according to instructions included with the listing or labeling

   _____ f. States that boxes specifically listed for floor installation must be used when installing a floor receptacle

   _____ g. States that raceways, cables, boxes, cabinets, and fittings must be securely fastened in place

   _____ h. States that any unused opening other than those for the operation of equipment or for mounting purposes must be closed

   _____ i. Tells an electrician where receptacle outlets must be installed in a dwelling unit

   _____ j. Allows outlet boxes to support luminaires weighing no more than 50 pounds

_____ k. Requires at least one communications outlet in all newly constructed dwelling units

_____ l. Permits a grounded conductor of 6 AWG or smaller to be identified by a white or gray color along its entire length

_____ m. States that bored holes must be made so there is a minimum distance of 1¼ inches from the edge of the hole to a cable

_____ n. Provides guidelines for the calculation of the maximum number of conductors allowed in an electrical box

_____ o. Requires boxes installed in a combustible wall or ceiling to be flush with the finished surface

REAL-WORLD SCENARIOS

1. Never throw away the instructions that come with any piece of electrical equipment. According to *NEC*® section _____, all electrical equipment must be installed according to the instructions that are included by the manufacturer of the equipment.

2. Many building contractors in charge of the actual framing of a house do not want electricians to "notch" framing members. Why is this?

   _____

   _____

   _____

   _____

3. An electrician will often use Type NM cable from a partially used roll. Typically, the package that the cable came in originally has been thrown away. How can you find out what the conductor size, maximum rated voltage, and letter type are for the cable?

   _____

   _____

   _____

   _____

Chapter 10

Electrical Box Installation

Outline

Objective: Select an appropriate electrical box type for a residential application.

1) Select an appropriate electrical box

- Selecting the appropriate electrical box type

  - To select the proper electrical box, check the electrical plans to determine what is being installed at a certain location.

  - If the symbol for a duplex receptacle is shown on the plans, a device box will need to be installed.

  - If the symbol for a ceiling-mounted incandescent lighting fixture is shown, an outlet box designed to accommodate a light fixture will have to be installed.

  - If the symbol for a junction box is shown, a box type that is designed to accommodate the proper number of conductors coming into that box will have to be installed.

  - Considerations when selecting an electrical device box. Does it need to be any of the following?

    - Single-gang

    - Two-gang

    - Three-gang or more

      - This information is found on the electrical plan.

  - Consider whether the box will be used in new or old work.

Objective: Size electrical boxes according to the *NEC*®.

2) Size electrical boxes

Key terms: pigtails, box fill

- Sizing electrical boxes

  - Once the proper box type is selected, make sure the box can accommodate the required number of conductors and other electrical equipment that will be placed in the box.

  - Become familiar with Tables 314.16(A) and 314.16(B).

  - Section 314.16 provides the requirements and identifies the allowances for the number of conductors permitted in a box.

  - Total box volume must be equal to or greater than total box fill.

  - Total box volume is determined by adding the individual volumes of the box components.

Objective: Demonstrate an understanding of the installation of metal and nonmetallic electrical device boxes in residential wiring situations.

3) Understand installing metal and nonmetallic electrical device boxes

Key terms: setout

- Installing nonmetallic device boxes

  o Most common method is to nail or screw them directly to a wood framing member.

  o Some styles come with a side-mounting bracket, which is nailed or screwed to a framing member.

- Installing metal device boxes

  o The best way to attach a metal device box with nails or screws is to keep the nails or screws outside of the box.

  o Some are attached to framing members using a side-mounting bracket.

  o There are times when installing a box directly to a stud, joist, or rafter does not put it in a location that is called for in the electrical plan. You may need to mount it so it is actually positioned between two studs, joists, or rafters.

    ▪ In this case, wood or metal strips can be placed between the framing members and the electrical box can be secured to them.

- Installing handy (utility) boxes

  o Box type classified by the *NEC*® as a device box.

  o Normally mounted directly on the surface of a framing member or existing wall.

  o Often used in places where a finished wall surface is not going to be used, like an unfinished basement, unfinished garage, or in some agricultural buildings.

Objective: Demonstrate an understanding of the installation of lighting outlet and junction boxes in residential wiring situations.

4) Understand installing lighting outlet and junction boxes

Key terms: lighting outlets, accessible

- o Installing lighting outlet and junction boxes

- o Lighting outlet boxes are round or octagonal.

- o Purely junction boxes are usually square and must always be accessible.

- o Lighting outlet boxes are mounted to the building framing members with side-mounting brackets or adjustable bar hangers.

- o Nails or screws are commonly used.

- o Make sure that a box for a ceiling-suspended paddle fan has been listed as suitable for this type of installation.

- o Use listed mounting kits when installing a lighting outlet in a suspended ceiling.

Objective: Demonstrate an understanding of the installation of electrical boxes in existing walls and ceilings.

5) Understand installing electrical boxes in existing walls and ceilings

Key terms: Madison hold-its

- Installing "old-work" boxes in existing walls and ceilings

  - To install "old-work" boxes, cut a hole into the wall or ceiling at the location where you wish to mount the box.

  - The plaster ears keep the boxes from going too far into the hole, but there has to be some other device that will keep the box from coming back out of the hole. Devices commonly used include:

    - Metal straps that spread open and do not allow the box to come back out of the hole.

    - Wallboard grips that are tightened to hold the box firmly in the wall or ceiling.

    - Madison hold-its (also known as Madison straps or Madison hangers), thin metal straps that are installed on each side of an "old-work" box to provide a secure attachment.

Quizzes

CODE/REGULATIONS/STANDARDS

1. When computing box fill, three equipment grounding wires are counted as _____ conductor(s) of the largest conductor in the box.

2. A _____ is a short length of wire used in electrical boxes to make connections to device terminals.

3. When computing box fill, a receptacle is counted as _____ conductors of the largest size wire connected to it.

4. When installing electrical boxes above a dropped ceiling, it is okay to have the boxes supported by the same wires that hold up the dropped ceiling. True or False?

5. When computing box fill, which of the following would count as two conductors?
   a. Single-pole switch
   b. External cable clamp
   c. Neither of the above

6. In which of the following *NEC®* tables would you find information on metal box fill?
   a. Table 220.12
   b. Table 314.16(A)
   c. Table 250.66

7. For each looped or coiled, unbroken length of conductor at least 12 inches long that passes through a box without splice or connection, the volume allowance is _____ of the actual conductor size.
   a. 1
   b. 2
   c. 3

8. A 3 × 2 × 3 ½-inch metal device box can hold a maximum of _____ 12 AWG conductors.
   a. 10
   b. 9
   c. 8

9. A plastic nail-on single-gang device box has a volume of 24 cubic inches. What is the maximum number of 14 AWG conductors it can hold?

    a. 12

    b. 14

    c. 18

TOOLS

1. A _____ saw is often used to cut holes in lath and plaster or plasterboard walls when installing old-work electrical boxes.

Chapter Quiz

1. Nonmetallic device boxes used in residential wiring typically have both their cubic inch volume and the maximum number of 14, 12, and 10 AWG conductors marked on the box. True or False?

2. The *NEC*® requires electrical boxes to be mounted so that they are flush with the finished surface of walls made of either combustible or noncombustible materials. True or False?

3. A wire that originates inside a device box and does not leave the device box, such as a pigtail, counts as one conductor when computing box fill. True or False?

4. When installing electrical boxes above a dropped ceiling, it is okay to have the boxes supported by the same wires that hold up the dropped ceiling. True or False?

5. The number of conductors permitted in a box is determined by _____.

    a. the area of the box

    b. the perimeter of the box

    c. the volume of the box

    d. Underwriters Laboratories

6. In walls or ceilings made of noncombustible materials, electrical boxes must be mounted so they are _____ the finished wall surface.

    a. set back no more than ⅛ inch from

    b. set back no more than ¼ inch from

    c. set back no more than ½ inch from

    d. flush with

7. In residential wiring, which of the following is an area where handy boxes would *not* be used?

    a. unfinished basements

    b. unfinished garages

    c. outdoors

    d. storage buildings

8. Nonmetallic electrical boxes used for lighting outlets in residential wiring are _____.

    a. rectangular

    b. square

    c. octagonal

    d. round

9. Ceiling-suspended paddle fans that weigh more than _____ pounds may be supported by an outlet box as long as the box has the maximum weight it can support marked on it.

10. The size of the opening of a standard metal device box used to mount a switch or receptacle in a wall is _____.

    a. 2×4 inches

    b. 2×2 inches

    c. 4×4 inches

    d. 3×2 inches

11. In which of the following *NEC®* tables would you find information on box fill?

    a. Table 220.12

    b. Table 314.16(A)

    c. Table 250.66

    d. Table 8 in Chapter 9

12. The actual volume for a $4 \times 1\frac{1}{2}$-inch square box with a 21-cubic-inch volume with a single-gang $4 \times \frac{3}{4}$-inch raised plaster ring with a marked volume of $4\frac{1}{2}$ cubic inches attached to it is _____ cubic inches.

13. For each looped or coiled, unbroken length of conductor at least 12 inches long that passes through a box without splice or connection, the volume allowance is _____ of the actual conductor size.

14. A $3 \times 2 \times 3\frac{1}{2}$-inch metal device box can hold a maximum of _____ 12 AWG conductors.

15. A $4 \times 1\frac{1}{2}$-inch square box can hold a maximum of _____ 14 AWG conductors.

16. A metal 4-inch square box has ten 12 AWG conductors, four 12 AWG equipment grounding conductors, and four external clamps. What is the minimum size square box you could use?

    a. $4 \times 1\frac{1}{4}$-inch

    b. $4 \times 1\frac{1}{2}$-inch

    c. $4 \times 2\frac{1}{8}$-inch

    d. $4\frac{11}{16} \times 1\frac{1}{4}$-inch

17. Explain why lighting outlet boxes are permitted to be used as junction boxes.

_____

_____

_____

_____

18. Explain what Madison hold-its are used for.

_____

_____

_____

_____

19. Explain what "ganging" electrical boxes means.

_____

_____

_____

_____

20. Explain why an electrician would use an adjustable bar hanger to mount a ceiling electrical box.

_____

_____

_____

_____

Troubleshooting/Quality Exercises

DISCUSSION QUESTIONS

1. Nonmetallic electrical boxes used in residential wiring usually have both their cubic inch volume marked on them as well as the maximum number of 14, 12, and 10 AWG conductors allowed in the box. How does this information help the installing electrician?

_____

_____

_____

_____

2. Describe how you could mount a box between two studs if you did not have a factory-made mounting bracket.

_____

_____

_____

_____

HANDS-ON APPLICATIONS

1. Using Table 314.16(A) in the *NEC*® or Figure 10-4 in the textbook, answer the following:

Box Size	Volume	Maximum # of 12 AWG Conductors
4 × 1½-inch octagon		
4 × 2⅛ × 2⅛-inch device		
4 × 1½-inch square		
3 × 2 × 3½-inch device		
4 × 2⅛-inch square		

2. Based on Table 314.16(B) in the *NEC*® or Figure 10-4 in the textbook, a 20-cubic-inch nonmetallic single-gang plastic box can hold a maximum of:

_____ 14 AWG conductors

_____ 12 AWG conductors

_____ 10 AWG conductors

2. Based on Section 314.16, Table 314.16(A), and Table 314.16(B) in the *NEC®*, a 4 × 1½-inch square box with a ½-inch single-gang raised plaster ring with 5 cubic inches marked on it can hold a maximum of:

_____ 14 AWG conductors

_____ 12 AWG conductors

_____ 10 AWG conductors

REAL-WORLD SCENARIOS

1. There is no *NEC®* rule for the height of electrical boxes in residential wiring. Name the common heights from the floor to the center of the box for the following:

Location	Box Height
Box for a receptacle in a living room	
Box for a single-pole switch in a bedroom	
Box for a switch or receptacle above a kitchen countertop	
Box for a receptacle in a garage	
Box for an inside wall-mounted lighting fixture	

2. List the seven steps of the procedure for installing outlet boxes with a side-mounting bracket in a house. The procedure is outlined at the end of Chapter 10 in the textbook.

Step 1:

Step 2:

Step 3:

Step 4:

Step 5:

Step 6:

Step 7:

Chapter 11   Cable Installation

Outline

Objective: Select an appropriate cable type for a residential application.

1) Select an appropriate cable type

- Depends on
  - ○ Cable type allowed by the authority having jurisdiction in your area
  - ○ Specific wiring application
- Many residential building plans will state the required cable type either on the electrical plan or in the electrical specifications.

Objective: State several *NEC*® requirements for the installation of the common cable types used in residential wiring.

2) List several *NEC*® requirements for cable installation

Key terms: running boards, jug handle, redhead

- Requirements for cable installation: Type NM
  - ○ Section 334.10(1) states that Type NM cable is allowed to be used in one- and two-family homes, including their attached or detached garages and storage buildings.
  - ○ Section 334.15 states that when Type NM cable is run exposed and not concealed in a wall or ceiling, the cable must be installed as follows.
    - Cable must closely follow the surface of the building finish or be secured to running boards.
    - Cable must be protected from physical damage where necessary by rigid metal conduit, electrical metallic tubing, Schedule 80 rigid PVC conduit, guard strips, or other means.
  - ○ Where the cable is run across the bottom of joists in unfinished basements and crawl spaces, it is permissible to secure cables not smaller than two 6 AWG conductors or three 8 AWG conductors directly to the bottom edge of the joists.
    - Smaller cables must be run either through bored holes in joists or on running boards.
  - ○ Section 334.17 covers running Type NM cable through or parallel to framing members.
    - If run through a hole that has an edge closer than 1¼ inches (32 millimeters) to the edge of the framing member, a metal plate at least ¹⁄₁₆ inch (1.6 millimeters) thick must be used to protect that spot from nails or screws that could penetrate the cable sheathing. A listed and marked steel plate can be less than ¹⁄₁₆ inch (1.6 millimeters) thick if it still provides equal or greater protection.
    - If the cable is placed in a notch in wood studs, joists, rafters, or other wood members, a steel plate must protect the cable and be at least ¹⁄₁₆ inch (1.6 millimeters) thick.

- If run along a framing member, where the cable is closer than 1¼ inches (32 millimeters) to the edge, metal protection plates at least ¹⁄₁₆ inch (1.6 millimeters) thick must be used to protect the cable.

o Installing Type NM cable in accessible attics is covered in Section 334.23.

- If Type NM cable is run across the top of floor joists, or within 7 feet (2.1 meters) of an attic floor or floor joists, cable must be protected by guard strips at least as high as the cable.

- Where the attic space is not accessible by permanent stairs or ladders, protection is required only within 6 feet (1.8 meters) of the nearest edge of a scuttle hole or attic entrance.

o Section 334.24 states that bends in nonmetallic sheathed cable must be made so that the cable will not be damaged.

- The radius of the curve of the inner edge of any bend must not be less than five times the diameter of the cable.

- When a Type NM cable is brought through a hole in a building framing member, the cable must be formed in a jug handle to ensure that the cable is not bent too sharply.

o Section 334.30 addresses securing and supporting Type NM cable.

- Nonmetallic sheathed cable must be secured by staples, cable ties, straps, hangers, or similar fittings designed and installed so as not to damage the cable at intervals not exceeding 4½ feet (1.4 meters) and within 12 inches (300 millimeters) of every cabinet, box, or fitting.

- Cable must be secured. Simply draping the cable over air ducts, timbers, joists, pipes, and ceiling grid members is not permitted.

o Section 334.30 also says that flat cables, like 14/2 or 12/2 Type NM cable, must not be stapled on the edge.

o Section 334.30(A) states Type NM cables that run horizontally through framing members spaced less than 4½ feet (1.4 meters) apart and passing through bored or punched holes in framing members without additional securing are considered supported by the framing members.

o Section 334.30(B) allows Type NM cables to not be supported under certain conditions.

- When fished between access points, where concealed in finished buildings and supporting is impracticable (as with remodel electrical work).

- When there is not more than 4½ feet (1.4 meters) from the last point of support for connections within an accessible ceiling to luminaires.

- Requirements for cable installation: Type UF

o Article 340 allows Type UF cable to be used in residential wiring for direct burial applications or for installation in the same wiring situations where you would use nonmetallic sheathed cable.

o When installing Type UF cable as nonmetallic sheathed cable, follow the installation requirements of Article 334.

o Section 340.24 states that bends in Type UF cable must be made so that the cable is not damaged.

- The radius of the curve of the inner edge of any bend must not be less than five times the diameter of the cable.

- Requirements for cable installation: Type AC

o Section 320.15 requires exposed runs of Type AC cable (also referred to as BX cable) to closely follow the surface of the building finish or be secured to running boards.

o Exposed runs are permitted on the underside of joists, as in a basement or crawl space, where they are supported at each joist and located so as not to be subject to physical damage.

○ Section 320.17 allows Type AC cable to be run through or along framing members in a house and must be protected in accordance with Section 300.4 (same requirements as for Type NM cable).

- If run through a hole that has an edge closer than 1¼ inches (32 millimeters) to the edge of the framing member, use a metal plate at least ¹⁄₁₆ inch (1.6 millimeters) thick to protect that spot from nails or screws that could penetrate the cable sheathing. A listed and marked steel plate can be less than ¹⁄₁₆ inch (1.6 millimeters) thick if it still provides equal or greater protection.

- If placed in a notch in wood studs, joists, rafters, or other wood members, use a steel plate at least ¹⁄₁₆ inch (1.6 millimeters) thick to protect the cable.

- If run along a framing member where the cable is closer than 1¼ inches (32 millimeters) to the edge, use metal protection plates at least ¹⁄₁₆ inch (1.6 millimeters) thick.

○ If installed in accessible attics, Section 320.23 requires Type AC cable to be installed exactly the same way as nonmetallic sheathed cable.

○ Section 320.24 states that, like Type NM cable, bends in Type AC cable must be made so that the cable will not be damaged.

- The radius of the curve of the inner edge of any bend must not be less than five times the diameter of the Type AC cable.

○ Secure Type AC cable with staples, cable ties, straps, hangers, or similar fittings designed and installed so as not to damage the cable at intervals not exceeding 4½ feet (1.4 meters) and within 12 inches (300 millimeters) of every outlet box, junction box, cabinet, or fitting according to Section 320.30.

○ Section 320.30(C) addresses Type AC being run horizontally through holes and notches in framing members.

○ In other than vertical runs, Type AC cables installed in accordance with Section 300.4 are to be considered supported and secured under the following conditions:

- Support intervals do not exceed 4½ feet (1.4 meters).

- Armored cable is securely fastened in place by an approved means within 12 inches (300 millimeters) of each box, cabinet, conduit body, or other armored cable termination.

○ According to 320.30(D), Type AC cable is permitted to be unsupported under the following conditions:

- Where cable is fished between access points, where concealed in finished buildings or structures and supporting is impractical, such as in remodel work in existing houses.

- Where cable is not more than 2 feet (600 millimeters) in length at terminals where flexibility is necessary, like a connection to an electric motor.

- Where cable is not more than 6 feet (1.8 meters) from the last point of support for connections within an accessible ceiling to a luminaire.

○ Section 320.40 states that at all points where the armor of Type AC cable terminates, a fitting must be provided to protect wires from abrasion, unless the design of the outlet boxes or fittings provides equivalent protection.

- The required insulating fitting is called many names by electricians in the field, including redhead or red devil.

- Requirements for cable installation: Type MC

○ Section 330.17 requires metal-clad cable be protected in accordance with Section 300.4 where installed through or parallel to framing members.

○ Section 330.23 requires the installation of Type MC cable in accessible attics or roof spaces to comply with Section 320.23.

- Section 330.24(B) states that the radius of the curve of the inner edge of any bend must not be less than seven times the external diameter of the metallic sheath for the style of Type MC used in residential wiring.

  - This style has an interlocking-type armor or corrugated sheathing and looks very similar on the outside to Type AC cable.

  - When installing Type MC cable, use bigger turns and "jug handles."

- Section 330.30 states that Type MC cable must be supported and secured by staples, cable ties, straps, or similar means and installed in a way that will not damage the cable.

- Section 330.30(C) covers horizontal installation through framing members and allows Type MC cables installed in accordance with Section 300.4 to be considered supported and secured where the framing members are no more than 6 feet (1.8 meters) apart.

  - Cable ties, or other securing methods, are not required as the cable passes through these members.

- Section 330.30(B) requires Type MC cable containing four or fewer 10 AWG or smaller conductors to be secured within 12 inches (300 millimeters) of each box, cabinet, or fitting and at intervals not exceeding 6 feet (1.8 meters).

- Section 330.30(D) allows Type MC cable to be installed unsupported under the following conditions:

  - Where fished between access points, or where concealed in finished buildings or structures and supporting is impractical.

  - Where not more than 6 feet (1.8 meters) from the last point of support for connections within an accessible ceiling to a luminaire.

- Requirements for cable installation: Types SEU and SER

  - Service entrance cable installation, when used as a service entrance wiring method, is covered in Chapter 8.

  - There are certain situations in residential wiring when service entrance cable is used as the branch-circuit wiring method for individual appliances.

    - Electric furnaces, electric ranges, electric clothes dryers.

    - As the wiring method for a feeder going to a subpanel.

  - Section 338.10(B)(1) allows us to use service entrance cable as branch circuit or feeder conductors.

    - Branch circuits using service entrance cable as a wiring method are permitted only if all circuit conductors within the cable are fully insulated with an insulation recognized by the *NEC*® and listed in Table 310.104(A).

  - Section 338.10(B)(2) states that service entrance cable containing an uninsulated conductor (Type SEU) is not permitted for new installations where it is used as a branch circuit to supply appliances such as ranges, wall-mounted ovens, counter-mounted cooking units, or clothes dryers. Install Type SER cable for these situations. However, the bare conductor can be used as an equipment grounding conductor.

  - Section 338.10(B)(4) says that in addition to the provisions of Article 338, Type SE cable used for interior wiring must comply with the installation requirements of Article 334 Nonmetallic Sheathed Cable.

    - This means that when Type SE for branch circuit or feeder runs inside a dwelling unit, install the Type SEU or Type SER service entrance cable the same way as for nonmetallic sheathed cable.

Objective: Demonstrate an understanding of the proper techniques for preparing, starting, and supporting a cable run in a residential wiring application.

3) Understand the techniques for preparing, starting, and supporting a cable run

Key terms: pulling in, reel

- Preparing the cable for installation
  - Know the correct procedure for setting up and then unrolling cables from a box, roll, or reel (spool).
  - When purchasing nonmetallic sheathed cable (Type NM) or underground feeder cable (Type UF) from the local electrical supplier, it is usually packaged in one of two ways.
    - 250-foot (75-meter) roll packaged in plastic or in a cardboard box.
    - On a 1000-foot (300-meter) reel.
  - When cables are in a roll, start unrolling them from the inside of the coil.
  - If the roll was purchased in a box, a round cutout on the box is provided that indicates the starting point for unrolling the cable.
  - If the cable is not unrolled properly, twists and kinks of the cable will develop that could damage the cable insulation.
  - Rolls of cable can also be unrolled easily with a device commonly called a spinner.
  - Installing the cable runs
  - Installing the cable in the framework of a house is often referred to as pulling in the cable.
  - Route the cables through or along studs, joists, and rafters to all of the receptacle, switch, or lighting outlet locations that make up a residential circuit.
  - Consider the following when installing a cable run.
    - Plan the cable route ahead of time and visually line it all up in your mind.
    - If holes have to be drilled in the building framing members, drill them all, making sure to maintain a minimum clearance of 1¼ inches (32 millimeters) from the edge of the hole to the edge of the framing member.
    - When drilling holes through consecutive framing members for longer cable runs, align the holes as closely as possible to make it easier to pull cables through the holes and make the installation neat and organized.
  - Once the cable has been pulled to all of the electrical boxes on the circuit run, double-check your work to make sure all locations on the circuit meet the following criteria:
    - Have the correct cable installed.
    - Cable is correctly secured and supported according to the *NEC*®.
    - There are no twists or kinks in the cable.
    - Cable turns have not been bent too tightly.
    - There is enough cable at each electrical box location for the required connections.
- Installing cables through wall studs
  - There are several ways to install the cable run through the wall studs of a house.
  - Four methods are commonly used.

- Drilling holes that are all a certain distance from the floor and in line, allowing for the easy pulling of the cables.
- When there is a need to make changes in elevation along the cable run, there are two ways to accomplish it:
  - Drill holes parallel to the floor at a certain height, keeping the holes in line, and then repeat the drilling process at a new elevation from the floor.
  - Drill holes that are gradually higher from one stud to the next.
- Drilling holes close to the bottom of the studs on outside walls, which keeps the cable run close to the bottom plate of a wall and allows for a better and easier fit for insulation.
- Notching the studs so that the cable is simply laid into the notches, and then placing a ¹⁄₁₆ inch (1.6 millimeters) or thicker metal plate over the notch both to protect the cables from screws and nails and to keep the cable from coming out of the notch.

  ○ Installing cables through ceiling or floor joists is similar to installing cables through wall studs and some of the methods previously described will work just fine. However, some cable installations in joists will require the use of one of the installation methods discussed below.

  - Drill holes through the joists, trying to line up the holes as you drill them so the cables will pull through easier.
  - Run the cables over the joists by drilling a hole through the top plate of a wall section and running the cable above the ceiling joists.
  - Running boards are sometimes used in a basement and are secured to the bottom of the joists in an end-to-end manner for the full length of the cable run. The cable is then stapled to the running boards at intervals consistent with *NEC®* requirements.

- Metal framing members
  ○ Several of the cable installation methods described previously will work with metal framing members; however, there are a few special considerations.
  ○ The following methods and materials for installation are recommended.

  - Line up drilled or punched holes in metal studs for easier cable pulling.
  - Install snap-in grommets or bushings made of an insulating material at each hole in the metal framing member to protect the cable sheathing from the sharp edges of the metal.
  - Mount electrical device boxes with side-mounting brackets or another box-mounting device and attach to the stud with sheet metal or self-drilling screws.
  - Since staples cannot be used to secure a cable to a metal stud within the required distance stated in the *NEC®*, use plastic tie wraps.
    - Install them by drilling two small holes into the stud at the proper location, passing the tie wrap through the holes, and then securing the cable in place by tightening the tie wrap.

- Starting the cable run
  ○ Although it is not a specific requirement, the best place to start a cable run for a branch circuit is at the service panel or subpanel.
  ○ This is considered to be the starting point of a circuit because it is the location on the circuit where electrical power is provided.
  ○ It is also where the circuit overcurrent protection devices are located.
  ○ The branch-circuit cable run that goes from the service panel or subpanel to the first electrical box of the circuit is called a home run.

Objective: Demonstrate an understanding of the proper installation techniques for securing the cable to an electrical box and preparing the cable for termination in the box.

4) Understand installation techniques for securing a cable to an electrical box

Key terms: secured, supported

- Securing the cable run at the device and outlet box

  ○ If you are using single-gang nonmetallic electrical device boxes, extend at least ¼ inch (6 millimeters) of the cable sheathing into the box.

  ○ Also, make sure to secure the cable to the building framing member no more than 8 inches (200 millimeters) from a single-gang nonmetallic box or 12 inches (300 millimeters) from a metal box with a staple or some other approved method.

  ○ Horizontal runs of cable through holes in building framing members are adequately supported as long as the distance between framing members does not exceed the *NEC®* support requirements for a particular wiring method.

    ■ For example, a Type NM cable installed through holes drilled in studs that are placed 16 inches O.C. (400 millimeters) means that the cable is adequately supported because the distance between supports is less than the 4½ feet (1.4 meters) requirement for supporting Type NM cable.

  ○ When the cable is run along the sides of a building framing member, support it according to the *NEC®* requirements.

    ■ For example, Type NM cable running along the side of a stud must be supported with staples or other approved means not more than every 4½ feet (1.4 meters).

Objective: Demonstrate an understanding of the common installation techniques for installing cable in existing walls and ceilings.

5) Understand installation techniques for installing cable in existing walls and ceilings

Key terms: fish

- Installing cable in existing walls and ceilings

  ○ This is not an easy task and usually will require two people to "fish" the cable from spot to spot.

  ○ Sometimes an electrician can remove the baseboard molding around a room, notch or drill holes in the exposed studs, and run cable through.

  ○ You can remove molding around doors and windows to expose a space where a cable could be run.

  ○ If the service entrance panel is located in a basement, new cable runs can start at the panel and be run along the exposed basement framing members until the cable is under the wall section in which you want to install an outlet.

  ○ Drill a hole from the basement up and into the wall cavity.

  ○ Drill at the proper angle to get from the basement into the wall.

  ○ Be prepared to drill through several items, including flooring, bottom wall plates, studs, joists, etc.

  ○ This same drilling technique can be used when running cable from an unfinished attic space into the wall cavities of the rooms below. The only difference is that you will be drilling down into the wall cavities from the attic, instead of drilling up from the basement.

- Be sure to measure accurately to avoid drilling into the wrong wall cavity or missing the wall cavity altogether and drilling out into the room.

- Once the holes have been drilled for the cable, use a fish tape to determine if the wall cavity is free from obstructions.

- Once the electrical outlet box hole has been cut, install the cable through the walls or ceiling to the box location (usually a two-person job).

- One electrician will push a fish tape through the bored holes toward the electrical box location.

- Another electrician will look for the fish tape at the electrical box location and will catch the end of the fish tape when it gets to the box hole.

- The electrician at the box hole will attach a cable to the end of the fish tape and feed it carefully through the box hole as the electrician at the other end carefully pulls the fish tape with the attached cable back into the basement or attic space.

- Then attach the installed cable to a junction box or to the service entrance panel.

Quizzes

CODE/REGULATIONS/STANDARDS

1. A "home run" may enter a loadcenter through knockouts in the back, sides, top, or bottom of the loadcenter. True or False?

2. The *NEC*® requires insulated staples to be used when securing and supporting Type NM cable. True or False?

3. An insulated anti-short fitting required to be installed in the ends of Type AC cable to protect the wires from abrasion is typically called a(n) _____.

    a. orangehead

    b. redhead

    c. blue devil

4. Type NM cable must be secured at no more than _____ inches from all electrical boxes except a nonmetallic single-gang box.

    a. 6 inches

    b. 8 inches

    c. 12 inches

5. The *NEC*® requires at least _____ of free conductor in electrical device boxes for connection to switches and receptacles.

    a. 6 inches

    b. 8 inches

    c. 10 inches

6. Manufacturers of Type NM cable are now color-coding their cables. The color typically used for 12 AWG cable is _____.

    a. red

    b. blue

    c. yellow

7. If Type AC is run through a hole in a framing member that has an edge closer than _____ inch(es) to the edge of the framing member, it must be protected by a metal plate at least $\frac{1}{16}$ inch thick.

8. Type AC cable must be supported at intervals not exceeding _____ feet along the run.

TOOLS

1. A _____ is used to pull cables into existing walls and ceilings.
   a. fish tape
   b. pair of pliers
   c. cable puller

2. Unrolling cable from a coil is very easily done with a device commonly called a _____.
   a. spinner
   b. roller
   c. reel jack

Chapter Quiz

1. Type NM Cable is considered to be supported when it passes through bored or punched holes in framing members as long as the framing members are no more than 4½ feet apart. True or False?

2. When service entrance cable is used for branch circuits, the only conductor permitted to be bare is the equipment grounding conductor. True or False?

3. A "home run" may enter a loadcenter through knockouts in the back, sides, top, or bottom of the loadcenter. True or False?

4. The *NEC*® requires insulated staples to be used when securing and supporting Type NM cable. True or False?

5. An insulated anti-short fitting required to be installed in the ends of Type AC cable to protect the wires from abrasion is typically called a(n) _____.
   a. orangehead
   b. redhead
   c. blue devil
   d. bluehead

6. Which of the following is not a type of nonmetallic sheathed cable?
   a. Type NM
   b. Type NM-B
   c. Type NMC
   d. Type NM-U

7. Cables run along the sides of framing members must be installed at least _____ inches from the nearest edge of studs, joists, or rafters.

8. In attics not accessible by permanent stairs or ladders, Type NM cable run across the top of floor joists within _____ feet of the scuttle hole must be protected by means such as guard strips.

9. Type NM cable must be secured at no more than _____ inches from all electrical boxes except a nonmetallic single-gang box.

10. Type NM cable installed in either a metal or plastic two-gang device box with interior cable clamps must be secured no more than _____ inches from the device box.

11. Type MC cable with an interlocked armor or corrugated sheath cannot be bent to a radius of less than _____ times the external diameter of the metallic sheath.

12. The *NEC*® requires at least _____ inches of free conductor in electrical device boxes for connection to switches and receptacles.

13. In single-gang nonmetallic electrical device boxes, at least —————— inch(es) of the cable sheathing must extend into the box.

14. When unrolling cable from a coil, the best way is to pull the cable out of the coil from the ——————.

    a. outside of the coil
    b. inside of the coil
    c. underside of the coil
    d. none of the above

15. Type NM cable is often purchased in —————— -foot rolls that are packaged in plastic wrapping or cardboard boxes.

16. When installing Type MC cable in a house, an anti-short bushing (redhead) is required at each end to provide protection for the conductors from the sharp edges of the cable armor. True or False?

17. Electricians commonly refer to armored-clad cable as —————— cable.

18. Type AC cable must be supported at intervals not exceeding —————— feet along the run.

19. Explain why the sheathing of nonmetallic sheathed cable should extend into a device box or beyond the cable clamp approximately ¼ inch.

    _____

    _____

    _____

    _____

20. Explain what a "jug handle" is when it is used on a cable wiring method.

    _____

    _____

    _____

    _____

Troubleshooting/Quality Exercises

DISCUSSION QUESTIONS

1. Most manufacturers color-code their nonmetallic sheathed cable so that an electrician can easily distinguish one size from another. Discuss the colors used for the outside sheathing of 14, 12, and 10 AWG nonmetallic sheathed cables.

    _____

    _____

    _____

    _____

2. Many electricians work with cables that come in 250-foot rolls. Discuss some of the more common techniques they use to unroll the cable so that it is not twisted.

    _____

    _____

    _____

3. List some advantages to drawing a cable diagram of the various cable runs before you start to actually install the cable.

_____

_____

_____

_____

4. Name the most important thing that you check before cutting the hole in a wall or ceiling for the installation of an old-work electrical box.

_____

_____

_____

_____

5. Some electricians place the cables in electrical boxes without first stripping the outside sheathing. Other electricians strip the outside sheathing first. Discuss the advantages and disadvantages of both techniques.

_____

_____

_____

_____

HANDS-ON APPLICATIONS

1. Match the following terms to their definitions. Write the correct number for the term in the blanks.

   1. fish
   2. home run
   3. jug handle
   4. pulling in
   5. supported
   6. reel
   7. running board
   8. secured

   _____ a. Held in place so a cable is not easily moved

   _____ b. The process of installing cables in an existing wall or ceiling

   _____ c. A drum having flanges on each end and used for cable storage

   _____ d. The type of bend used with certain types of cable to ensure that the cable is not bent too tightly

   _____ e. Pieces of lumber nailed to the joists in an attic or basement for securing cables

   _____ f. Fastened in place so a cable cannot move

   _____ g. The process of installing cables through the framework of a house

   _____ h. The part of the branch-circuit wiring that originates in a loadcenter and terminates at the first electrical box in the circuit

REAL-WORLD SCENARIOS

1. When running a home run, the end that is left at the loadcenter needs to be identified so that the electrician trimming out the loadcenter will know where it goes. Name some methods commonly used to identify the cable.

_____

_____

_____

_____

2. Cables can enter a loadcenter from the top, bottom, back, or sides. Indicate where you would typically attach the cables to a loadcenter in the following situations by placing the correct letter of the location in the blank. There could be more than one answer and answers can be used more than once.

   a. Top
   b. Bottom
   c. Back
   d. Sides

   _____ Loadcenter in a basement

   _____ Loadcenter flush mounted in a wall

   _____ Loadcenter surface mounted on a wall

   _____ Weatherproof loadcenter on the outside of a house

Chapter 12 Raceway Installation

Outline

Objective: Select an appropriate raceway size and type for a residential application.

1) Select an appropriate raceway size and type

Key terms: raceway, conduit, thin-wall

- Selecting the appropriate raceway type and size

  - ○ Branch-circuit installation using a raceway wiring method is seldom used.

  - ○ Cable wiring methods are less expensive and easier to install than raceway wiring methods. This is the main reason why most houses are wired using as little conduit as possible.

  - ○ However, some areas of the country require that all wiring in a house be installed in a raceway wiring method.

  - ○ Use individual conductors when installing a circuit in a raceway wiring method.

  - ○ It is common wiring practice to install a green insulated equipment grounding conductor in every raceway.

  - ○ The most common raceway type used for branch-circuit installation is EMT.

    - ■ EMT is relatively easy to bend and connect.

    - ■ EMT is much less expensive than other metal raceways.

- Raceway type installation requirements: Type RMC

  - ○ Article 344 covers the installation requirements for rigid metal conduit (RMC).

  - ○ RMC is a threadable raceway of circular cross-section designed for the physical protection and routing of conductors and cables and for use as an equipment grounding conductor when installed with appropriate fittings.

  - ○ RMC is generally made of steel with a protective galvanized coating.

    - ■ It can be used in all atmospheric conditions and occupancies.

  - ○ Section 344.22 states that the number of conductors or cables allowed in RMC must not exceed that permitted by the percentage fill specified in Table 1, Chapter 9 in the *NEC*®.

    - ■ Table 1 of Chapter 9 specifies the maximum fill percentage of a conduit or tubing.

    - ■ For an application where only two conductors are installed, the conduit cannot be filled to more than 31% of the conduit's cross-sectional area.

    - ■ If three or more conductors are to be installed (a common practice), the conduit cannot be filled to more than 40% of its cross-sectional area.

  - ○ Informative Annex C is located in the back of the *NEC*® and, through 12 sets of tables, indicates the maximum number of conductors that are the same size and have the same insulation permitted in a conduit or tubing.

- The maximum number of conductors allowed in a conduit, according to Informative Annex C, takes into account the fill percentage requirements of Table 1, Chapter 9.

- No additional calculations are necessary.

- Use Table C8 for RMC.

o To determine the maximum number of conductors of different sizes and different insulations in a conduit.

- Use Table 4, Chapter 9 to get the usable area of a specific size and type of conduit or tubing.

- Use Table 5, Chapter 9 to get the required area of each conductor size with a specific insulation type used in the conduit.

o Section 344.24 requires that when bending RMC, the bends must be made so that the conduit is not damaged and the internal diameter of the conduit is not effectively reduced.

o Section 344.26 limits the number of bends in one conduit run from one box to another to no more than 360° total.

o When cutting RMC, Section 344.28 requires all cut ends to be reamed or otherwise finished to remove rough edges.

o Install RMC as a complete system as required in Section 300.18(A) and securely fasten it in place and support it in accordance with Section 344.30.

o Section 344.46 states that where a rigid metal conduit enters a box, fitting, or other enclosure, a bushing must be provided to protect the wire from abrasion unless the design of the box, fitting, or enclosure is made so it already gives protection.

- Section 300.4(G) requires the use of an insulated bushing (like plastic) for the protection of conductors sizes 4 AWG and larger that are installed in conduit.

o Section 344.60, RMC is permitted as an equipment grounding conductor.

- Raceway type installation requirements: Type IMC

o Article 342 covers the installation requirements for intermediate metal conduit (IMC).

o IMC is a steel threadable raceway of circular cross-section designed for the physical protection and routing of conductors and cables.

o It is used as an equipment grounding conductor when installed with associated couplings and appropriate fittings.

o IMC is a thinner-walled version of rigid metal conduit used in all locations where rigid metal conduit is permitted to be used.

o Section 342.22 covers the requirements for determining the maximum number of conductors allowed in a specific size of IMC.

o The procedures are the same as for RMC with the exception of using Table C4 in Informative Annex C when determining the maximum number of conductors that are all the same size and with the same insulation type.

o Also, use the IMC section of Table 4 in Chapter 9 when determining the minimum size IMC for conductors of different sizes with different insulation types.

o All other *NEC*® installation requirements for IMC, including support, are the same as for RMC.

- Raceway type installation requirements: Type EMT

o Article 358 covers the installation requirements for electrical metallic tubing (EMT).

o EMT is an unthreaded thin-wall raceway of circular cross-section designed for the physical protection and routing of conductors and cables.

- ○ It can be used as an equipment grounding conductor when installed using appropriate fittings.

- ○ Section 358.22 covers the requirements for determining the maximum number of conductors allowed in a specific size of EMT.

- ○ The procedures used are exactly the same as for RMC and IMC with the exception of using Table C1 in Informative Annex C when determining the maximum number of conductors that are all the same size and with the same insulation type.

- ○ Also, use the EMT section of Table 4 in Chapter 9 when determining the minimum size EMT for conductors of different sizes that have different insulation types.

- ○ Section 358.24 requires bends in EMT to be made so that the tubing is not damaged and the internal diameter of the tubing is not effectively reduced.

- ○ Section 358.26 does not allow more than the equivalent of 360° total bending between termination points of EMT.

- ○ Section 358.28 requires all cut ends of EMT to be reamed or otherwise finished to remove rough edges.

- ○ Section 358.30 requires EMT to be installed as a complete system and to be securely fastened in place and supported.

  - ■ EMT must be securely fastened in place at least every 10 feet (3 meters).

  - ■ Between termination points, EMT must be securely fastened within 3 feet (900 millimeters) of each outlet box, junction box, device box, cabinet, conduit body, or other tubing termination.

  - ■ *Exception No. 2* to Section 358.30(A) states that for concealed work in finished buildings or pre-finished wall panels where such securing is impracticable, unbroken lengths (without a coupling) of EMT is permitted to be fished.

  - ■ Section 358.30(B) allows horizontal runs of EMT supported by openings through framing members at intervals not greater than 10 feet (3 meters) and securely fastened within 3 feet (900 millimeters) of termination points.

  - ■ Section 358.30(C) allows unbroken lengths of EMT to be unsupported when lengths of 18 inches (450 millimeters) or less are used between boxes or enclosures.

- ○ Section 358.42 requires couplings and connectors used with EMT to be made up tight.

- ○ Section 358.46 states that where electrical metallic tubing enters a box, fitting, or other enclosure, a bushing must be provided to protect the wire from abrasion unless the design of the box, fitting, or enclosure is made so it already gives protection.

  - ■ Section 300.4(G) requires the use of an insulated bushing (like plastic) for the protection of conductors sizes 4 AWG and larger that are installed in conduit.

- ○ Section 358.60, EMT is also permitted as an equipment grounding conductor.

- • Raceway type installation requirements: Type FMC

  - ○ Article 348 covers the installation requirements for flexible metal conduit (FMC).

  - ○ FMC is a raceway of circular cross-section made of helically wound, formed, interlocked metal strip.

  - ○ Many electricians often refer to this raceway type as Greenfield.

  - ○ FMC is appropriate for use indoors where a need for flexibility at the connection points is required.

  - ○ Section 348.22 covers the requirements for determining the maximum number of conductors allowed in a specific size of FMC.

    - ■ The procedures used are exactly the same as for the raceways we previously discussed, with the exception of using Table C3 in Informative Annex C when determining the maximum number of conductors that are all the same size, with the same insulation type.

- Also, use the FMC section of Table 4 in Chapter 9 when determining the minimum size FMC for conductors of different sizes that have different insulation types.

- Use Table 348.22 when determining the maximum number of conductors in a ⅜-inch trade size FMC.

o Section 348.24 and Section 348.26 state that a run of FMC installed between boxes, conduit bodies, and other electrical equipment is not permitted to contain more than the equivalent of 360° total.

o Proper shaping and support of this flexible wiring method will ensure that conductors can be easily installed or taken out at any time.

o Section 348.28 requires all cut ends to be trimmed or otherwise finished to remove rough edges, except where fittings that thread into the convolutions (so-called inside fittings) are used.

o Section 348.30 gives the securing and supporting requirements.

- They are basically the same as for Type AC cable and Type NM cable. However, the supporting and securing rules do not have to be followed when FMC is fished. The rules also do not apply at terminals where flexibility is required when lengths do not exceed 3 feet (900 millimeters) for trade sizes ½–1¼ inches; 4 feet (1200 millimeters) for trade sizes 1½–2 inches; and 5 feet (1500 millimeters) for trade sizes 2½ inches and larger.

o According to Section 348.60, FMC can be used as a grounding means if it is not longer than 6 feet (2 meters). See Section 250.118(5).

o Any installation of FMC over 6 feet (2 meters) in length will require an installed equipment grounding conductor.

o An additional equipment grounding conductor is always required where FMC is used for flexibility.

- Raceway type installation requirements: Type LFMC

  o Article 350 covers the installation requirements for liquid-tight flexible metal conduit (LFMC).

  o Raceway of circular cross-section having an outer liquid-tight, nonmetallic, sunlight-resistant jacket over an inner flexible metal core with associated couplings, connectors, and fittings for the installation of electric conductors.

  o Intended for use in wet locations for connections to equipment located outdoors, such as air-conditioning equipment.

  o Section 350.22 covers the requirements for determining the maximum number of conductors allowed in a specific size of LFMC.

  o The procedures are exactly the same as for the raceways we have previously discussed, with the exception of using Table C7 in Informative Annex C when determining the maximum number of conductors that are all the same size and with the same insulation type.

  o Use the LFMC section of Table 4 in Chapter 9 when determining the minimum size LFMC for conductors of different sizes that have different insulation types.

  - Use Table 348.22 when determining the maximum number of conductors in a ⅜-inch trade size LFMC.

  o Sections 350.24 and 350.26 state that a run of LFMC installed between boxes, conduit bodies, and other electrical equipment is not permitted to contain more than the equivalent of 360° total.

  o The securing and supporting requirements are given in Section 350.30.

  - Exactly the same as for FMC.

    o According to Section 350.60, LFMC can be used as a grounding means if it is not longer than 6 feet (2 meters). See Section 250.118(6).

        ■ Any installation of LFMC over 6 feet (2 meters) in length requires an installed equipment grounding conductor.

        ■ An additional equipment grounding conductor is always required where LFMC is used for flexibility.

- Raceway type installation requirements: Type PVC

    o Article 352 covers the installation requirements for rigid polyvinyl chloride conduit (PVC).

    o Nonmetallic raceway of circular cross-section, with integral or associated couplings, connectors, and fittings for the installation of electrical conductors.

    o Two types commonly used:

        ■ Schedule 40

        ■ Schedule 80

    o Schedule 40 PVC

        ■ Suitable for underground use by direct burial or encasement in concrete.

        ■ Suitable for aboveground use indoors or outdoors exposed to sunlight and weather where not subject to physical damage (unless marked "Underground Use Only" or equivalent wording).

    o Schedule 80 PVC

        ■ Suitable for use wherever Schedule 40 conduit may be used.

        ■ Marking "Schedule 80" identifies conduit as suitable for use where exposed to physical damage.

    o Section 352.22 covers requirements for determining maximum number of conductors allowed in a specific size of PVC.

    o The procedures used are the same as for the other solid-length conduits discussed with the exception of using Table C9 in Informative Annex C when determining the maximum number of conductors that are all the same size with the same insulation type.

    o Use the PVC section of Table 4 in Chapter 9 when determining the minimum size PVC for conductors of different sizes with different insulation types.

    o Section 352.24 requires that when bending PVC, you must make bends so the conduit is not damaged and the internal diameter of the conduit is not effectively reduced.

        ■ Field bends must be made only with bending equipment identified for the purpose.

    o Section 352.26 limits the number of bends in one conduit run from one box to another to no more than 360°.

    o When cutting PVC, Section 352.28 requires all cut ends to be reamed or otherwise finished to remove rough edges.

    o Section 352.30 requires PVC to be installed as a complete system and to be fastened to permit movement from thermal expansion or contraction.

        ■ PVC must be securely fastened and supported.

        ■ Must be fastened within 3 feet (900 millimeters) of each outlet box, junction box, device box, conduit body, or other conduit termination.

        ■ Must be supported as required in Table 352.30.

        ■ Horizontal runs of PVC are permitted when supported by openings through framing members at intervals not exceeding those in Table 352.30 and securely fastened within 3 feet (900 millimeters) of termination points.

- Section 352.30(C) allows unbroken lengths of PVC to be unsupported when lengths of 18 inches (450 millimeters) or less are used between boxes or enclosures.

○ Expansion fittings for PVC are covered in Section 352.44.

  - They are required to compensate for thermal expansion and contraction where the length change, in accordance with Table 352.44, is expected to be ¼ inch (6 millimeters) or greater in a straight run between securely mounted items such as boxes, cabinets, elbows, or other conduit terminations.

  - Expansion fittings are generally provided in exposed runs of rigid PVC conduit in the following instances.

    • The run is long.

    • The run is subjected to large temperature variations during or after installation.

    • Expansion and contraction measures are provided for the building or other structures.

○ Section 352.46 states that where PVC enters a box, fitting, or other enclosure, a bushing must be provided to protect the wire from abrasion unless the design of the box, fitting, or enclosure affords equivalent protection.

  - Section 300.4(G) states that an insulated bushing (like plastic) must be used for the protection of conductors sizes 4 AWG and larger that are installed in conduit.

○ Section 352.60 requires a separate equipment grounding conductor to be installed in the conduit.

- Raceway type installation requirements: Type ENT

○ Article 362 covers the installation requirements for electrical nonmetallic tubing (ENT).

○ ENT is a nonmetallic pliable corrugated raceway of circular cross-section with integral or associated couplings, connectors, and fittings for the installation of electric conductors.

○ It is composed of a material that is resistant to moisture and chemical atmospheres and is also flame-retardant.

○ Section 362.22 covers the requirements for determining the maximum number of conductors allowed in a specific size of ENT.

  - The procedures are exactly the same as for electrical metallic tubing (EMT) with the exception of using Table C2 in Informative Annex C when determining the maximum number of conductors that are all the same size and with the same insulation type.

  - Use the ENT section of Table 4 in Chapter 9 to determine minimum size ENT for conductors of different sizes with different insulation types.

○ Section 362.24 requires bends in ENT to be made so that the tubing is not damaged and the internal diameter of the tubing is not effectively reduced.

○ Section 362.26 does not allow more than the equivalent of 360° total bending between termination points.

○ Section 362.28 requires all cut ends of ENT to be reamed or otherwise finished to remove rough edges.

○ Section 362.30 covers securing and supporting of ENT.

  - ENT must be installed as a complete system and be securely fastened in place and supported.

  - ENT must be securely fastened at intervals not exceeding 3 feet (900 millimeters).

  - ENT must be securely fastened in place within 3 feet (900 millimeters) of each outlet box, device box, junction box, cabinet, or fitting where it terminates.

- Where ENT is run on the surface of framing members, it is required to be fastened to the framing member every 3 feet (900 millimeters) and within 3 feet (900 millimeters) of every box.

- Horizontal runs of ENT are permitted when:

  - Supported by openings in framing members at intervals not exceeding 3 feet (900 millimeters) and securely fastened within 3 feet (900 millimeters) of termination points.

  o Section 362.46 states that where ENT enters a box, fitting, or other enclosure, a bushing must be provided to protect the wire from abrasion unless the design of the box, fitting, or enclosure is such as to afford equivalent protection.

    - Section 300.4(G) requires the use of an insulated bushing (like plastic) for the protection of conductors sizes 4 AWG and larger that is installed in conduit.

  o Section 362.60 requires a separate equipment grounding conductor to always be installed in the conduit.

Objective: Demonstrate an understanding of the proper techniques for cutting, threading, and bending electrical conduit for residential applications.

2) Understand the proper techniques for cutting, threading, and bending conduit

Key terms: take-up, back-to-back bend, box offset bend, field bend, offset bend, saddle bend, stub-up bend

- Cutting conduit

  o Solid-length metal conduit, like rigid metal conduit, intermediate metal conduit, and electrical metallic tubing, can be cut to length by using one of the following tools.

    - Pipe cutter

    - Tubing cutter

    - Portable bandsaw

    - Hacksaw

  o The ends of cut lengths of RMC and IMC will also have to be threaded.

  o PVC conduit can be cut using a hacksaw or with saws equipped with special blades for PVC.

  o FMC and LFMC are usually cut with a hacksaw.

  o ENT can also be cut with a hacksaw, but special nonmetallic tubing cutters that look like big scissors are available.

- Threading conduit

  o Either hand threaders or power threading equipment can be used.

  o When cutting new threads on a conduit end, the following items should always be observed:

    - Wear safety glasses as well as proper foot and hand protection and observe all applicable safety rules.

    - Secure the conduit in a pipe vise before you attempt to thread it.

    - Choose the proper threading die for the size of conduit you are threading.

    - Always use plenty of cutting oil during the threading process.

- Bending conduit

○ Bending conduit is definitely a skill that improves with practice.

○ The most common electrical conduit installed in houses is EMT and, for this reason, the discussion that follows on conduit bending will focus on EMT. The bending techniques described will also apply to the other types of circular metal raceway, such as RMC and IMC.

- 90° bend or stub-up

  - Most common type.

- Back-to-back bend

  - Distance is measured between the outside diameters of two sections of the pipe.

- Offset bend

  - Two equal bends result in the direction of the conduit being changed to avoid an obstruction blocking the conduit run.

- Saddle bend

  - Similar to offset bend.

  - Allows conduit run to go over the obstruction.

  - Two styles:

    ○ Three-point saddle

    ○ Four-point saddle

- Box offset bend

  - Smaller version of regular offset bend.

  - Used when conduits enter a box or other electrical enclosure that is surface mounted.

○ EMT is bent in the field by using either a hand bender, a hydraulic bender, or an electric bender.

- Bend sizes ½ inch through 1¼ inches with a hand bender.

- Use a hydraulic or electric power bender for larger sizes.

Objective: Demonstrate an understanding of the proper installation techniques for common raceway types used in residential wiring.

3) Understand the installation techniques for common raceway types

- Raceway installation using EMT

  ○ When EMT is used, use only metal electrical boxes.

  ○ Connect the EMT to the boxes with approved fittings, called connectors.

  ○ The set-screw type of connector is used most often because of its ease of installation and lower cost as compared to a compression connector.

  ○ When lengths of EMT need to be coupled together, use only approved couplings.

  - Just like the connectors, couplings are available in a set-screw or a compression type.

  ○ EMT is an approved grounding method, and as such, does not always require an equipment grounding conductor to be run in the raceway with the other circuit conductors.

  ○ Most electricians will run a green insulated equipment grounding conductor in the raceway, though it is not required.

  ○ Electricians choose the conductor insulation color for the conductors they install in the raceway.

- For a 120-volt branch circuit, use a white insulated wire and a black insulated wire.

- For a straight 240-volt circuit (like an electric water heater), use two black conductors or a black and a red conductor.

- If the circuit is a 120/240-volt circuit (like an electric clothes dryer), run a white insulated wire, a black insulated wire, and a red insulated wire.

o EMT can be installed:

- Though drilled holes in studs, joists, and rafters like Type NM cable.

- On a wall or a ceiling's finished surface.

Objective: Demonstrate an understanding of the common installation techniques for installing conductors in an installed raceway system.

4) Understand installation techniques for installing conductors in raceways

- Raceway conductor installation

  o Conductors are usually pulled into the conduit, but in shorter runs between electrical boxes, conductors may be pushed through the raceway.

  o Conductors are taken off spools in a way that results in the conductors coming off the spools easily and not becoming tangled with each other.

  o One of the easiest ways to do this is to use a wire cart that allows several spools of wire to be put on at one time.

  o If the length of conduit between boxes is fairly long, use a fish tape.

  o Pull the fish tape out of its reel. Insert it into a raceway and push it through until it comes out at a box location.

  o The fish tape will have a hook on the end of it. Attach the conductors to the fish tape end.

  o While one electrician pulls the conductors slowly off the spools, another electrician will pull the fish tape with the attached conductors back through the raceway.

Quizzes

SAFETY

1. When trimming and reaming the cut ends of PVC conduit with a knife, electricians should wear
   _____.
   a. an apron
   b. leather or Kevlar gloves
   c. goggles

2. When threading conduit you should wear safety glasses and proper foot and hand protection. True or False?

3. Use a _____ fish tape when pulling or pushing wires in a conduit that is connected to an energized loadcenter.
   a. conductive
   b. round
   c. nonconductive

CODE/REGULATIONS/STANDARDS

1. To find the maximum number of conductors of the same size and insulation type permitted in various types and sizes of conduit, you would look in ———————— of the *NEC®*.

   a. Chapter 8

   b. Informative Annex C

   c. Informative Annex D

2. The *NEC®* defines ———————— as an unthreaded thin-wall raceway of circular cross-section designed for the physical protection and routing of conductors and cables.

   a. EMT

   b. IMC

   c. PVC

3. Rigid metal conduit must be secured within ———————— of outlet boxes, junction boxes, cabinets, and conduit bodies. (general rule)

   a. 12 inches

   b. 24 inches

   c. 36 inches

4. The *NEC®* requires that a(n) ———————— be used for the protection of conductors sized 4 AWG and larger installed in conduit.

   a. antioxidant

   b. coupling

   c. insulated bushing

TOOLS

1. EMT is bent in the field by using either a ———————— or a hydraulic or electric bender.

   a. hand bender

   b. pair of pliers

   c. power threader

2. PVC conduit can be cut with special PVC cutting saws or with a ————————.

   a. right-angle drill

   b. keyhole saw

   c. hacksaw

3. A ———————— can hold several spools of wire at one time and allows the wire to be pulled off easily.

   a. wire box

   b. wire cart

   c. piece of wood

4. A ———————— can be used to pull wires through a length of conduit.

   a. fish tape

   b. pair of pliers

   c. wire puller

Chapter Quiz

1. EMT must be secured within 30 inches of every outlet box, junction box, cabinet, and conduit body. True or False?

2. The *NEC®* considers horizontal runs of EMT supported by openings through framing members at intervals not greater than 10 feet and securely fastened within 3 feet of termination points to be properly supported. True or False?

3. Schedule 80 rigid PVC conduit has a larger inside diameter than Schedule 40 rigid PVC conduit. True or False?

4. A(n) ———————— bend is a conduit bend formed by two 90° bends with a straight length of conduit between the two bends.
   a. back-to-back
   b. field
   c. offset
   d. saddle

5. A(n) ———————— bend is a conduit bend that is made with two equal-degree bends in such a way so the conduit changes elevation and avoids an obstruction.
   a. back-to-back
   b. field
   c. offset
   d. saddle

6. To find the maximum number of conductors of the same size and insulation type permitted in various types and sizes of conduit, you would look in ———————— of the *NEC®*.
   a. Chapter 8
   b. Informative Annex C
   c. Informative Annex D
   d. the index

7. Rigid metal conduit must be supported at least every ———————— feet along a conduit run. (general rule)

8. Rigid metal conduit must be secured within ———————— feet of outlet boxes, junction boxes, cabinets, and conduit bodies. (general rule)

9. The *NEC®* requires that a(n) ———————— be used for the protection of conductors sized 4 AWG and larger installed in conduit.
   a. antioxidant
   b. coupling
   c. locknut
   d. insulated bushing

10. EMT must be supported at least every ———————— feet. (general rule)

11. According to Underwriters Laboratories, lengths of flexible metal conduit longer than ———————— feet have not been judged suitable for grounding purposes.

12. Expansion fittings are generally provided in exposed runs of rigid PVC conduit where ————————.
    a. the run is long
    b. the run is subject to large temperature variations during or after installation
    c. expansion and contraction measures are provided for the building or other structures
    d. all of the above

13. The metric designator for 6-inch trade size rigid metal conduit is —————————.
    a. 100
    b. 155
    c. 175
    d. 208

14. A(n) ————————— plastic cap on the ends of rigid metal conduit designates inch trade sizes.
    a. black
    b. red
    c. blue
    d. none of the above

15. RMC, IMC, EMT, and PVC conduits come in standard ————————— lengths.

16. The maximum number of 14 AWG conductors with THHN insulation that you could install in a ⅜-inch trade size FMC is ————————— if the connector fittings are inside or ————————— if the connector fittings are outside. (Assume the use of a grounding conductor.)
    a. 1, 2
    b. 3, 4
    c. 4, 5
    d. 5, 6

17. The ————————— on a conduit hand bender head is where you will place the bending mark when doing a back-to-back bend.
    a. arrow
    b. rim notch
    c. star point
    d. hook

18. The ————————— on a conduit hand bender head is where you will place the bending mark when doing a stub-up bend.
    a. arrow
    b. rim notch
    c. star point
    d. hook

19. The ————————— on a conduit hand bender head is where you will place the middle bending mark when doing a three-point saddle bend.
    a. arrow
    b. rim notch
    c. star point
    d. hook

20. When installing more than two conductors in a conduit or tubing, the maximum percentage of fill allowed by the *NEC®* is —————————.
    a. 25%
    b. 31%
    c. 40%
    d. 60%

Troubleshooting/Quality Exercises

DISCUSSION QUESTIONS

1. Rigid metal conduit (RMC) and intermediate metal conduit (IMC) come from the manufacturer in standard 10-foot lengths with a coupling on one end and colored plastic caps on the other end that protect the threads. Name the colors of the caps and what conduit size(s) each represents.

2. Where would you look in the *NEC®* to determine the maximum number of conductors allowed in a conduit that are the same size and have the same insulation type?

3. Discuss how the cut ends of Schedule 40 or Schedule 80 PVC conduit are typically trimmed and reamed.

4. Some electricians cut flexible metal conduit (FMC) by bending it so tightly that it actually breaks apart. A pair of cutting pliers is then used to complete cutting off the length needed. Explain why this method of cutting FMC is not recommended.

HANDS-ON APPLICATIONS

1. Match the following terms to their definitions. Write the correct number for the term in the blanks.

   1. saddle bend
   2. thin-wall
   3. back-to-back bend
   4. conduit
   5. offset bend
   6. stub-up bend
   7. box offset bend
   8. raceway
   9. take-up
   10. field bend

   _____ a. A type of conduit bend that results in a 90° change of direction

   _____ b. A type of conduit bend that is formed by two 90° bends with a straight length of conduit between them

   _____ c. An enclosed channel of metal or nonmetallic material designed for holding wires or cables

   _____ d. Any bend or offset made by electricians during the installation of a conduit system

_____ e. A type of conduit bend that uses two equal bends to cause a slight change of direction of the conduit where it is attached to an electrical box

_____ d. A trade name often used for electrical metallic tubing

_____ g. The amount that must be subtracted from the desired stub-up height so the bend will come out right

_____ h. A type of conduit bend that results in a conduit run going over an object that is blocking its path

_____ i. A type of conduit bend that is made with two equal degree bends in such a way that the conduit changes elevation and avoids an obstruction

_____ j. A raceway with a circular cross-section

2. When bending an offset bend there is a certain distance between the two bends that needs to be marked on the conduit before starting the offset bend. This distance is based on the height of the offset and the degrees of the offset bends. Assuming the use of 30° bends, write in the distance between the two bends for the following offset heights.

Offset Height	Distance Between Bends
4 inches	
5 inches	
6 inches	
8 inches	
10 inches	

REAL-WORLD SCENARIOS

1. Many residential electrical contractors do not install conduit systems on a regular basis. Because of this they probably do not have a commercially available wire cart. If you were getting ready to pull conductors through a conduit system and did not have a wire cart to hold the wire spools, what could you do so that the wire on the spools could be pulled off easily?

_____

_____

_____

_____

2. Explain why it is a good idea to use a pencil or permanent marker to mark the locations on a length of conduit before you bend it.

_____

_____

_____

_____

3. The bends you are making with a certain conduit bender always seem to come out ¼-inch shorter than what you want. What adjustment can you make when marking the conduit so that the finished bend will come out closer to the desired length?

_____

_____

_____

_____

Chapter 13 Switching Circuit Installation

Outline

Objective: Select an appropriate switch type for a specific residential switching situation.

Objective: List several *NEC*® requirements that apply to switches.

1) Select an appropriate switch type

2) List several *NEC*® requirements that apply to switches

Key terms: single-pole switch, double-pole switch, three-way switch, four-way switch, dimmer switch, combination switch

- Selecting the appropriate switch type
    - Single-pole switch
        - Most common switch type used.
        - Used in 120-volt residential circuits to control a lighting load from one specific location.
    - Double-pole switch
        - Used to control 240-volt load such as an electric water heater.
        - Like the single-pole switch, the load can be controlled from only one specific location.
    - Three-way switch
        - Used to control lighting loads from two separate locations, such as at the top and bottom of a stairway.
    - Four-way switch
        - Allows for control of lighting loads from three or more locations, such as in a room that has three doorways with a wiring plan that calls for switches controlling the room lighting to be located at each doorway.
    - Dimmer switch
        - Found in both a single-pole and a three-way configuration.
        - Used to brighten or dim a lamp or lamps in a lighting fixture.
    - Electricians also install timers, motion sensors, and occupancy sensors in residential electrical systems.
    - These devices are not only convenient and provide added safety, but they also reduce electricity consumption.
    - Combination switch
        - Consists of two switches on one strap or yoke.
        - Allows placement of multiple switches in a single-gang electrical box.

○ When selecting the proper switch type for a residential lighting application, there are many factors to consider.

■ Base switch selection on the voltage and current rating of the circuit in which the switch is being used.

■ According to *NEC®* Section 404.15(A), each switch must be marked with the current and voltage rating for the switch.

■ Switch rating must be matched to the voltage and current you encounter with the circuit on which you are using the switch.

○ Many residential lighting circuits are wired with 14 AWG conductors protected with a 15-amp circuit breaker and will require switches with at least a 15-amp, 120-volt rating.

■ This switch rating is the most common found in residential wiring.

○ Consider the number of switching locations for the lighting load on the circuit.

■ This allows you to choose the appropriate type of switch for the load.

○ All switching is done in the ungrounded circuit conductor.

■ No switching in the grounded (neutral) conductor is allowed.

■ Section 404.2(B) says switches must not disconnect the grounded conductor of a circuit.

■ There is no need to connect a white insulated grounded conductor to any switch in a residential switching circuit.

○ The *NEC®* and UL refer to the switches used to control lighting outlets as snap switches.

○ Most electricians refer to them as toggle switches or as simply switches.

○ Switches that are used to control lighting circuits are classified as general-use snap switches and the requirements for these switches are the same for both the *NEC®* and UL.

Objective: Select a switch with the proper rating for a specific switching application.

3) Select a switch with the proper rating

- Section 404.14(A) and (B) cover switch requirements and recognize two distinct switch categories.

  ○ Section 404.14(A) categorizes one switch type as an alternating current general-use snap switch.

  ○ Section 404.14(B) categorizes the other switch type as an alternating current or direct current general-use snap switch.

- Alternating current general-use snap switches are suitable only for use on alternating current circuits for controlling the following:

  ○ Resistive and inductive loads, including fluorescent lamps, not exceeding the ampere rating of the switch at the voltage involved.

  ○ Tungsten-filament lamp loads (incandescent lamps) not exceeding the ampere rating of the switch at 120 volts.

  ○ Motor loads not exceeding 80% of the ampere rating of the switch at its rated voltage.

- Alternating-current or direct-current general-use snap switches can be used on either alternating current or direct current circuits for controlling the following:

  ○ Resistive loads not exceeding the ampere rating of the switch at the voltage applied.

  ○ Inductive loads not exceeding 50% of the ampere rating of the switch at the applied voltage.

  ○ Tungsten-filament lamp loads not exceeding the ampere rating of the switch at the applied voltage if T-rated.

Objective: Demonstrate an understanding of the proper installation techniques for single-pole, three-way, and four-way switches.

4) Understand the proper installation techniques for single-pole, three-way, and four-way switches

- Remember three things when making the switch and load connections.
  - At all electrical box locations in the switching circuit, connect the circuit grounding conductors to:
    - Metal electrical boxes with a green grounding screw.
    - Green screw on the switching devices.
    - Lighting fixture ground screw or grounding conductor.
  - At the lighting outlet locations, connect the white insulated grounded conductor to the silver screw terminal or wire identified as the grounded wire on the lighting fixture. Then, connect the "hot" ungrounded conductor at the lighting fixture to the brass screw or the wire identified as the ungrounded conductor on the lighting fixture.
  - At the switch locations, determine the conductor connections to the switch depending on the specific switching situation and make those connections.
- Single-pole switches
  - Single-pole switching is the most common switching found in residential wiring.
  - On a single-pole switch, two wires will be connected to the two terminal screws on the switch.
  - Both wires will be considered "hot" ungrounded conductors.
- Installing single-pole switches using a switch loop
  - It is very common for residential electricians to run the power source to the lighting outlet first and then to run a two-wire cable to the single-pole switching location.
    - Called a switch loop.
    - Requires the white wire in the cable to be used as a "hot" ungrounded conductor.
    - The white wire is typical reidentified as a "hot" conductor by wrapping some black electrical tape around the conductor, although another color tape (like red) may also be used.
    - Section 404.2(C) in the 2011 NEC requires a grounded conductor be installed at the switch location unless there is a way to install a grounded conductor in the future if one is needed.
    - Switch boxes installed on framing members with a basement under or an attic above would qualify as areas where the grounded conductor would not have to be installed.
- Three-way switches
  - Three-way switches are used to control a lighting load from two different locations.
  - Beginning electricians often find the connections for three-way switches confusing.
  - Learning some common rules will make the process much easier no matter if you wire three-way switches all of the time or only once in awhile.
- Common rules for installing three-way switches
  - Three-way switches must always be installed in pairs.
  - A three-wire cable must always be installed between the two three-way switches.
  - If you are wiring with conduit, three separate wires must be pulled into the conduit between the two three-way switches.

- The black common terminal on a three-way switch should always have a black insulated wire attached to it.

- One three-way switch will have a black "hot" feed conductor attached to it.

- The other three-way switch will have the black insulated conductor that will be going to the lighting load attached to it.

- Assuming the use of a nonmetallic sheathed cable:

  - When the power source feed is brought to the first three-way switch, the traveler wires that interconnect the traveler terminals of both switches will be black and red.

  - When the power source feed is brought to the lighting outlet first, the traveler wires will be red and white.

  - Reidentify the white traveler conductors with black tape at each switch location.

  o There is no marking for the ON or OFF position of the toggle on a three-way switch.

    - It does not make any difference which way it is positioned in the electrical device box.

- Common rules for installing four-way switches

  o Always install four-way switches between two three-way switches.

  o Always install a three-wire cable between all four-way and three-way switches.

  o If you are wiring with conduit, pull three separate wires into the conduit between all of the four-ways and three-ways.

  o Assuming the use of a nonmetallic sheathed cable:

    - When the power source feed is brought to the first three-way switch in the circuit, the traveler wires that interconnect the traveler terminals of all four-way and three-way switches will be black and red.

    - When the power source feed is brought to the lighting outlet first, the traveler wires will be red and white.

    - The white traveler conductors will need to be reidentified with black tape at each switch location.

  o When a four-way switch is positioned in a vertical position, the top two screws have the same color and are a traveler terminal pair. The bottom two screws are the other traveler terminal pair and have the same color.

    - The colors of each traveler pair are different.

  o As with a three-way switch, there is no marking for the ON or OFF position of the toggle on the four-way switch.

  o It makes no difference which way it is positioned in the electrical device box.

Objective: Demonstrate an understanding of the proper installation techniques for switched duplex receptacles, combination switches, and double-pole switches.

5) Understand the proper installation techniques for switched duplex receptacles, combination switches, and double-pole switches

- Switched duplex receptacles

  o Switched receptacles are often found in areas such as bedrooms, living rooms, and family rooms.

○ Switching of receptacles can be done so the whole receptacle is switched ON or OFF, or the wiring can be installed so half of a duplex receptacle is energized with the switch while the other half remains "hot" at all times.

○ This is done by splitting a duplex receptacle.

■ Splitting a duplex receptacle means removing the tab between the two brass screw terminals on the ungrounded side of a duplex receptacle.

• Double-pole switches

○ Double-pole switches (also called two-pole switches) are used for wiring situations needing to control a 240-volt load.

○ The double-pole switch has four brass terminals on it and, at first glance, looks like a four-way switch.

○ However, unlike a four-way (or three-way) switch, the toggle on the double-pole switch does have the words ON and OFF written on it.

■ This means that as with a single-pole switch, there is a correct mounting position for the switch so that when it is ON, the toggle will indicate it.

○ The double-pole switch also has markings that usually indicate the "load" and the "line" sides of the switch.

Objective: Demonstrate an understanding of the proper installation techniques for installing single-pole and three-way dimmer switches.

6) Understand the proper installation techniques for single-pole and three-way dimmer switches

• Dimmer switches

○ Available in both single-pole and three-way models

○ Differ from regular switches

■ They typically do not have terminal screws.

■ They usually have colored insulated pigtail wires coming off the switch installed by the manufacturer.

○ Connect the dimmer switch pigtails to the appropriate circuit conductor with a wirenut.

○ Connect single-pole or three-way dimmer switches in a switching circuit exactly as you would regular single-pole and three-way switches.

• Rules for connecting dimmer switches

○ Single-pole dimmers have two black insulated conductors coming off them.

○ Like the two brass screw terminals on a regular single-pole toggle switch, it does not really make any difference which of the two wires is connected to the incoming "hot" feed wire or the outgoing switch leg.

○ Three-way dimmer switches typically come from the manufacturer with two red insulated pigtails and one black insulated pigtail.

■ The two red conductor pigtails are the traveler connections.

■ The black conductor pigtail is the common connection.

- When three-way dimmers are used, it is common wiring practice to use one three-way dimmer and one three-way regular toggle switch when controlling and dimming a lighting load from two locations.
- Some dimmer switch manufacturers make "digital" dimmers that will require the use of two three-way dimmer switches when wiring a switching circuit that controls a lighting load from two locations.
  - Always check the manufacturer's instructions to find out for sure whether two three-way dimmer switches must be used.
- Both single-pole and three-way dimmer switches will also have a green insulated grounding pigtail.
- Connect this pigtail to the circuit equipment grounding conductor.
- Never connect dimmer switches to an energized electrical circuit.
  - The solid state electronics that allow a dimmer switch to operate will be severely damaged.
  - Always de-energize the electrical circuit first before installing a dimmer switch.

Objective: Demonstrate an understanding of the proper installation techniques for installing ceiling-suspended paddle fan/light switches.

7) Understand the proper installation techniques for ceiling-suspended paddle fan/light switches.

- Ceiling paddle-fan switches
  - Several styles of paddle-fan control switches are available.
    - A rotary switch is used to control the paddle fan.
    - It has three specific speed settings: low, medium, and high.
  - A single-pole switch controls the attached lighting fixture.
  - A two-gang electrical device box must be installed for this switch combination.
  - Another common switching installation to control a ceiling-suspended paddle fan/light is to use two single-pole switches.
    - One switch controls the paddle fan.
    - One switch controls the lighting fixture.
    - Set the speed of the paddle fan by a switch located on the fan itself in this switch arrangement.

Quizzes

SAFETY

1. When installing the wiring for switching circuits, electricians should wear safety glasses and observe all applicable safety rules. True or False?

CODE/REGULATIONS/STANDARDS

1. The *National Electrical Code®* requires that all switching be done in the _____ circuit conductor.
   a. grounded
   b. grounding
   c. ungrounded

2. The white wire used in a switch loop wired with a two-wire cable like Type NM must be reidentified. This is normally done with black tape. True or False?

3. When a house is wired with conduit, the *NEC®* permits a white insulated conductor to be used as a "hot" conductor in a switch loop as long as it is reidentified as such. True or False?

4. When a white wire is used as a traveler conductor for three-way and four-way switching, it is required to be reidentified as an ungrounded conductor. True or False?

5. Motor loads not exceeding 80% of the ampere rating of the switch at its rated voltage are permitted to be controlled by alternating-current general-use snap switches. True or False?

Chapter Quiz

1. There is no need to connect a white insulated grounded conductor to any single-pole, double-pole, three-way, or four-way switch in a residential switching circuit. True or False?

2. If the circuit grounding conductor is connected to the green grounding terminal on the switch, there is no need to ground the metal device box in which the switch is installed. True or False?

3. When a house is wired using conduit, the *NEC®* permits a white insulated conductor to be used as a "hot" conductor in a switch loop as long as it is reidentified as such. True or False?

4. Switch loops are not permitted in three-way switch circuits. True or False?

5. Like single-pole switches, dimmer switches always come from the factory with screw terminals. True or False?

6. A double-pole switch can be used to control two 120-volt circuits. True or False?

7. A permanent black marker can be used to reidentify a white insulated conductor as a "hot" conductor in a cable. True or False?

8. When you are reidentifying a white insulated conductor as an ungrounded conductor, the identifying mark must completely encircle the conductor. True or False?

9. The *National Electrical Code®* requires that all switching be done in the _____ circuit conductor.
   a. grounded
   b. grounding
   c. switch loop
   d. ungrounded

10. The black common terminal on a three-way switch should always have a _____ insulated wire connected to it.
    a. white
    b. black
    c. red
    d. blue

11. To control a lighting outlet from four locations, a circuit requires ——————

    a. four single-pole switches

    b. a four-way switch

    c. two three-way switches and one four-way switch

    d. two three-way switches and two four-way switches

12. Which of the following switches tend to generate the most heat?

    a. double-pole switches

    b. dimmer switches

    c. three-way switches

    d. four-way switches

13. When installing a ceiling-suspended paddle fan with a lighting kit and you want to have separate control of the fan and light, a ——————— must be run from the switching location to the paddle fan location.

    a. two-wire cable

    b. three-wire cable

    c. four-wire cable

    d. five-wire cable

14. A switch type that raises or lowers the lamp brightness of a lighting fixture is a ———————— switch.

    a. two-way

    b. dimmer

    c. pilot light

    d. combination

15. Three-way switches are always installed in ———————

    a. pairs

    b. threes

    c. fours

    d. none of the above

16. Four-way switches are always installed between two ——————— switches.

    a. single-pole

    b. double-pole

    c. three-way

    d. dimmer

17. A(n) ——————— switch is used to control a 240-volt load, such as an electric water heater, in a residential wiring system.

18. ——————— switch devices consist of two switches on one yoke or strap.

19. Explain how to tell the difference between a four-way switch and a double-pole switch.

    ———————————————————————————————————

    ———————————————————————————————————

    ———————————————————————————————————

    ———————————————————————————————————

20. Explain why one dimmer switch and one regular three-way switch are used in three-way circuits instead of two three-way dimmer switches unless the switches are digital.

_____

_____

_____

_____

Troubleshooting/Quality Exercises

DISCUSSION QUESTION

1. When installing switches in a metal or nonmetallic electrical box, a connection must be made between the switch and the branch-circuit equipment grounding conductor. Discuss how this connection can be made.

_____

_____

_____

_____

HANDS-ON APPLICATIONS

1. Match the following terms to their definitions. Write the correct number for the term in the blanks.

   1. combination switch   5. double-pole switch
   2. three-way switch   6. single-pole switch
   3. dimmer switch   7. four-way switch
   4. switch loop

   _____ a. A switch that when used with two three-way switches will allow control of a lighting load from more than two locations

   _____ b. A switch type used to control a lighting load from one location

   _____ c. A switching arrangement where the feed is brought to the lighting outlet first and a two-wire loop is run from the lighting outlet to the switch

   _____ d. A switch type used to control a lighting load from two locations

   _____ e. A device with more than one switch type on the same strap

   _____ f. A switch type that raises or lowers a lamp's brightness

   _____ g. A switch type used to control two separate 120-volt loads or one 240-volt load from one location

REAL-WORLD SCENARIOS

1. Because a residential electrical system uses 120/240-volt alternating current, electricians use alternating current general-use snap switches. If you encounter a direct current switching situation, what kind of switch rating would you need to use?

_____

_____

_____

_____

2. When a house is wired with a conduit wiring method, a white insulated conductor cannot be used in a switch loop as a "hot" conductor because *NEC*® Section 200.7(C)(2) allows the reidentification of a white conductor as a "hot" conductor only in a cable assembly. As this is the case, describe what you could use to wire the switch loop in a conduit system.

_____

_____

_____

_____

3. Some four-way switches are made with the traveler terminal pairs on each side rather than with a traveler pair on the top and one on the bottom. They are not common but if encountered can confuse an electrician making connections at four-way switch locations. How can you tell if a four-way switch is configured this way?

_____

_____

_____

_____

Chapter 14 Branch-Circuit Installation

Outline

Objective: Demonstrate an understanding of the installation of general lighting branch circuits.

1) Understand the installation of general lighting branch circuits

- Installing general lighting branch circuits
  - Wire general lighting branch circuits with:
    - 14 AWG copper conductor protected with a 15-ampere fuse or circuit breaker or
    - 12 AWG copper conductor protected with a 20-ampere fuse or circuit breaker.
  - The circuit will originate in the service entrance panel or a subpanel and will be routed to the various lighting outlets, switching points for the lighting outlets, and receptacle outlets on the circuit.
  - Follow these steps when roughing in general lighting branch circuits.
    - Determine whether you will be installing 15- or 20-amp-rated general lighting branch circuits.
    - Use 14 AWG wire for 15-amp-rated branch circuits and 12 AWG wire for 20-amp-rated branch circuits.
    - Determine the number of lighting and receptacle outlets to be included on the general lighting branch circuit.
    - If they have not been installed previously, install the required lighting outlet boxes, switching location boxes, and receptacle outlet boxes at the appropriate locations.
    - Starting at the electrical panel where the circuit overcurrent protection device is located, install the wiring to the first electrical box on the circuit.
    - Continue roughing in the wiring for the rest of the circuit, following the installation practices described in previous chapters.
    - Double-check to make sure all circuit wiring for the general lighting branch circuit you are installing has been roughed in.

Objective: Demonstrate an understanding of the installation of small-appliance branch circuits.

2) Understand the installation of small-appliance branch circuits

- Installing small-appliance branch circuits
  - Small-appliance branch circuits are wired with 12 AWG copper conductors.
    - Could use larger wire such as a 10 AWG to compensate for voltage drop problems when the distance back to the electrical panel is excessively long.
  - Small-appliance branch circuits originate in the service entrance panel or a subpanel and are routed to the various receptacle outlet boxes on the circuit.

143

o There is no maximum number of receptacles that can exist on a small-appliance branch circuit.

   ▪ It is a good idea to limit the number of receptacles on them to 13.

      • (20 amps/1.5 amps per receptacle = 13)

o Follow these steps when roughing in small-appliance branch circuits.

   ▪ Use 12 AWG wire and 20-amp overcurrent protection devices for all small-appliance branch circuits.

   ▪ Install a minimum of two small-appliance branch circuits.

   ▪ Determine the number of receptacle outlets to be included on each small-appliance branch circuit.

   ▪ If they have not been installed previously, install the required receptacle outlet boxes at the appropriate locations.

   ▪ Starting at the electrical panel where the overcurrent protection device is located, install the wiring to the first electrical box on the circuit.

   ▪ Continue roughing in the wiring for the rest of the circuit, following the installation practices described in previous chapters.

   ▪ Double-check to make sure all circuit wiring for the small-appliance branch circuit you are installing has been roughed in.

Objective: Demonstrate an understanding of the installation of electric range branch circuits.

3) Understand the installation of electric range branch circuits

Key terms: electric range, cord-and-plug connection, hardwired

• Installing electric range branch circuits

o To install the branch circuit for an electric range, first determine the minimum size of the wire and maximum overcurrent protection device required.

   ▪ This was shown in Chapter 6.

o The branch circuit will be rated at 120/240 volts and will need a conductor size that will handle, at a minimum, the calculated load.

o Electric ranges are usually connected to the electrical system in a house through a cord-and-plug connection.

o Use an 8/3 copper cable with ground protected by a 40-ampere circuit breaker, or a 6/3 copper cable with ground protected by a 50-ampere circuit breaker.

o Use nonmetallic sheathed cable, Type AC cable, Type MC cable, or Type SER service entrance cable for the branch-circuit wiring method.

o Use three conductors (with a fourth grounding conductor) of the proper size installed in a raceway.

o Install the wiring method between the electrical panel and the location of the receptacle for the range.

   ▪ The receptacle will be either surface-mounted or flush-mounted and is located close to the floor at the rear of the final position of the electric range.

o The wiring method is attached to either the electrical box for flush mounting or the receptacle body itself for surface mounting.

o Assuming the use of a cable wiring method, connect the wires of the branch circuit to the receptacle terminals.

- Connect the white grounded neutral conductor to the terminal marked "W."
- Connect the black ungrounded conductor to the terminal marked "Y."
- Connect the red ungrounded conductor to the terminal marked "X."
- Connect the green or bare equipment grounding conductor to the terminal marked "G."
  - Sometimes the electrical installation calls for the connection to a range to be hardwired.
  - Bring the branch-circuit wiring method directly to the back of a range and make the terminal connections on the terminal block on the range.
  - Assuming a cable wiring method, make the connections as follows:
    - Connect the white grounded neutral conductor to the silver screw terminal on the range terminal block.
    - Connect the black ungrounded conductor to the brass screw on the range terminal block.
    - Connect the red ungrounded conductor to the other brass screw on the range terminal block.
    - Connect the green or bare equipment grounding conductor to the green grounding screw on the range terminal block.

Objective: Demonstrate an understanding of the installation of countertop cook unit and wall-mounted oven branch circuits.

4) Understand the installation of countertop cook unit and wall-mounted oven branch circuits

Key terms: counter-mounted cooktop, wall-mounted oven

- Installing counter-mounted cooktop and wall-mounted oven branch circuits
  - Three common methods are used to install the branch circuit and make the necessary connections to counter-mounted cooktops and wall-mounted ovens.
    - Running a separate branch circuit from the electrical panel to both the countertop cook unit and the wall-mounted oven.
    - Running one larger size branch circuit from the electrical panel to a metal junction box located close to both appliances.
    - Installing individual branch-circuit wiring to a receptacle outlet box located in a cabinet near the cooking appliance.

Objective: Demonstrate an understanding of the installation of a garbage disposal branch circuit.

5) Understand the installation of a garbage disposal branch circuit

- Installing the garbage disposal branch circuit
  - Two installation methods used
    - Running the wiring method to a single-pole switch and then hard-wiring directly to the garbage disposal.
      - Switch is usually located above the kitchen countertop near the sink.
    - Installing a receptacle outlet under the sink during the rough-in stage that will be controlled by a single-pole switch located above the countertop and near the sink.

○ Assuming the use of a cable wiring method, make the wiring connections in the disposal junction box as follows:

■ Loosen the screw that secures the blank cover to the junction box and remove the cover.

■ Strip the outside sheathing back about 6 inches and attach the cable to the disposal junction box though the ½-inch KO located on the unit with an approved connector.

■ Connect the incoming equipment grounding conductor to the green screw or green grounding jumper.

■ Wirenut the white incoming grounded conductor to the white pigtail.

■ Wirenut the black incoming ungrounded wire to the black pigtail.

■ Carefully push the wire connections into the junction box area and put the blank cover back on.

Objective: Demonstrate an understanding of the installation of a dishwasher branch circuit.

6) Understand the installation of a dishwasher branch circuit

- Installing the dishwasher branch circuit

  ○ There are two common ways to electrically connect a dishwasher:

  ■ Hardwire it in the branch circuit.

  ■ Cord-and-plug connect it.

  ○ When wiring the dishwasher, install the branch-circuit wiring from the electrical panel to the dishwasher location.

  ○ Make the electrical connections in the dishwasher junction box normally located on the lower right side of the front of the appliance.

  ○ Remove the bottom "skirt" on the dishwasher to gain access to the junction box.

  ○ Assuming the use of a cable wiring method, make the wiring connections as follows.

  ■ Remove the junction box cover and take out the ½-inch KO on the junction box.

  ■ Attach the wiring method used to the dishwasher junction box though the ½-inch KO with an approved connector.

  ■ Connect the incoming equipment grounding conductor to the green screw or green grounding jumper.

  ■ Wirenut the white incoming grounded conductor to the white pigtail.

  ■ Wirenut the black incoming ungrounded wire to the black pigtail.

  ■ Carefully push the wire connections into the junction box area and put the blank cover back on.

  ○ If a dishwasher is to be cord-and-plug connected, the installation is similar to that of a cord-and-plug connected garbage disposal.

  ■ However, there is no need to switch the receptacle that has the dishwasher cord plugged into it.

Objective: Demonstrate an understanding of the installation of the laundry branch circuit.

7) Understand the installation of the laundry branch circuit

- Installing the laundry area branch circuit

  ○ In Section 210.11(C)(2), the *NEC®* requires the laundry branch circuit to be rated at 20 amps, resulting in wiring it with 12 AWG wire.

○ In Section 210.50(C), the *NEC®* requires the receptacle for the clothes washer to be located within 6 feet (1.8 meters) from the appliance location.

○ When wiring the laundry circuit, install the branch-circuit wiring from the electrical panel to the receptacle outlet box installed at the clothes washer location.

Objective: Demonstrate an understanding of the installation of an electric clothes dryer branch circuit.

8) Understand the installation of an electric clothes dryer branch circuit

- Installing the electric dryer branch circuit

  ○ Located in the laundry area of a house.

    ▪ Not supplied by the 20-ampere laundry circuit.

    ▪ Supplied by an individual 120/240-volt branch circuit installed similar to the branch circuit for an electric range.

  ○ Usually connected to the electrical system in a house through a cord-and-plug connection.

  ○ Assuming the use of a cable wiring method, make the wiring connections to the receptacle terminals as follows.

    ▪ Connect the white grounded neutral conductor to the terminal marked "W."

    ▪ Connect the black ungrounded conductor to the terminal marked "Y."

    ▪ Connect the red ungrounded conductor to the terminal marked "X."

    ▪ Connect the green or bare equipment grounding conductor to the terminal marked "G."

  ○ Sometimes the electrical installation calls for the connection to a dryer to be hard-wired.

    ▪ Done by bringing the branch-circuit wiring method directly to the back of the clothes dryer and making the terminal connections on the terminal block on the dryer.

  ○ Assuming the use of a 10/3 cable wiring method, make the connections as follows:

    ▪ Connect the white grounded neutral conductor to the silver screw terminal on the dryer terminal block.

    ▪ Connect the black ungrounded conductor to the brass screw on the dryer terminal block.

    ▪ Connect the red ungrounded conductor to the other brass screw on the dryer terminal block.

    ▪ Connect the green or bare equipment grounding conductor to the green grounding screw on the dryer terminal block.

Objective: Demonstrate an understanding of the installation of branch circuits in a bathroom.

9) Understand the installation of branch circuits in a bathroom

- Installing the bathroom branch circuit

  ○ The *NEC®* requires at least one 20-amp-rated branch circuit be installed to supply the bathroom receptacle or receptacles.

  ○ Each 15- and 20-amp, 125-volt-rated receptacle in a bathroom must be GFCI protected.

  ○ When roughing in bathroom branch circuits, follow these steps:

    ▪ Use 12 AWG wire and 20-amp overcurrent protection devices for all bathroom branch circuits.

- Install at least one bathroom branch circuit for each bathroom.

- Determine the number of receptacle outlets to be included on each bathroom branch circuit.

- If they have not been installed previously, install the required lighting outlet and receptacle outlet boxes at the appropriate locations.

- Starting at the electrical panel where the circuit overcurrent protection device is located, install the wiring to the first electrical box on the circuit.

- Mark the cable or other wiring method with the circuit number, type, and area served at the panel.

- Continue roughing in the wiring for the rest of the circuit, making sure to follow the installation practices described in previous chapters.

- Double-check to make sure all circuit wiring for the bathroom branch circuit you are installing has been roughed in.

Objective: Demonstrate an understanding of the installation of a water pump branch circuit.

10) Understand the installation of a water pump branch circuit

Key terms: dual-element time-delay fuse, inverse time circuit breakers, submersible pump

- Installing a water pump branch circuit

  - Houses built in a town or city generally get their supply of water from a city water system.

  - In rural areas where water system piping is not available, a well is used to supply water to the house.

  - A pump brings the water from the well to the house.

  - Two types of water pumps are commonly used:

    - Deep-well jet pump

    - Submersible pump

  - Installing a water pump branch circuit: jet pump

    - The jet pump and the electric motor that turns it are usually located in a basement or crawl space area, although other locations may be used.

    - The electric motors are usually around 1 horsepower, single-phase, capacitor start, and operate on either 115 or 230 volts.

      - Called a dual-voltage electric motor.

      - When the motor is connected at the higher voltage (230 volts), it draws half as much current as when it is connected to operate on the lower voltage (115 volts).

      - It is common wiring practice to connect water pump motors to the higher voltage.

      - At the higher voltage of 230 volts (actually 240 volts from the circuit breaker panel), a two-pole circuit breaker is required.

  - Installing a water pump branch circuit: submersible pump

    - A submersible pump has both the electric motor and pump enclosed in the same housing.

    - The housing is lowered into a well casing and placed below the water level in the well.

    - Wired using a special submersible pump cable buried in the ground, usually in the same trench with the incoming water pipe from the well.

Objective: Demonstrate an understanding of the installation of an electric water heater branch circuit.

11) Understand the installation of an electric water heater branch circuit

- Installing an electric water heater branch circuit
  - ○ Electric water heaters come in several sizes.
    - ▪ 40- and 42-gallon are probably the most common sizes for residential applications.
  - ○ Electric water heaters used in homes usually operate on 240 volts.
  - ○ They normally require a 10 AWG conductor with a 30-ampere overcurrent protection device.
  - ○ Some smaller electric water heaters may require 120 volts and will be wired with a 12 AWG conductor with a 20-ampere overcurrent protection device.
  - ○ Section 422.10 requires the branch-circuit rating to not be less than 125% of the nameplate rating of the water heater.
    - ▪ Rating of the branch circuit is based on the size of the overcurrent protection device.
  - ○ To calculate the size of the overcurrent protection device, refer to Section 422.11(E), which says to size the overcurrent device no larger than the protective device rating on the nameplate.
  - ○ Assuming the use of a 10/2 nonmetallic sheathed cable wiring method, wire the connections as follows:
    - ▪ Remove the junction box cover and take out the KO on the junction box.
    - ▪ Attach the wiring method used to the water heater junction box though the KO with an approved connector.
    - ▪ Connect the incoming equipment grounding conductor to the green screw or green grounding jumper.
    - ▪ Wirenut the white incoming conductor to one of the black pigtail wires in the junction box.
    - ▪ Reidentify the white wire as a "hot" conductor with black tape.
    - ▪ Wirenut the black incoming ungrounded wire to the other black pigtail.
    - ▪ Carefully push the wire connections into the junction box area and put the blank cover back on.

Objective: Demonstrate an understanding of the installation of branch circuits for electric heating.

12) Understand the installation of branch circuits for electric heating

Key term: thermostats

- Installing branch circuits for electric heating
  - ○ Many different styles of electric heating are available.
  - ○ Electric furnaces are used to heat an entire house.
  - ○ Individually controlled baseboard electric heaters are used to heat specific rooms.
  - ○ Unit heaters are used to heat a specific area like a basement or garage.
  - ○ Installing an electric furnace
    - ▪ Install the electric furnace on a separate circuit as specified by Section 422.12.
    - ▪ Determine the size and type of the individual branch circuit required.
    - ▪ Determine the size and location of the electric furnace disconnecting means.
      - • It is common practice to mount the disconnecting means on the side of the electric furnace or on a wall space adjacent to the furnace.

- Determine the location of the thermostat and install the Class 2 wiring from the furnace controller box to the thermostat.
- Installing an electric baseboard heater
    - Determine whether a line voltage thermostat or a low-voltage thermostat will be used to control the electric baseboard heating.
    - The most common thermostat type used is the line-voltage type.
        - It works at the full 240 volts required for the baseboard heating units.
        - It is connected directly in the electric baseboard heater branch circuit.
    - Determine the size and type of the individual branch circuit required.
    - Most baseboard installations are done on 240-volt branch circuits wired with 12 AWG wire and protected with a 20-amp-rated two-pole circuit breaker.
    - Starting at the location where the branch-circuit overcurrent protection device is located, install the wiring to the electrical box on the circuit that will contain the line-voltage thermostat.
    - Install the wiring from the electrical box containing the line-voltage thermostat to the electric baseboard heater unit.
    - Double-check to make sure all circuit wiring for the electric baseboard heating branch circuit you are installing has been roughed in.

Objective: Demonstrate an understanding of the installation of branch circuits for heating and air-conditioning.

13) Understand the installation of branch circuits for heating and air-conditioning

Key terms: hydronic system, heat pump, transformers, nameplate

- Installing the branch circuits for air-conditioning
    - Air-conditioning is accomplished in one of two ways
        - A central air-conditioning system that blows cooled air from one centrally located unit into various areas of a house through a series of ducts.
        - Individual room air conditioners located in different rooms throughout a house.
- Installing the branch circuits for air-conditioning: central air
    - Install branch-circuit wiring from the electrical panel to the main air-handling unit located inside the home.
    - The air-handling unit contains the fan that blows the cool air throughout a house and will need either a 120-volt or 240-volt circuit.
    - Install another branch circuit from the electrical panel to the disconnect switch of the compressor unit located outside of the house.
    - The circuit supplying the compressor unit will require a 240-volt, two-wire circuit with a grounding conductor.
    - If the disconnect located outside next to the compressor contains fuses or a circuit breaker, the circuit supplying it is a feeder.
    - If the disconnect has no overcurrent protection devices in it, the circuit supplying it is considered a branch circuit.

○ Article 440 contains most of the installation requirements for air-conditioning systems.

○ When selecting the type of overcurrent protection device for an air-conditioning circuit (or heating circuit), consult the nameplate on the unit.

- If it says "maximum size fuse," the overcurrent protection device MUST be a fuse.

- If it says "maximum size fuse or HACR-type circuit breaker," either a fuse *or* an HACR circuit breaker can be used.

  • HACR is an acronym for heating, air-conditioning, and refrigeration.

○ A thermostat is required for the system control.

- Run thermostat wire from the compressor and the air handler to the thermostat.

- This low-voltage wiring is called a Class 2 circuit. (Article 725 covers Class 2 wiring.)

- Thermostat wire comes in a cable form and is usually 18 AWG or 20 AWG solid copper wire.

- Class 2 circuit power is limited because of the transformers that supply the low voltage on which they operate.

- Class 2 circuits are considered to be safe from electric shock and do not present much, if any, fire hazard.

- The transformers used to supply power to Class 2 circuits are normally marked "Class 2 Transformer."

- Conductors do not need separate fuses or circuit breakers to protect them because they are inherently current limited by the transformers from which they originate.

- Wiring can be run exposed as a cable or in a raceway.

- Article 725 states that Class 2 wiring cannot be run in the same raceway, cable, compartment, or electrical box with regular light and power conductors.

  • Room air conditioners are available in 120-volt and 240-volt models.

○ They typically fit into a window opening.

○ They can also be installed in a more permanent way directly in an outside wall.

○ They are cord-and-plug connected.

○ Circuits that power them must be sized and installed according to the installation requirements as outlined in Article 440.

○ Sections 440.60 through 440.65 cover room air conditioners.

- They must be grounded and connected with a cord-and-plug set.

- They may not have a rating greater than 40 amps at 250 volts.

- The rating of the branch-circuit overcurrent protection device must not exceed the branch-circuit conductor rating *or* the receptacle rating, whichever is less.

- On an individual branch circuit where no other items are served by the circuit, the air conditioner load cannot exceed 80% of the branch-circuit ampacity.

- On a branch circuit on which other loads are also served, the air conditioner load cannot exceed 50% of the branch-circuit ampacity.

- The plug on the end of the cord can serve as the air conditioner's disconnecting means.

- The maximum length of the cord on a 120-volt room air conditioner is 10 feet (3 meters).

- The maximum length of the cord on a 240-volt room air conditioner is 6 feet (1.8 meters).

Objective: Demonstrate an understanding of the installation of a branch circuit for smoke detectors.

14) Understand the installation of a branch circuit for smoke detectors

Key term: interconnecting

- Installing the smoke detector branch circuit

- The National Fire Protection Association (NFPA) publishes the *National Fire Alarm Code,* called *NFPA 72.*

  o Chapter 2 of NFPA 72 covers residential fire alarm systems.

  o It defines a household fire alarm system as a system of devices that produces an alarm signal in the house for the purpose of notifying the occupants of the house of a fire so that they will evacuate the house.

  o The most common fire warning device used in a house is a smoke detector.

  o Should be placed:

    ▪ In each bedroom

    ▪ In the area just outside of the bedroom areas

    ▪ On each level of a house

- The smoke detectors should be interconnecting and installed so that when one detector is operated, all other detectors in the house will also operate.

- Follow these installation requirements:

  o Install the smoke detectors on the ceiling at a location where there is no dead airspace.

  o Install smoke detectors at the top of a stairway instead of the bottom since smoke will rise.

  o An exception is in the basement, where it is better to install a smoke detector on the ceiling, but close to the stairway to the first floor.

  o Install smoke detectors in new houses that are hard-wired directly to a 120-volt circuit.

  o Wire the smoke detectors so that they are interconnected so when one goes off, they all go off.

  o Do not install smoke detectors to branch-circuit wiring controlled by a wall switch since the electrical power to the smoke detectors must be on at all times.

  o Do not install smoke detectors on circuits that are GFCI protected because if the GFCI trips off, you have lost power to the smoke detectors.

Objective: Demonstrate an understanding of the installation of a branch circuit for carbon monoxide detectors.

15) Understand the installation of a carbon monoxide detector branch circuit.

  o Installing carbon monoxide detectors

  o Statistics indicate that the leading cause of accidental death in homes is from carbon monoxide (CO) poisoning.

  o CO is an invisible, odorless, tasteless, and non-irritating gas that is completely undetectable to your senses.

  o Carbon monoxide alarms are now required in most areas of the country to alert people living in the home in case there is a dangerously high level of CO.

○ CO alarms should be mounted in or near bedrooms and living areas.

○ It is recommended that you install a CO alarm on each level of a home.

○ Carbon monoxide detectors are hardwired into a 120-volt branch circuit.

○ Install the wiring for interconnected CO alarms similar to the way interconnected smoke detectors are wired.

Objective: Demonstrate an understanding of the installation of a low-voltage chime circuit.

16) Understand the installation of a low-voltage chime circuit

- Installing the low-voltage chime circuit:

  ○ A chime system is used in homes to signal when somebody is at the front or rear doors.

  ○ Consists of:

    ▪ Chime

    ▪ Momentary contact switch buttons

    ▪ Transformer

    ▪ Wire used to connect the system together

  ○ The signal from the door buttons is delivered to the chime itself through low-voltage Class 2 wiring.

  ○ Chimes are available in a variety of styles.

  ○ If surface mounted, it is common wiring practice to just bring the low-voltage wiring into the back of the chime enclosure and use hollow wall anchors to hold the chime onto the wall.

  ○ If flush mounted, the installation requires adequate backing and an electrical box for attachment of the chime enclosure to the wall.

  ○ Chime buttons are of the momentary contact type.

    ▪ When the button is pushed, contacts are closed and current can pass through the switch to the chime, causing a tone to be sounded.

    ▪ When pressure on the button is taken away, springs cause the contacts to come apart and the switch will not pass current. This de-energizes the chime and the tone stops.

  ○ The chime transformer is used to transform the normal residential electrical system voltage of 120 volts down to the value on which a chime system will work, usually around 16 volts.

    ▪ These transformers have built-in thermal overload protection and no additional overcurrent protection is required for them to be installed.

    ▪ Install the transformer in a separate metal electrical box or right at the service entrance panel or subpanel.

  ○ The wire used to connect the buttons, chime, and transformer is often called bell wire or simply thermostat wire since it is the same type as that used to wire heating and cooling system thermostats.

    ▪ This wire is usually 18 AWG or 20 AWG solid copper with an insulation type that limits it to use on circuits of 30 volts or less.

- Running the cable
  - ○ Article 725 tells us to not install Class 2 wiring in the same raceway or electrical box as the regular power and lighting conductors.
  - ○ Article 725 tells us to keep the Class 2 bell wire at least 2 inches (50 millimeters) from light and power wiring.
    - However, when the wiring method is NMSC, Type AC (alternating current), Type MC (multi-contact), or a raceway wiring method, you are allowed to run the bell wire right next to these wiring methods.
  - ○ Two common wiring schemes
    - Running a two-wire thermostat cable from each doorbell button location and from the transformer location to the chime.
    - Installing a two-wire thermostat cable from each doorbell button location to the transformer location.

Quizzes

SAFETY

1. When installing the wiring for branch circuits, electricians should wear safety glasses and observe all applicable safety rules. True or False?

CODE/REGULATIONS/STANDARDS

1. The *NEC*® requires that small-appliance branch circuits be protected with _____ ampere overcurrent protection devices.

2. The *NEC*® requires a laundry circuit to be rated at _____ amperes.

3. The *NEC*® requires at least one 15- or 20-amp, 125-volt receptacle in a bathroom and it must be located no more than _____ inches from the edge of the basin.
   a. 12 inches
   b. 24 inches
   c. 36 inches

4. The minimum number of small-appliance circuits required by the *NEC*® in each house is _____.
   a. 1
   b. 2
   c. 3

5. The *NEC*® permits low-voltage Class 2 doorbell cable to be run next to 120- or 240-volt branch circuits wired with nonmetallic sheathed cable. True or False?

6. A receptacle located under a kitchen sink for a cord-and-plug-connected garbage disposal must be GFCI protected. True or False?

Chapter Quiz

1. A receptacle for a gas-fired range is permitted to be connected to a small-appliance branch-circuit. True or False?

2. The receptacle installed for a clothes washing machine located in a laundry area of an unfinished basement must be GFCI protected. True or False?

3. The *NEC®* permits low-voltage Class 2 doorbell cable to be run next to 120- or 240-volt branch circuits wired with nonmetallic sheathed cable. True or False?

4. HACR is an acronym for heating, air-conditioning, and ventilation. True or False?

5. A(n) _____ circuit breaker has a trip time that gets faster as the fault current flowing through it gets larger.
   a. dual-element
   b. inverse time
   c. plug fuse
   d. Type S

6. If a counter-mounted cooktop unit has a nameplate rating of 7300 watts at 240 volts, the current draw will be _____ amperes.

7. In the *NEC®*, a bathroom is defined as an area that has a basin and a _____.
   a. toilet
   b. tub
   c. shower
   d. Any of the above

8. The *NEC®* requires at least one 15- or 20-amp, 125-volt receptacle in a bathroom and it must be located no more than _____ inches from the edge of the basin.

9. The size of the circuit breaker and the circuit conductors is determined by the _____ of the motor.
   a. horsepower rating
   b. full load current
   c. voltage rating
   d. all of the above

10. The *NEC®* requires the vast majority of motor applications to be protected by overload devices sized not more than _____ of the motor's nameplate current rating.
   a. 100%
   b. 115%
   c. 125%
   d. 150%

11. Most residential electric water heaters operate on 240 volts, are wired with 10 AWG conductors, and are protected with a _____-ampere overcurrent protection device.

12. On an individual branch circuit where no other items are served by the circuit, an air conditioner load cannot exceed _____ percent of the branch-circuit ampacity.

13. A hot water heating system is sometimes referred to as a(n) _____ heating system.
   a. forced air
   b. gas
   c. electric
   d. hydronic

14. A(n) _____ switch cover is used to identify an oil burner safety switch.

    a. orange

    b. red

    c. blue

    d. yellow

15. Who usually decides whether 15- or 20-amp-rated general lighting circuits will be installed in a residential electrical system?

    a. Home owner

    b. Architect

    c. Electrical contractor

    d. Plumber

16. Instead of installing the required bathroom receptacle on a wall or partition adjacent to the basin, it may be installed on the side or face of the basin cabinet, as long as it is located no more than _____ inches below the countertop.

17. The *NEC*® requires a laundry branch-circuit to be rated at _____ amperes.

18. If an electric clothes dryer has a rating of 6 kW at 120/240 volts, it will draw _____ amperes.

19. A(n) _____ is a fuse type that has a time-delay feature built into it and is often used as an overcurrent protection device for electric motor branch-circuits.

    a. dual-element fuse

    b. inverse time fuse

    c. plug fuse

    d. Type S fuse

20. When an installation technique is used that brings the circuit conductors directly to an electrical appliance and then terminates them at the appliance, the appliance is said to be _____.

Troubleshooting/Quality Exercises

DISCUSSION QUESTIONS

1. Discuss why it is a good idea to split up some lighting outlets and receptacle outlets in a room and put them on different branch circuits.

   _____

   _____

   _____

   _____

2. Why do receptacles located under a sink for a cord-and-plug-connected garbage disposal or a dishwasher not have to be GFCI protected?

   _____

   _____

   _____

   _____

3. Discuss why local oil burner codes typically require a safety switch to be installed at the head of a basement stairway or at the entrance to a room or crawl space that houses an oil-fired furnace.

_____

_____

_____

_____

HANDS-ON APPLICATIONS

1. Match the following terms to their definitions. Write the correct number for the term in the blanks.

   1. nameplate
   2. thermostat
   3. submersible pump
   4. dual-element time-delay fuse
   5. jet pump
   6. hard-wired

   7. inverse time circuit breaker
   8. heat pump
   9. interconnecting
   10. cord-and-plug connection
   11. carbon monoxide detector

   _____ a. A device used with a heating or cooling system to establish a set temperature

   _____ b. A type of water pump where the pump and motor are enclosed in the same housing and are lowered down a well casing below the water level

   _____ c. The label located on an appliance that contains information such as amperage and voltage

   _____ d. A type of water pump in which the pump and motor are separate items and are located away from the well

   _____ e. A type of circuit breaker that has a trip time that gets faster as the fault current gets larger

   _____ f. The process of connecting together smoke detectors so that if one is activated, they will all be activated

   _____ g. A reversible air-conditioning system

   _____ h. An installation technique where the circuit conductors are brought directly to an electrical appliance and terminated at the appliance

   _____ i. A device that detects the presence of carbon monoxide (CO) gas in order to prevent carbon monoxide poisoning

   _____ j. A fuse type that has a time-delay feature

   _____ k. An installation technique in which electrical appliances are connected to a branch circuit with a flexible cord

2. Using Table 430.248 in the 2011 *NEC*® or Figure 14-22 of the House Wiring textbook, list the full load currents for the following motors:

Horsepower	115 Volts	230 Volts
¼		
½		
¾		
1		
2		

REAL-WORLD SCENARIOS

1. Who usually decides whether 15- or 20-amp-rated general lighting circuits will be installed in a residential electrical system?

   _____

   _____

   _____

   _____

2. Wallboard installers typically cover all of your electrical boxes with wallboard and then cut out the electrical box openings. They often use a tool called a RotoZip® with a zip bit. What kind of problem can this installation practice present for an electrician and what can the electrician do to lessen the chance that this problem will occur?

   _____

   _____

   _____

   _____

3. A trash compactor is to be installed in the house you are wiring. The branch-circuit installation and connection procedure is the same as that of what other appliance?

   _____

   _____

   _____

   _____

Chapter 15 · Special Residential Wiring Situations

Outline

Objective: Demonstrate an understanding of the installation of garage feeders and branch circuits.

1) Understand the installation of garage feeders and branch circuits

- Installing attached garage feeders and branch circuits

  ○ If a garage is attached, *NEC®* Section 210.52(G) requires at least one 15- or 20-amp, 125-volt-rated receptacle be located there.

  ○ Section 210.70(A)(2)(a) requires at least one wall-switch-controlled lighting outlet in the attached garage.

  ○ There is no *NEC®* requirement concerning the maximum distance between receptacle outlets in a garage.

  ○ The branch circuits supplying the receptacle(s) and lighting outlet(s) may be on the same branch circuit and can be rated either 15-amp or 20-amp.

  ○ Good wiring practice is to install the lighting outlets on a separate 15-amp or 20-amp branch circuit and to install the receptacle(s) on a separate 20-amp-rated branch circuit.

  ○ Section 210.8 (A)(2) requires all 15- or 20-amp, 125-volt rated receptacles installed in a garage to be GFCI protected.

  ○ Installing lighting outlets

    ■ Two lighting outlets minimum in a one-car garage (one on each side of the vehicle parking area).

    ■ Three lighting outlets minimum in a two-car garage (one located in the middle and one to the outside of each vehicle parking location).

    ■ Four lighting outlets minimum in a three-car garage (located where light shines down on each side of the vehicle parking areas).

- Installing detached garage feeders and branch circuits

  ○ The *NEC®* does not require that electrical power be brought to a detached garage. However, if electrical power is brought to a detached garage, the same rules apply as those for an attached garage.

  ○ Supplying a detached garage with electrical power

    ■ Install overhead conductors from the main house to the detached garage.

    ■ Install underground wiring from the main house to the detached garage.

  ○ Usually the electrical loads required in a detached garage require more electrical power than a single branch circuit can supply.

  ○ Install feeder wiring from the main service entrance panel to a subpanel located in the garage.

     ○ Several different branch circuits can originate in the garage subpanel and supply the electrical load requirements of the garage.

     ○ Bring the feeder to the detached garage as either a cable, typically Type UF cable, or a raceway wiring method, typically rigid PVC conduit (PVC).

     ○ A multiwire branch circuit could be installed using a three-wire Type UF cable or three wires and an equipment grounding conductor in a raceway.

Objective: Demonstrate an understanding of the installation of branch circuits for a swimming pool.

2) Understand the installation of branch circuits for a swimming pool

Key terms: dry-niche luminaire, exothermic welding, forming shell, hydromassage bathtub, no-niche luminaire, permanently installed swimming pool, pool cover, electrically operated, spa or hot tub, self-contained spa or hot tub, storable swimming pool, wet location, wet-niche luminaire

- Installing branch-circuit wiring for a swimming pool

     ○ Article 680 covers swimming pools, hydromassage bathtubs, hot tubs, and spas.

     ○ Section 680.2 has many terms and definitions that relate specifically to the items covered in Article 680.

          ■ A spa or hot tub is a hydromassage pool or tub designed for immersion of users and usually has a filter, heater, and motor-driven pump.

               ● Can be installed indoors or outdoors, on the ground, or in a supporting structure.

          ■ A self-contained spa or hot tub is a factory-fabricated unit consisting of a spa or hot tub vessel having integrated water-circulating, heating, and control equipment.

          ■ A hydromassage bathtub is a permanently installed bathtub with recirculating piping, pump, and associated equipment.

          ■ A wet-niche luminaire is a type of lighting fixture intended for installation in the wall of a pool.

               ● It is accessible by removing the lens from the forming shell.

               ● It is designed so that water completely surrounds the fixture inside the forming shell.

          ■ A dry-niche luminaire is intended for installation in the wall of a pool and goes in a niche.

               ● It has a fixed lens that seals against water entering the niche and surrounding the lighting fixture.

               ● A no-niche luminaire is intended for above or below the water level installation.

               ● It does not have a forming shell.

               ● It sits on the surface of the pool wall.

               ● It can be located above or below the waterline.

          ■ A permanently installed swimming pool is one constructed totally or partially in the ground with a water-depth capacity of greater than 42 inches (1 meter).

          ■ A storable swimming pool is constructed on or above the ground and has a maximum water-depth capacity of 42 inches (1 meter).

               ● Also can be a pool with nonmetallic, molded polymeric walls (or inflatable fabric walls) regardless of size or water-depth capacity.

          ■ A pool cover, electrically operated, is a motor-driven piece of equipment designed to cover and uncover the water surface.

- A forming shell is the support structure designed and used with a wet-niche lighting fixture.
    - It is installed in the wall of a pool.
- Installing branch-circuit wiring for a permanently installed swimming pool
    - The *NEC®* rules for permanently installed pools are located in Part II of Article 680.
    - Part II provides installation rules for clearances of overhead lighting, underwater lighting, receptacles, switching, associated equipment, bonding of metal parts, grounding, and electric heaters.
    - Probably the most important installation practice is the proper method of bonding together all metal parts in and around the pool, as required in Section 680.26.
        - Use an 8 AWG or larger solid copper conductor to make the bonding connections.
        - Make the connection to the metal parts with exothermic welding or by pressure connectors labeled for the purpose and made of stainless steel, brass, copper, or a copper alloy.
        - Exothermic welding is a process using specially designed connectors, a form, a metal disk, and explosive powder.
            - Process is sometimes called CADWELD®.
    - Other methods acceptable to the *NEC®* for bonding swimming pool metal parts:
        - Structural reinforcing steel of a concrete pool where the rods are bonded together by steel tie wires or the equivalent.
        - Walls of a bolted-together or welded-together metal pool.
        - Brass rigid metal conduit (RMC) or other identified corrosion-resistant metal.
    - Swimming pool electrical installation usually includes installing one or more receptacles.
        - Section 680.22(A) includes the installation requirements for receptacles that must be followed closely.
        - Receptacles installed for a pump motor or other loads directly related to the circulation and sanitation system for the pool can be located between 6 feet (1.83 meters) and 10 feet (3 meters) from the inside walls.
        - Receptacles used and located accordingly must be:
            - A single receptacle of the proper voltage and amperage rating.
            - Locking and grounding type.
            - GFCI protected.
            - Section 680.22(B) requires outlets supplying pool motors and that are rated 15- or 20-amp, 120 through 240 volts to be GFCI protected whether they are cord and plug connected or hardwired.
        - A permanently installed swimming pool must have at least one 125-volt, 15- or 20-ampere-rated receptacle located a minimum of 6 feet (1.83 meters), but no more than 20 feet (6 meters) from the pool's inside wall.
        - Section 406.9(B) requires any 15- or 20-amp, 125- or 250-volt receptacle installed in a wet location (indoors or outdoors) to have a type of cover or enclosure that keeps it weatherproof at all times whether a plug is inserted in it or not.
        - The receptacle must also be a listed weather-resistant type.

- ○ Some permanently installed pool installations will require luminaires or ceiling-suspended paddle fans.

  - Section 680.22(C) lists the installation of luminaires over the pool.

  - Some swimming pools will require the installation of luminaires underwater in the walls of the pool.

- ○ Follow these rules when installing the junction boxes used with underwater lighting.

  - Locate the junction box at least 4 feet (1.2 meters) from the pool's inside wall, unless separated from the pool by a solid fence, wall, or other permanent barrier.

  - The junction box must be made of copper, brass, suitable plastic, or other approved corrosion-resistant material.

  - Measured from the inside of the bottom of the junction box, locate the box at least 8 inches (200 millimeters) above the maximum water level of the pool.

  - The junction box must be equipped with threaded entries or a specifically listed nonmetallic hub.

  - Locate the junction box no less than 4 inches (100 millimeters) above the ground level or pool deck.

- ○ The junction box support must comply with Section 314.23(E), which requires the box to be supported by at least two or more conduits threaded into the box.

- ○ Section 680.23(B) requires that the metal conduit connection from the junction box to the forming shell to be made with a brass rigid metal conduit or other identified corrosion resistant metal.

- ○ PVC conduit is what most electricians install.

- Installing branch-circuit wiring for storage pools

  - ○ Most storable pools have no lighting fixtures installed in or on them.

  - ○ However, Section 680.33(A) says that if lighting fixtures are installed, they must be cord-and-plug connected and be a listed assembly.

  - ○ The lighting fixture assembly must be properly listed for use with storage pools and must have the following construction features:

    - No exposed metal parts.

    - A lamp that operates at no more than 15 volts.

    - Impact-resistant polymeric lens, lighting fixture body, and transformer enclosure.

    - A transformer (to drop the voltage from 120 volts to 15 volts) with a primary voltage rating of not over 150 volts.

  - ○ Lighting fixtures without a transformer can also be used.

- Section 680.33(B) says that they must operate at 150 volts or less and can be cord-and-plug connected.

  - ○ The lighting fixture assembly must have the following construction features:

    - No exposed metal parts.

    - An impact-resistant polymeric lens and fixture body.

    - GFCI with open neutral conductor protection as an integral part of the assembly.

  - ○ When installing the wiring for a storable pool, install a receptacle in a location that allows the filter system pump to be plugged in.

    - Section 680.32 states that it must be GFCI protected.

- Section 406.9(B) requires the receptacle to have a cover or enclosure that maintains its weatherproof capability, whether a pump cord is plugged in or not.

- Must also be a listed weather-resistant receptacle.

- Section 680.31 states that the cord-and-plug-connected pool filter pump must incorporate an approved double-insulated system and must have a means for grounding the appliance's internal and non-accessible noncurrent-carrying metal parts.

  - It also requires that cord-connected pool filter pumps be provided with a GFCI that is built into the attachment plug or located in the power supply cord within 12 inches (300 millimeters) of the attachment plug.

- Installing branch-circuit wiring for spas and hot tubs

  o Electrical installation requirements for spas and hot tubs are found in Part IV of Article 680.

  o Section 680.42 specifies the wiring methods allowed for outdoor installations.

  - They include flexible connections using flexible raceway or cord and plug connections.

  - A spa or hot tub assembly can be connected using regular wiring methods recognized in Chapter 3 of the *NEC®* as long as the wiring method has a copper equipment grounding conductor not smaller than 12 AWG.

  o Section 680.43 covers indoor installation requirements.

  o The following installation requirements apply to receptacle installation.

  - At least one 125-volt, 15- or 20-ampere-rated receptacle connected to a general purpose branch circuit must be installed at least 6 feet (1.83 meters), but not more than 10 feet (3 meters) from the inside wall of the spa or hot tub.

  - All receptacles installed in the area of the spa or hot tub must be located at least 6 feet (1.83 meters), measured horizontally, from the inside wall of the spa or hot tub.

  - All 125-volt, 15-, 20-, or 30-amp-rated receptacles located within 10 feet (3 meters) of the inside walls of the spa or hot tub must be GFCI protected.

  - The receptacle that supplied power to the spa or hot tub must be GFCI protected.

  o The following installation requirements apply to luminaire installations or ceiling-suspended paddle-fan installations around a spa or hot tub.

  - They must not be installed less than 7 feet (3.7 meters) above the spa or hot tub unless GFCI protected.

  - They must not be installed less than 7 feet 6 inches (2.3 meters) above the spa or hot tub, even when GFCI protected.

    - If located less than 7 feet 6 inches (2.3 meters) above the spa or hot tub, they must be suitable for a damp location.

  - A recessed luminaire with a glass or plastic lens and a nonmetallic or isolated metal trim.

  - A surface-mounted luminaire with a glass or plastic globe, a nonmetallic body, or a metal body that is isolated from contact.

  - If underwater lighting is installed for a spa or hot tub, the same rules apply as for underwater lighting in a swimming pool.

  - When installing wall switches for the lighting fixtures or ceiling-suspended paddle fans, keep them a minimum of 5 feet (1.5 meters) from the inside edge of the spa or hot tub.

- Installing branch-circuit wiring for hydromassage bathtubs
  - ○ Part VII of Article 680 covers the installation requirements for hydromassage bathtubs usually installed in a bathroom of a home.
  - ○ Remember that Section 210.8(A)(1) requires all bathroom receptacles (125-volt, 15- or 20-ampere-rated) to be GFCI protected.
  - ○ Section 680.71 also states that all hydromassage bathtubs and any associated electrical equipment must be on an individual branch circuit and be GFCI protected.
  - ○ Section 680.74 requires:
    - ■ All metal piping, electrical equipment metal parts, and pump motors associated with the hydromassage bathtub must be bonded together using an 8 AWG or larger solid copper conductor.
    - ■ Electricians usually run a separate circuit to a hydromassage bathtub.
    - ■ It is normally either a 15-amp circuit using a 14 AWG conductor in a cable or raceway, or a 20-amp circuit using a 12 AWG conductor.

Objective: Demonstrate an understanding of the installation of branch circuits in outdoor situations.

3) Understand the installation of branch circuits in outdoor situations

- Installing outdoor branch-circuit wiring
  - ○ Outdoor electrical wiring in residential situations includes installing the wiring and equipment for lighting and power equipment located outside of the house.
  - ○ Wiring may be installed overhead or underground.
- Installing outdoor branch-circuit underground wiring
  - ○ Most underground receptacle and lighting circuits installed in residential wiring are done using Type UF Cable (Article 340 covers Type UF Cable).
  - ○ When determining the size of Type UF Cable to use for a particular wiring situation, find the ampacity of the cable using the 60°C column in Table 310.15(B)(16), just like nonmetallic sheathed cable.
  - ○ According to Article 340, Type UF cable:
    - ■ Must be marked as underground feeder cable.
    - ■ Is available from 14 AWG through 4/0 AWG copper and from 12 AWG through 4/0 aluminum.
    - ■ Can be used outdoors in direct exposure to the sun only if listed as being sunlight-resistant with a sunlight-resistant marking on the cable sheathing.
    - ■ Can be used with the same fittings as used with nonmetallic sheathed cable.
    - ■ Can be buried directly in the ground and installed according to Section 300.5 and Table 300.5.
    - ■ Must be installed according to the same rules as for nonmetallic sheathed cable when used as an interior wiring method.
    - ■ Contains an equipment grounding conductor (bare) used to ground equipment fed by the Type UF cable.
  - ○ Any wiring installed in the underground conduits must have a "W" in its insulation designation, such as "THWN" or "XHHW."
    - ■ The "W" means that the conductor insulation is suitable for installation in a wet location.

○ Use rigid PVC conduit (PVC) for underground conduit installation.

○ The minimum burial depths for both Type UF cable and for any of the conduit wiring methods are shown in Table 300.5 of the *NEC®*.

- Installing outdoor receptacles

  ○ You must install receptacle outlets located outdoors in weatherproof enclosures.

  ○ The electrical boxes are usually made of metal and are of the "FS" or "FD" type.

    ▪ They have threaded openings or hubs that allow attachment to the box with conduit or a cable connector.

    ▪ Each of these boxes comes from the factory with a few threaded plugs that are used in any unused threaded openings to make the box truly weatherproof.

    ▪ These boxes can be mounted on the surface of an outside wall or on some other structural support such as a wooden post driven into the ground.

    ▪ They are often installed with underground wiring and supported by conduits coming up out of the ground.

    ▪ Section 314.23(E) and (F) covers the support rules for electrical boxes when they are fed using an underground wiring method and supported by conduit.

    ▪ According to Section 408.9(B), when a receptacle is installed outdoors, the enclosure and cover combination must maintain its weatherproof characteristics whether a cord plug is inserted into the receptacle or not.

      • Done by installing a self-closing cover that is deep enough to also cover the attached plug cap on a cord.

    ▪ Section 404.8(A) and (B) requires all 15- and 20-ampere, 125- and 250-volt non-locking receptacles installed in damp or wet locations to be a listed weather-resistant type.

- Installing outdoor lighting

  ○ Outdoor lighting can be mounted on the side of building structures, on poles, or even on trees.

  ○ Any luminaire installed outdoors and exposed to the weather must be listed as suitable for the location.

  ○ Must have a label with a marking that states "suitable for wet locations."

  ○ If a luminaire is to be installed under a canopy or under an open porch, it is considered a 'damp' location and the fixture only needs a label that states "suitable for damp locations."

  ○ Section 410.36(G) allows outdoor lighting fixtures to be mounted on trees.

    ▪ However, Sections 225.26 and 590.4(J) state that overhead conductor spans cannot be supported by trees or other living or dead vegetation.

    ▪ So, if installing wiring to a tree-mounted lighting fixture, use an underground wiring method.

  ○ A pole light is a common wiring installation in residential wiring.

  ○ Photoelectric sensors, often called a "photocell" can be used to switch outside lighting fixtures on when it gets dark outside and then switch them off when the sun comes up in the morning.

Objective: Demonstrate an understanding of the installation of a standby power system.

4) Understand the installation of a standby power system

> **Key terms: standby power system, twistlock receptacle, critical loads, transfer switch, generator**

- Installing the wiring for a standby power system
  - Many home owners have opted to have a standby power system installed so they will not need to be without electrical power for prolonged periods of time.
  - Most common standby power system installations use a portable generator and the associated equipment necessary to safely feed branch circuits already installed as part of the residential wiring system.
    - Permanently installed generators can also be used.
  - Generator ratings can range from around 3000 watts up to 16,000 watts or more.
  - Before installing the standby power system, a home owner must determine which electrical loads are critical and which loads are not.
  - Some generator manufacturers suggest that after all of the critical loads are added up, another 20% should be added for any future loads that may be included.
  - The installation for the standby power system that is being presented in this chapter includes a generator that is cord-and-plug connected to an outdoor power inlet and then to a transfer switch.
    - When the transfer switch is in the "normal" position, electrical power from the electric utility is routed to the branch circuits in the critical load panel.
    - When the transfer switch is in the "standby generator" position, it disconnects the critical load branch circuits from the incoming electric utility power.
    - The transfer switch eliminates the possibility of the generator voltage being applied to the secondary of the electric utility's supply transformer, resulting in a very high voltage being put back on the "dead" electric utility wiring from the primary of the home supply transformer.
      - This is commonly called backfeeding and could be a very dangerous situation when utility linemen working on the power lines believe them to be de-energized, but they are really "live" with a high voltage from the home owner's generator.

Quizzes

SAFETY

1. When installing the wiring for special residential wiring situations, electricians do not need to wear safety glasses and observe all applicable safety rules. True or False?

CODE/REGULATIONS/STANDARDS

1. When used as an interior wiring method, Type UF cable must be installed according to the same rules as Type NM cable. True or False?

2. Overhead conductor spans are permitted to be supported by trees. True or False?

3. Any 15- or 20-amp, 125- through 250-volt receptacle installed in a wet location must have a cover or enclosure that keeps it weatherproof, but only when there is not anything plugged into it. True or False?

4. The maximum distance between receptacles in a garage is _____ feet.

    a. 6

    b. 12

    c. There is no *NEC*® requirement.

5. For a permanently installed swimming pool, any 15- or 20-amp, 125- through 250-volt-rated receptacle located within _____ feet of a pool's inside wall must be GFCI protected.

    a. 3

    b. 10

    c. 20

6. The minimum burial depth for type UF cable can be found in Table _____ of the *NEC*®.

    a. 110.26

    b. 220.12

    c. 300.5

7. Assuming a 20-amp circuit breaker and GFCI protection, the minimum burial depth for a Type UF cable running to a 120-volt pole light at a residence is _____ inches.

    a. 6

    b. 12

    c. 18

8. Article _____ in the *NEC*® covers the requirements for swimming pools, hot tubs, and hydromassage tubs.

    a. 110

    b. 250

    c. 680

Chapter Quiz

1. Rigid PVC conduit is not permitted as the only means of support for a junction box used with underwater lighting. True or False?

2. A GFCI receptacle is an acceptable means to provide the required GFCI protection to a hydromassage bathtub. True or False?

3. Overhead conductor spans are permitted to be supported by trees. True or False?

4. A detached garage must always have at least one 125-volt, 15- or 20-amp receptacle and at least one wall switch-controlled lighting outlet. True or False?

5. In attached garages, the *NEC*® requires at least _____.

    a. one single 125-volt receptacle

    b. one duplex 125-volt receptacle

    c. one wall switch-controlled lighting outlet

    d. one 125-volt, 15- or 20-amp receptacle and one wall switch-controlled lighting outlet

6. The maximum distance between receptacles in a garage is _____ feet.

    a. 4

    b. 6

    c. 12

    d. There is no *NEC*® requirement.

7. According to the *NEC®*, which of the following 125-volt, 15- or 20-amp duplex receptacles installed in a garage to serve cord-and-plug appliances, such as a refrigerator or freezer, requires GFCI protection?

   a. Duplex receptacles located beside the appliance

   b. Duplex receptacles located behind the appliance

   c. Duplex receptacles located on the wall and just above the appliance

   d. All of the above

8. The *NEC®* requires that if a feeder supplying a detached garage from the service in the main house has an equipment grounding conductor run with the feeder, the grounded conductor _____ be connected to the equipment grounding conductor or to the grounding electrode system in the garage.

   a. may

   b. shall

   c. shall not

   d. could

9. A permanently installed swimming pool must have at least one 125-volt, 15- or 20-amp receptacle located a minimum of _____ feet from the inside wall of the pool.

10. In an outdoor pool area, the *NEC®* requires luminaires, ceiling-suspended paddle fans, or lighting outlets not to be installed over the pool or an area extending _____ feet horizontally from the inside wall of the pool.

11. The conductors that provide power to pool lighting fixtures must be GFCI protected if the voltage is above _____ volts.

12. For spas and hot tubs, at least one 125-volt, 15- or 20-ampere-rated receptacle must be installed at least _____ from the inside wall of the spa or hot tub.

13. Type UF cable can be used outdoors in direct exposure to the sun only if _____.

   a. it is listed as being sunlight resistant

   b. it has a sunlight-resistant marking on the cable sheathing

   c. it is rated for less than 300 volts

   d. both A and B

14. The minimum burial depth for Type UF cable can be found in Table _____ of the *NEC®*.

15. Pool motors can be cord-and-plug connected but if they are, the cord cannot be more than _____ feet long and must contain an equipment grounding conductor sized according to Table 250.122.

16. According to the *NEC®*, the minimum burial depth for a direct buried cable like Type UF is _____ inches. (general rule)

17. Article _____ in the *NEC®* covers the requirements for swimming pools, hot tubs, and hydromassage tubs.

18. Without a special ballast, fluorescent lamps installed in a garage usually will not start when exposed to temperatures below _____ °F.

19. A(n) _____ switch eliminates the possibility of a generator voltage being applied to the secondary of the utility's supply transformer.

20. A circuit rated at 6600 watts and 240 volts will draw _____ amperes.

Troubleshooting/Quality Exercises

DISCUSSION QUESTIONS

1. A storage shed is a structure often found on a home owner's property and is used to store things such as lawnmowers, gardening tools, and kid's toys. Discuss the wiring requirements for a storage shed.

_____

_____

_____

_____

HANDS-ON APPLICATIONS

1. Match the following terms to their definitions. Write the correct number for the term in the blanks.

   1. critical load
   2. wet location
   3. exothermic welding
   4. twistlock receptacle
   5. generator
   6. transfer switch
   7. permanently installed swimming pool
   8. storable swimming pool

   _____ a. Installations underground or in concrete slabs in direct contact with the earth

   _____ b. An electrical load that is determined to require power from a standby power generator

   _____ c. A swimming pool constructed on or above the ground with a maximum water depth of 42 inches

   _____ d. A switching device for transferring one or more load conductor connections from one power source to another

   _____ e. A rotating machine used to convert mechanical energy into electrical energy

   _____ f. A process for making bonding connections on the binding grid for a permanently installed swimming pool

   _____ g. A swimming pool constructed totally or partially in the ground with a water depth of greater than 42 inches

   _____ h. A type of receptacle that requires the attachment plug to be inserted and then turned in a clockwise direction to lock the plug in place

2. The outside sheathing of a Type UF cable used in underground wiring is much more difficult to remove than the outside sheathing of a Type NM cable. List the five steps used to remove the outside sheathing of a Type UF as suggested in Chapter 15 of the *House Wiring* textbook.

   Step 1:

   Step 2:

   Step 3:

   Step 4:

   Step 5:

REAL-WORLD SCENARIOS

1. When installing a standby power system for a home, a transfer switch is required that will eliminate the chance of electrical power from a generator being fed back onto the electric utility lines. In an effort to save money, many electrical contractors use two double-pole circuit breakers in the main loadcenter that are mechanically interlocked as the transfer switch. Explain how this arrangement works.

_____

_____

_____

_____

2. When a standby power system includes a portable generator that is cord-and-plug connected to the transfer switch, you should follow a specific procedure to safely connect the generator power to the critical load branch circuits when electrical power from the utility is lost. List the seven steps that you should follow to connect a generator's electrical power to the critical load branch circuits as suggested in the *House Wiring* textbook.

Step 1:

Step 2:

Step 3:

Step 4:

Step 5:

Step 6:

Step 7:

Chapter 16

Video, Voice, and Data Wiring Installation

Outline

Objective: List several common terms and definitions used in video, voice, and data cable installations.

1) List common terms and definitions for video, voice, and data cable installations

Key terms: attenuation, bandwidth, category, coaxial cable, EIA/TIA, EIA/TIA 570-B, F-Type connector, structured cabling, insulation displacement connection (IDC), unshielded twisted pair, RG-6 (series 6), RG-59, RJ-11, RJ-45, jack, punch-down, horizontal cabling, work area outlet, patch cord, service center, megahertz, megabits per second, work area outlet

- Terms used in structured cabling
  - Structured cabling
    - A system for video, voice, and data communications cabling specified by EIA/TIA.
    - Used as a voluntary standard by manufacturers to ensure compatibility.
  - EIA/TIA
    - Abbreviation for the Electronic Industry Association (EIA) and the Telecommunications Industry Association (TIA).
  - Bandwidth
    - The amount of data that can be sent on a given cable.
    - Measured in hertz (Hz) or megahertz (MHz).
  - Insulation displacement connection (IDC)
    - Type of termination where the wire is "punched down" into a metal holder with a punch-down tool.
  - Unshielded twisted pair (UTP)
    - Type of cable normally used to install voice and data communication wiring in a house.
    - Consists of four pairs of copper conductors and is graded for bandwidth as "categories."
  - Category
    - Rating, based on the bandwidth performance, of UTP cable.
  - Shielded twisted pair (STP)
    - Cable resembles UTP but has a foil shield over all four pairs of copper conductors.
    - Used for better high-frequency performance and less electromagnetic interference (EMI).

171

- Coaxial cable
  - The center signal-carrying conductor is centered within an outer shield and separated from the conductor by a dielectric.
  - Used to install video signal wiring or to carry a high-speed Internet signal.
- RG-6 (Series 6)
  - Type of coaxial cable that is "quad shielded."
  - Used in residential structured cabling systems to carry video signals such as cable and satellite TV.
- RG-59
  - Type of coaxial cable typically used for residential video applications.
  - EIA/TIA-570-B recommends using RG-6 coaxial cable instead of RG-59 because RG-6 cable has better performance characteristics.
- F-Type connector
  - 75-ohm coaxial cable connector that can fit RG-6 and RG-59 cables.
  - Used for terminating video system cables.
- RJ-11
  - Popular name given to a six-position connector or jack.
  - RJ stands for "registered jack."
- RJ-45
  - Popular name given to an eight-pin connector or jack used to terminate UTP cable.
- Jack
  - The receptacle device that accepts an RJ-11 or RJ-45 plug.
- Punch-down block
  - Connecting block that terminates UTP cables directly.
  - 110 blocks most popular for residential applications.
  - Require a 110-block punch-down tool for making the terminations.
- Horizontal cabling
  - Identifies cables that run from a service center serving as the "hub" for the structured cabling system to the work area outlet.
- Work area outlet
  - Jack on the wall connected to a desktop computer by a patch cord.
- Service center
  - The hub of a structured wiring system with telecommunications, video, and data communications installed.
- Patch cord
  - Short length of cable with an RJ-45 plug on either end.
  - Used to connect a home computer to the work area outlet or to interconnect various punch-down blocks in the service center.

- ○ Megahertz (MHz)

  - ▪ Upper frequency band on the ratings of a cabling system.

- ○ Megabits per second (Mbps)

  - ▪ Rate at which digital bits (1s and 0s) are sent between two pieces of digital electronic equipment.

  - ▪ Attenuation

- ○ The decrease in the power of a signal as it passes through a cable measured in decibels

Objective: Demonstrate an understanding of EIA/TIA 570-B standards for the installation of video, voice, and data wiring in residential applications.

2) Understand EIA/TIA 570

- • Introduction to EIA/TIA 570-B standards

- ○ EIA/TIA 570-A was developed by the committee and published in 1999 to establish a standard for a generic cabling system that can accommodate many different applications.

- ○ In 2004 EIA/TIA 570-B was published and included updates to the original 570-A standard.

- ○ In 2008 EIA/TIA 570-B-1 was published with an addendum that included some new guidelines for coaxial cable broadband installations.

- ○ In today's homes, there is often a need for a structured cabling system to be installed.

  - ▪ Completely separate from the residential electrical power system.

  - ▪ Includes the wiring and other necessary components for providing video, voice, and data signals throughout a house.

  - ▪ Wired with special cables:

    - • Unshielded twisted pair cable (UTP)

    - • Shielded twisted pair cable (STP)

    - • Coaxial cables

Objective: Identify common materials and equipment used in video, voice, and data wiring.

3) Identify materials and equipment used in video, voice, and data wiring

Key terms: UTP, STP, tip, ring, punch-down block

- • Cable categories

- ○ Category 1

  - ▪ Four wires, not twisted.

  - ▪ Called "quad wire" by electricians.

  - ▪ Okay for audio and low-speed data transmission.

  - ▪ Not recommended for use in residential applications.

- ○ Category 2

  - ▪ Four pairs, with a slight twist to each pair.

  - ▪ Okay for audio and low-speed data transmission.

  - ▪ Not recommended for use in residential applications.

- ○ Category 3
  - 16-MHz bandwidth.
  - Supports applications up to 10 Mbps.
  - 100-ohm UTP-rated Category 3.
  - Declining in popularity, but can be used in residential telephone applications.
- ○ Category 4
  - 20-MHz bandwidth.
  - Supports applications up to 16 Mbps.
  - 100-ohm UTP-rated Category 4.
  - Basically obsolete.
- ○ Category 5
  - 100-MHz bandwidth.
  - Supports applications up to 100 Mbps.
  - 100-ohm UTP-rated Category 5.
  - Often-used UTP cable for residential voice and data applications.
- ○ Category 5e
  - 100-MHz bandwidth.
  - Supports applications up to 100 Mbps.
  - 100-ohm UTP-rated Category 5e.
  - Higher performance over a minimally compliant Category 5 installation by following Category 5e technical specifications.
  - The preferred cable type for both voice and data in residential applications.
- ○ Category 6
  - 250-MHz bandwidth.
  - Supports applications over 100 Mbps.
  - 100-ohm UTP-rated Category 6.
  - The favorite in commercial applications and is recommended for use in residential applications where the high bandwidth of Cat 6 is needed to provide the speed that many newer applications require.
- ○ Category 7
  - 750-MHz bandwidth.
  - Has a metal shield around the conductors.
  - Not used in residential applications at this time.
- Cable characteristics
  - ○ The voice and data cable recommended by EIA/TIA 570-B for use in residential applications has the following characteristics:
    - Carries both voice and data.
    - Normally 22 or 24 AWG copper.

- Always described and connected in pairs.
- The wire pairs should be twisted together to preserve signal quality.
  ○ Each pair of wires in a voice/data structured cabling system cable consists of a tip and a ring wire
    - Carry-over from the old days in the telephone industry.
    - Use the tip wire as the positive (+) conductor and the ring as the negative (–) conductor.
  ○ Cable color coding
    - The standard color coding for a four-pair UTP cable.
      • Pair 1: tip is white/blue; ring is blue.
      • Pair 2: tip is white/orange; ring is orange.
      • Pair 3: tip is white/green; ring is green.
      • Pair 4: tip is white/brown; ring is brown.
      • Easy way to remember the color coding for a four-pair UTP cable is to use the acronym BLOGB.
        • BL stands for blue.
        • O stands for orange.
        • G stands for green.
        • B stands for brown.

Objective: Demonstrate an understanding of the installation of video, voice, and data wiring in a residential application.

Objective: Install crimp-on and compression style F-Type coaxial cable connectors.

Objective: Install RJ-45 jacks and plugs on Category 5e and Category 6 unshielded twisted pair cable.

4) Understand the installation of video, voice, and data wiring

5) Install F-Type coaxial cable connectors

6) Install RJ-45 jacks and plugs

- Installing residential video/voice/data circuits
  ○ Install a service center that serves as the origination point for all video, voice, and data systems in a house.
  ○ Locate the service center (sometimes called a distribution center) in the basement, garage, or some other utility area of the house.
    - There must be readily available electrical power for the service center and access to the service grounding electrode system.
- Installing a voice wiring system
  ○ Use four-pair 100-ohm UTP cable.
  ○ Category 3 cable is the minimum performance category that should be installed.
  ○ Use at least a Category 5e in all new voice installations.

- At each wall outlet location, an eight-position RJ-45 jack with T568A wiring should be used.

  - T568B wiring may also be used.

  - Whichever wiring scheme is used, make sure all jacks in the house are wired the same way.

- There should be a minimum of one voice jack per outlet location and each voice outlet location should have a separate "home run" back to the service center.

- Terminate the "home runs" at the service center by punching down the cable to the proper 110 terminal blocks.

- The telephone company will terminate their wiring to a telephone network interface (TNI) box mounted on the outside of the house.

- The point where the telephone company's wiring ends and the home owner's interior wiring begins is called the demarcation point.

- You will need to install telephone wiring from the demarcation point to the service center.

- Installing a data wiring system

  - When wiring for the data wiring system, use a four-pair 100-ohm UTP cable.

  - It is recommended you use a Category 5e as the minimum category rated cable for this type of installation.

  - Category 6 cable is often specified in anticipation of faster data transmission speed requirements in the future.

  - At each wall outlet location, use an eight-position RJ-45 jack with T568A wiring.

    - T568B wiring may also be used.

    - Whichever wiring scheme is used, make sure all jacks in the house are wired the same way.

  - There should be a minimum of one data jack at each wall outlet location and each data outlet location should have a separate "home run" back to the service center.

  - Terminate the "home runs" at the service center by punching down the cable to the proper 110 terminal blocks.

- Installing a video wiring system

  - Use a RG-6, 75-ohm coaxial cable.

  - Install two runs of coaxial cable from the service center to each TV outlet location.

    - This allows for video distribution from any video source as well as distribution to a TV at each outlet location.

    - Extra cable could be used to serve as a data transmission line for high-speed Internet for a computer located close to the TV outlet location.

  - Two runs are also recommended to be run to a convenient attic or basement location in case a satellite television system will be installed at a later date.

  - At each end of the coaxial cable, install F-Type connectors.

  - Threaded F-type connectors that are crimped on or compressed on using a special tool are recommended to reduce signal interference.

  - At the wall-mounted TV outlet end, use a female-to-female F-Type coupler.

  - At the service center end, thread the F-Type connector onto the proper fitting.

  - At the TV outlet end, connect a video device using a 75-ohm RG-6 coaxial patch cord.

  - Sometimes, because of the higher cost associated with the EIA/TIA recommended method, home owners choose to install the more traditional coaxial wiring technique of running just one cable to each location.

- Voice/data/video installation safety considerations
    - Never install or connect telephone wiring during an electrical storm.
    - Never install jacks where a person could use a telephone (hard-wired) while in a bathtub, hot tub, or swimming pool.
    - Do not run open communications wiring between structures where it may be exposed to lightning.
    - Avoid telecommunications wiring in or near damp locations.
    - Never place telephone wires near bare power wires or lightning rods.
    - Never place voice/data wiring in any conduit, box, or other enclosure that contains power conductors.
    - Always maintain adequate separation between voice/data wiring and electrical wiring according to the *NEC®*.
    - 50 to 60 volts of direct current is normally present on an idle telephone tip and ring pair.
    - When a call comes in, there are 90 volts of alternating current that can cause a shock under the right conditions.
    - Always disconnect the dial tone service from the house when working on an existing phone system.
        - If you cannot disconnect, simply take the receiver off the hook and the DC value will drop.
        - The 90-volt AC ring will not be available.
- Common voice/data/video installation practices
    - Follow these steps when installing the structured wiring cables.
        - Keep the cable runs as short as possible.
        - Do not splice wires on the cable runs.
        - Run the cables as one continuous length.
        - Do not pull the wire with more than 25 pounds of pulling tension (four-pair).
        - Do not run the wire too close to electrical power wiring.
        - Do not bend too sharply.
        - Do not install the cable with kinks or knots.
        - There are no *NEC®* or EIA/TIA 570-B requirements for the maximum distance between supports for the cables.
        - There is no requirement for the minimum distance from a box or panel that the cables have to be secured.
        - Provide adequate support so that the cables follow the building framing members closely and so that there are no lengths of the cable that sag excessively.
        - Use insulated rounded or depth-stop plastic staples to secure the cables to the building framing members.
        - Use tie-wraps (secured loosely) when you have a bundle of several cables to support.
        - Maintain polarity and match color coding throughout the house.
        - To provide compatibility with two-line telephones, wire up the two inner pairs of an RJ-45 jack.
            - Use the T568B wiring scheme.
        - If conduit is installed, leave a pull string so the voice/data and video cable can be pulled in later.

- Never run voice/data or video wiring in the same conduit with power wires.
- Use inner structural walls instead of outer walls for cable runs whenever possible.
- Do not run the cables through bored holes with power wires.
- Keep the cables away from heat sources like hot water pipes, furnaces, etc.
- Avoid running exposed cables whenever possible.
- Leave about 24 inches of wire at outlets and connection points.
- Always check for shorts, opens, and grounds when the rough in is complete.
- Testing a Residential Structured Cabling System
  - Certification refers to the process of making measurements and then comparing the results obtained to pre-defined standards, so that a pass/fail determination can be made.
  - Rather than being "certified," most home wiring systems are "verified."
  - In verifying residential cabling, the most important measurement is often called wire-mapping.

Quizzes

SAFETY

1. When installing the wiring for video, voice, and data wiring, electricians need to wear safety glasses and observe all applicable safety rules. True or False?

2. Electricians cannot get an electrical shock from a telephone line. True or False?

3. Telephone jacks should not be installed where a person could use a hard-wired telephone while in a _____.

   a. bathtub
   b. swimming pool
   c. both of the above

CODE/REGULATIONS/STANDARDS

1. EIA/TIA 570-B was developed to establish a standard for a _____ structured cabling system that can accommodate many different applications.

   a. communications
   b. voice
   c. generic

2. The letters CMR found on a UTP cable is an abbreviation for _____.

   a. communications plenum cable
   b. communications riser cable
   c. communications general purpose cable

3. When installing voice, video, and data wiring in a house, Articles _____ and _____ of the *NEC*® contain some requirements that must be followed.

   a. 100, 300
   b. 430, 440
   c. 800, 820

TOOLS

1. A _____ with a 110 blade is used to terminate voice and data wires to a 110 terminal block.
   a. screwdriver
   b. punch-down tool
   c. wire stripper

2. A _____ is used to strip coaxial cable in preparation for installing an F-Type connector.
   a. coaxial cable stripper
   b. T-stripper
   c. knife

Chapter Quiz

1. A standard network interface box is used when the telephone company terminates their wiring outside the house. True or False?

2. A suspended ceiling grid cannot be used to support structured cabling systems. True or False?

3. An RJ-45 is a popular six position UTP connector. True or False?

4. When the prefix "mega" is used in measurement, it means _____.
   a. hundred
   b. thousand
   c. million
   d. billion

5. A(n) _____-Type connector is a 75-ohm coaxial cable connector that can fit RG-6 and RG-59 cables and is used for terminating coaxial cables.

6. Category _____ cable is the preferred UTP cable for residential voice and data applications.

7. Which of the following is not a recommended installation method for structured wiring cables?
   a. Keep the cable runs as short as possible
   b. Keep splices to a minimum and make sure they are properly made
   c. Do not pull wire with more than 25 pounds of pulling tension
   d. Do not run wires too close to electrical power wiring

8. It is recommended that UTP cable be bent so that the radius of the bend is not less than _____ times the diameter of the cable.

9. The letters CMR found on a UTP cable is an abbreviation for _____.
   a. communications plenum cable
   b. communications riser cable
   c. communications general purpose cable
   d. communications radio cable

10. Coaxial cable limited to use in dwelling units or raceways is labeled _____.
    a. CATV
    b. CATVP

    c. CATVR

    d. CATVX

11. Coaxial cable must be separated from electric light and power circuits by at least _____ inches. (general rule)

12. The popular name given to an eight-pin connector or jack used to terminate UTP cable is

    _____.

    a. F-Type

    b. RG-59

    c. RJ-11

    d. RJ-45

13. A _____ is the hub of a structured wiring system and is usually located in the basement or garage. It is sometimes called a "distribution center."

14. The color coding for a four-pair UTP cable can be remembered by using the acronym BLOGB. Using this as a reminder, the colors used for pair #3 are _____ and _____.

    a. white/blue stripe; blue

    b. white/green stripe; green

    c. white/orange stripe; orange

    d. white/brown stripe; brown

15. It is a good idea to use 75-ohm F-Type _____ _____ in each unused television outlet and at each unused splitter port throughout a house.

16. When installing voice, video, and data jacks throughout a house, the preferred piece of equipment used to mount the jacks in a wall is a _____.

    a. metal device box

    b. mud ring (plaster ring)

    c. special voice, video, and data box

    d. plastic device box

17. When installing voice, video, and data wiring in a house, Articles _____ and _____ of the *NEC*® contain some requirements that must be followed.

18. The rating on the bandwidth performance of UTP cable is called the _____ and includes 3, 4, 5, 5e, and 6.

19. A(n) _____ is the smallest unit of measure in the binary system.

20. The minimum bending radius for a coaxial cable is recommended to be no less than _____ times the cable diameter.

Troubleshooting/Quality Exercises

DISCUSSION QUESTIONS

1. In many areas of the country a compression F-Type coaxial cable connector has become the coaxial cable connector of choice. Discuss why this type of coaxial cable connector has become so popular.

_____

_____

_____

_____

2. Wireless computer networks are becoming increasingly popular. Discuss some of the advantages and disadvantages of a wireless system.

_____

_____

_____

_____

3. When roughing in the wiring for a residential structured cabling system in a house with more than one floor, it is a good idea to install a couple of empty conduits from the basement, or from where the distribution center is located, to the attic. Why?

_____

_____

_____

_____

HANDS-ON APPLICATIONS

1. Match the following terms to their definitions. Write the correct number for the term in the blanks.

   1. bandwidth
   2. category
   3. coaxial cable
   4. F-Type connector
   5. IDC
   6. UTP
   7. punch-down block
   8. RG-6
   9. RJ-45
   10. structured cabling
   11. attenuation

   _____ a. An acronym for "unshielded twisted pair" cable

   _____ b. An architecture for communications cabling specified by the EIA/TIA

   _____ c. Identifies the amount of data that can be sent on a given cable

   _____ d. An acronym for "insulation displacement connection"

   _____ e. The popular name for the modular eight-pin connector used to terminate UTP cable

   _____ f. The connecting block that terminates cables directly

   _____ g. Ratings on the bandwidth performance of UTP cable

   _____ h. A cable in which the center signal-carrying conductor is centered within an outer shield and separated from the conductor with a dielectric

   _____ i. The decrease in the power of a signal as it passes through a cable; it is measured in decibels

   _____ j. A type of coaxial cable that is "quad-shielded"

   _____ k. A type of 75 ohm coaxial cable connector

REAL-WORLD SCENARIOS

1. An easy way to remember the color coding for a four-pair UTP cable is to use the acronym BLOGB. Explain how this acronym will help you remember the color coding of the conductor pairs.

   _____

   _____

   _____

   _____

2. State two reasons why Category 5e UTP cables are the preferred cable category types for wiring a residential structured cabling system.

   1. _____

   _____

   _____

   _____

   2. _____

   _____

   _____

   _____

3. Name the item that should be installed at each *unused* television outlet and splitter port throughout a house.

   _____

   _____

   _____

   _____

Chapter 17   Lighting Fixture Installation

Outline

Objective: Demonstrate an understanding of lighting basics.

Objective: Demonstrate an understanding of common lamp and lighting fixture terminology.

1) Understand lighting basics

2) Understand lamp and light fixture terminology

Key terms: **color temperature, lumen, luminaire, color rendition, lamp efficacy**

- Lighting basics
  - Luminaire:
    - Complete lighting unit consisting of a lamp or lamps together with the parts designed to distribute the light, to position and protect the lamps and ballast (where applicable), and to connect the lamps to the power supply.
  - The overall performance of a lighting system is a combination of the quantity and quality of light the lamps produce.
  - Light is the visible portion of the electromagnetic spectrum.
  - Lamp manufacturers are concerned with three factors:
    - Color temperature.
    - Color rendering.
    - Lamp efficiency.
  - Color temperature of a light source.
    - A measurement of its color appearance.
    - It is measured in degrees Kelvin (°K).
    - Light at higher-temperature wavelengths is referred to as "cool."
      - This light is whiter in color.
    - Light from lower-temperature wavelengths is referred to as "warm."
      - This light is more yellow in color.
    - Warm light sources are ideal for residential applications.
  - Many people think that a higher-wattage lamp will produce more light.
    - They are confusing light output with the amount of energy a lamp uses.
  - Light output is measured in lumens.
    - A lumen is the unit of light emitted from a light source.

o The amount of energy used by a lamp type is measured in watts.

■ For example, a 20-watt compact fluorescent lamp will produce as much usable light as a 75-watt incandescent lamp and use much less energy.

o Efficacy

■ The best indicator of a lamp's performance.

■ Rated in LPW (lumens per watt) units.

■ A ratio of the number of lumens a lamp produces to each watt of power it uses.

■ The higher the LPW, the more efficient the light source.

Objective: Demonstrate an understanding of the four different lamp types used in residential wiring applications: incandescent, LED, fluorescent, and high-intensity discharge.

3) Understand incandescent, fluorescent, and high-intensity discharge lamp types

Key terms: ballast, Class P ballast, fluorescent lamp, high-intensity discharge lamp, incandescent lamp, LED lamp

- Incandescent lighting

  o First type of electric lamp.

  o Same basic technology for over 100 years.

  o Light is produced when a tungsten filament, placed inside a glass enclosure, has an electric current passed through it.

  o The resistance of the filament causes it to heat up, giving off light.

  o Incandescent lamps are the most common lamp type for several reasons.

    ■ They come in a wide variety of sizes and styles.

    ■ They have the lowest initial cost of any lamp type.

    ■ They produce a warm light that has excellent color tones.

    ■ They are easily controlled with dimmers.

  o They are very inefficient and produce more heat than light.

    ■ Approximately 11% of electricity is transformed to light; the rest to heat.

    ■ By the end of 2014 the old style (non-halogen) tungsten filament incandescent lamps as we know them will be phased out in the United States because of their low efficiency.

    ■ 100-watt incandescent lamps will be the first to be phased out in 2012 and 40-watt incandescent lamps will be the last size to be phased out in 2014

- LED lighting

  o LEDs are a viable light source for homes and can produce usable amounts of light using very small amounts of electricity and with very little or no production of heat when in use.

  o The money savings because of the reduced amount of electricity used to light a home is significant.

  o LED lamps have many advantages over traditional incandescent lamps, including:

    ■ Better energy efficiency with an energy savings of 80% to 90%.

    ■ A very long operating life (up to 100,000 hours).

- An ability to operate at many different voltages, including 12V AC/DC or 85- to 240-volt AC.

- A cool light output with little or no heat energy produced.

- Available with narrow, medium and wide-angle lenses for a variety of light directions.

- Available in a variety of lighting colors including Warm White (2700K to 4300k), Daylight white (5000K to 6500K), Cool white (up to 8000K), Amber, Orange, Pink, Red, Cyan, Blue, Royal Blue, and Green.

- Fluorescent lighting
  - Referred to as electric discharge lamps by the *NEC*®.
  - Produced when an electric current is passed through tungsten cathodes at each end of a sealed glass tube.
  - Tube is filled with an inert gas like argon or krypton, as well as a very small amount of mercury.
  - Electrons emitted from the cathodes strike particles of mercury vapor, resulting in the production of ultraviolet radiation.
    - Causes a phosphor coating on the inside of the glass tube to glow.
  - Two electrical requirements
    - High-voltage source needed to start the lamp.
    - Once the lamp is started, mercury vapor offers a decreasing amount of resistance to the current. To prevent the lamp from drawing more and more current and quickly burning itself out, the current flow must be regulated.
  - Ballast
    - Provides both the voltage surge needed to start the lamp and the current control that allows the lamp to operate efficiently.
  - Two common types of ballasts
    - Magnetic.
    - Electronic.
  - Electronic ballast offers several advantages over the magnetic ballast.
  - Electronic ballasts:
    - Produce their full light output using 25–40% less energy.
    - Are more reliable.
    - Last longer.
    - Produce constant, flicker-free light.
    - Are more expensive to buy.
  - Three types of circuitry for fluorescent ballasts
    - Preheat.
    - Rapid start.
    - Instant start.
  - Preheat ballasts are easily identified because they have a "starter."
    - When the ballast receives power, the starter causes the cathodes to glow, or preheat, for a few seconds before the arc is established to produce light.

- Lamps used with preheat ballasts have two pins (bi-pin) on each end of the lamp.

- Preheat lamps and ballasts cannot be used with dimmers.

- The preheat type is not used much anymore.

o Rapid-start ballasts are the most common type.

- Rather than a starter, the filaments remain energized by a low-voltage circuit from the ballast; this allows the lamps to come to full power in less than one second.

- To start the lamps properly, ground the ballast and the fluorescent fixture case to the supply circuit grounding conductor.

- Rapid-start ballast lamps are bi-pin and may be dimmed with a special dimming ballast.

o Instant-start ballasts provide a high-voltage surge to start the lamp instantly.

- They require special lamps that do not require preheating of the cathodes.

- This system shortens lamp life by 20–40%.

- Single pin on each end of the lamp.

- Cannot be used with dimmers.

o The fastest-growing application of fluorescent lighting is in compact fluorescent lamps (CFLs).

- Consists of a narrow fluorescent tube doubled back on itself with both ends terminated in a plastic base.

- Base also contains the ballast.

- Provide better value in the long term than incandescent lamps.

- Longer lamp life and more efficient operation.

- High energy discharge (HID) lamps

o A type of gaseous discharge lamp

o Two main types used

- Sodium vapor

- Metal halide

o Arc is established between two cathodes in a glass tube filled with a metallic gas such as mercury, halide, or sodium.

o Arc causes the metallic gas to produce radiant energy, resulting in light.

o Electrodes are only a few inches (or less) apart in the lamp arc tube, not at opposite ends of a glass tube like a fluorescent lamp.

o Arc generates extremely high temperatures, allowing the metallic elements of the gas to release large amounts of visible energy.

o Main disadvantages:

- Even a very brief power interruption can cause the lamp to restart its arc and have to warm up again.

- Process usually takes several minutes.

o Not generally used inside residences.

o Used for outdoor security and area lighting applications.

Objective: Select a lighting fixture for a specific residential living area.

4) Select a lighting fixture for a specific area

Key terms: Type IC, Type Non-IC

- Selecting the appropriate lighting fixture
    - Installation requirements for light fixtures are contained in Article 410 of the *NEC*®.
    - Each fixture comes with specific installation instructions provided by the manufacturer.
    - Each fixture also comes with labeling listing installation restrictions pertaining to location and wiring methods.
    - Common information found on a label:
        - For wall mount only or ceiling mount only.
        - Maximum lamp wattage.
        - Lamp type.
        - Suitable for operation in an ambient temperature not exceeding ___°F (°C).
        - Suitable for use in suspended ceilings, damp locations, or wet locations.
        - Suitable for mounting on low-density cellulose fiberboard.
        - For supply connections, use wire rated at least ___ °F (°C).
        - Thermally protected.
        - Type IC or Type Non-IC.
    - Residential lighting divided into four separate groups:
        - General lighting.
        - Accent lighting.
        - Task lighting.
        - Security lighting.
    - Factors to consider when selecting light fixtures:
        - Match both the lamp and the fixture to the desired application.
        - How important is color rendition to the lighting application?
        - Consider the energy efficiency of the lamp and fixture.
        - How long will a particular lamp last?

Objective: Demonstrate an understanding of the installation of common residential lighting fixtures.

5) Understand the installation of lighting fixtures

Key terms: sconce, troffer, track lighting

- Recessed luminaires
    - Fixtures installed above the ceiling.
    - Designed so little or none of the luminaire body, lamp, lens, or trim extends below the level of the ceiling.

- o   Both fluorescent and incandescent recessed luminaires are available.
- o   Recessed incandescent luminaires are rated either Type IC (insulated ceiling) or Type Non-IC.
- o   Type Non-IC
  - ▪   Installed so the insulation is no closer than 3 inches (75 millimeters) to any part of the fixture.
- o   Type IC
  - ▪   Designed to be in direct contact with thermal insulation.
- • Surface-mounted luminaires
  - o   Make up the majority of light fixtures installed in a residence
  - o   Wall-mounted lighting fixtures
    - ▪   Called sconces.
    - ▪   Usually connected to either metallic or nonmetallic lighting outlet boxes after the wallboard is installed and finished.
  - o   Ceiling-mounted fixtures
    - ▪   Connected to metallic or nonmetallic lighting outlet boxes after the ceiling has been installed and finished.
  - o   Three basic types of ceiling-mounted fixtures
    - ▪   Direct mount.
    - ▪   Chandelier.
    - ▪   Pendant.
- • Track lighting
  - o   Used to illuminate a specific work area or to highlight special items like a painting hanging on a wall.
  - o   Available in 12-volt or 120-volt models and consists of a track with a number of lighting "heads" attached to it.
  - o   The lighting heads can be adjusted to direct their light wherever a home owner may want it.
- • Installing common residential lighting fixtures
  - o   Consider these factors:
    - ▪   Light fixtures are very fragile and are easily damaged or broken.
    - ▪   Handle and store with care.
    - ▪   Read the manufacturer's instructions that come with each lighting fixture.
    - ▪   Connect them to the electrical system with the proper polarity.
- • Direct connection to a lighting outlet box
  - o   Not many incandescent fixtures are mounted directly to the outlet box.
  - o   Two main types are the porcelain or plastic lamp holder, which is available as a:
    - ▪   Keyless fixture.
    - ▪   Pull-chain fixture.

- Direct connection to the ceiling
  - Surface-mounted fluorescent fixtures are installed and connected to the electrical system in a slightly different manner because the fixture is not attached directly to the lighting outlet box.
  - Fixture is attached directly to the ceiling.
  - Electrical wiring is brought into the fixture from an electrical lighting outlet box.
  - There are two ways that a surface-mounted fluorescent fixture can be connected to the lighting branch-circuit wiring.
    - Have the wiring method used, such as a Type NM cable, connected directly to the lighting fixture.
    - Mount fixture over previously installed lighting outlet box and run wiring from the box into the fixture.
- Strap to lighting outlet box
  - Very similar to the direct connection method discussed previously.
  - Main difference is that a metal strap is connected to the lighting outlet box and the fixture is attached to the metal strap with headless bolts and decorative nuts.
- Stud and strap connection to a lighting outlet box
  - There are larger and heavier types of light fixtures that use the stud and strap method of installation.
  - Hanging fixtures, like a chandelier or pendant fixture, often require extra mounting support when compared to smaller light fixtures.
  - The electrical connections for each installation are the same as discussed previously.
  - Read and follow the manufacturer's instructions because there are some slight variations with this type of installation.
- Recessed luminaire installation
  - Recessed fixtures are very easy to finish out.
  - Recessed fixture rough in frame has already been installed and connected to the electrical system.
  - All that remains is to install the lamps and the trim rings or lenses.
- Luminaire installation in a suspended ceiling
  - Make sure the fixture is listed for that type of installation.
  - Look for these two listings:
    - "Suitable for use in suspended ceilings."
    - "Suitable for mounting on low density cellulose fiberboard."
  - Fluorescent lighting fixtures installed in a dropped ceiling are often referred to as a troffer.
- Outside luminaire installation
  - Consider whether the fixture is listed for the desired application.
  - Outdoor fixtures are listed for use in either damp or wet locations.
  - Wet location fixtures may be installed in damp locations, but damp location fixtures cannot be installed in wet locations.

Objective: Demonstrate an understanding of the installation of ceiling-suspended paddle fans.

6) Understand the installation of ceiling-suspended paddle fans

Key term: ceiling-suspended paddle fan

- Installing a ceiling-suspended paddle fan:
  - Requires a special electrical box for proper support.
  - Many ceiling-suspended paddle fans have a light kit attachment.
  - When the ceiling is low, mount close to ceiling.
  - For high or sloped ceilings, extension rods are used.
  - Before you start installing a ceiling-suspended paddle fan:
    - Locate the manufacturer's instructions and follow them.
    - Inspect the contents of the carton.
    - Install all wiring according to the *NEC*® and appropriate local codes.
    - Make sure the outlet box is listed for the weight of the paddle fan.

Quizzes

SAFETY

1. Electricians should wear safety glasses and observe all applicable safety rules when installing lighting fixtures. True or False?

2. When installing recessed incandescent fixtures where thermal insulation will be placed over and around the fixture, Type _____ fixtures must be used, otherwise the fixture could overheat.
   a. Non-IC
   b. IC
   c. INS

CODE/REGULATIONS/STANDARDS

1. The *NEC*® requires that all fluorescent ballasts installed indoors, for both new and replacement applications, must have thermal protection built into the unit by the manufacturer. True or False?

2. A Type IC recessed incandescent light fixture may be completely covered by thermal insulation. True or False?

3. A _____ is defined by the *NEC*® as a complete lighting unit consisting of a lamp or lamps together with the parts designed to distribute the light, to position and protect the lamps and ballast, and to connect the lamps to the power supply.
   a. fixture
   b. sconce
   c. luminaire

4. Installation requirements for lighting fixtures are contained in Article _____ of the *NEC*®.
   a. 210
   b. 310
   c. 410

5. _____ lamps are referred to as "electric discharge lamps" by the *NEC*®.
   a. Fluorescent
   b. Metal halide
   c. Incandescent

Chapter Quiz

1. A Type IC recessed incandescent light fixture must be ___ thermal insulation.
   a. kept at least 2 inches from
   b. kept at least 3 inches from
   c. kept at least ½ inch from
   d. none of the above

2. A _____ is defined by the *NEC*® as a complete lighting unit consisting of a lamp or lamps together with the parts designed to distribute the light, to position and protect the lamps and ballast, and to connect the lamps to the power supply.

3. There are _____ basic colors in the visible portion of the electromagnetic spectrum.

4. Lamp manufacturers are concerned with which of the following factors?
   a. color temperature
   b. color rendering
   c. lamp efficiency
   d. all of the above

5. The amount of electrical power used by a lamp is measured in _____

6. A(n) _____ lamp has a cool light output with little or no heat energy produced.

7. Which of the following lamp types is the most inefficient?
   a. fluorescent
   b. incandescent
   c. metal halide
   d. high pressure sodium

8. Which of the following ballasts may be used with fluorescent dimmers?
   a. preheat
   b. rapid start
   c. instant start
   d. none of the above

9. Which of the following lamp types require the use of phosphors to produce light?
   a. metal halide
   b. sodium vapor
   c. incandescent
   d. fluorescent

10. The most energy-efficient source of white light on the market today is the _____ lamp.

11. Installation requirements for lighting fixtures are contained in Article _____ of the NEC®.

12. Which of the following is not considered one of the three main independent testing laboratories for electrical lighting fixtures?

    a. National Electrical Manufacturers Association

    b. Underwriters Laboratories

    c. Canadian Standards Association

    d. Electrical Testing Laboratories

13. The measure of the color appearance of a light source that helps describe the apparent "warmth" or "coolness" of the light source is called _____.

14. A measure of a lamp's ability to show colors accurately is called _____.

15. Lamp diameters are measured in _____ of an inch.

16. An incandescent lamp with a designation of G40 means that the lamp is a globe shape and _____ inches in diameter.

17. A T-8 fluorescent tube is _____ inch(es) in diameter.

18. The fastest-growing application of fluorescent lighting is in _____.

    a. plant-growing lamps

    b. germicidal lamps

    c. compact fluorescent lamps

    d. u-shaped lamps

19. A type of lighting fixture that has adjustable "heads" used to illuminate a specific work area or to highlight special items like a painting hanging on a wall is called _____.

20. When installing ceiling-suspended paddle fans in a low ceiling, make sure the fan blades are at least _____ feet above the floor.

Troubleshooting/Quality Exercises

DISCUSSION QUESTION

1. Discuss why fluorescent lighting in basements, garages, and other types of outdoor buildings may not be the best choice in colder climates.

_____

_____

_____

_____

HANDS-ON APPLICATIONS

1. Match the following terms to their definitions. Write the correct number for the term in the blanks.

1. ballast	9. incandescent lamps
2. LED lamp	10. lamp efficacy
3. Class P ballast	11. lumen
4. color temperature	12. luminaire
5. color rendition	13. sconce
6. fluorescent lamp	14. troffer
7. track lighting	15. Type IC
8. high intensity discharge	

_____ a. A light fixture designation that allows the fixture to be completely covered by thermal insulation

_____ b. A type of lighting fixture that consists of a surface-mounted or suspended track with several lighting heads attached to it

_____ c. A component that controls the voltage and current to a fluorescent lamp

_____ d. The unit of light energy emitted from a light source

_____ e. A measure of the color appearance of a light source that helps describe the apparent warmth or coolness of that light source

_____ f. A solid-state lamp that uses light-emitting diodes (LEDs) as the source of light

_____ g. A wall-mounted lighting fixture

_____ h. The original electric lamp type

_____ i. A gaseous discharge light source

_____ j. A gaseous discharge lamp that does not use phosphors

_____ k. A term commonly used by electricians to refer to a fluorescent lighting fixture installed in the grid of a suspended ceiling

_____ l. A measure of a lamp's ability to show colors accurately

_____ m. A measure used to compare light output to energy consumption

_____ n. A complete lighting unit

_____ o. A ballast with a thermal protection unit built in by the manufacturer

REAL-WORLD SCENARIOS

1. When considering the different lighting fixture types that are permitted to be installed in a clothes closet, most electricians install surface-mounted fluorescent strips. Why?

_____

_____

_____

_____

2. On some lighting fixtures, especially the ones hung with a decorative chain, the lighting fixture wiring is not color coded. Explain how you can distinguish the ungrounded conductor from the grounded conductor in this situation.

_____

_____

_____

_____

Chapter 18 Device Installation

Outline

Objective: Demonstrate an understanding of the proper way to terminate circuit conductors to a switch or receptacle device.

1) Understand the proper way to terminate conductors to a switch or receptacle

- Connecting circuit conductors to a receptacle or switch.
    - Form terminal loops in the wire.
        - Place the loop around the proper terminal in a clockwise manner and tighten the terminal screw.
    - A terminal screw that is not tightened properly or a wire not looped properly around a screw will typically be the cause of future problems.
- Back-wiring can be used.
- Two ways to back-wire:
    - Screw connected
        - Found on higher-quality switches and receptacles.
    - Push-in
- For back-wired switches and receptacles that are screw connected, strip the insulation from the conductor, insert the end of the conductor into the proper hole, and tighten the terminal screw.
- For back-wired switches and receptacles that are of the push-in type, strip the insulation from the conductor and push the conductor into the proper hole.
    - The conductor is held in place by a thin copper strip.
    - The push-in method is not recommended.

Objective: Select the proper receptacle for a specific residential application.

2) Select the proper receptacle for a specific residential application

Key terms: receptacle, single receptacle, duplex receptacle, strap (yoke), split-wired receptacles

- Selecting the appropriate receptacle type
    - Chapter 2 of the textbook presented an overview of receptacle devices.
        - Review the information that covers:
            - Single receptacles.
            - Duplex receptacles.
            - Multiple receptacles.

195

- Split-wired receptacles.
- Ground fault circuit interrupter (GFCI) receptacles.
  - NEMA configurations.
  - Section 406.12 in the *NEC®* requires tamper-resistant receptacles to be used in all non-locking 125-volt, 15- and 20-ampere receptacle locations specified in Section 210.52. Exceptions include:
    - Receptacles located more than 5½ feet (1.7 m) above the floor
    - Receptacles that are part of a luminaire or appliance
    - Receptacles located within dedicated space for appliances where only the appliance will be plugged into the receptacle

Objective: Demonstrate an understanding of the proper installation techniques for receptacles.

3) Understand proper installation techniques for receptacles

Key term: split-wired receptacle

- Installing receptacles
  - The more circuit testing and wire marking that occurs during the rough in stage, the easier receptacle installation will be at the trim-out stage.
  - Always check circuits before the wallboard is installed.
  - Circuit tracing and, if necessary, cable or raceway replacement is much easier during the rough in stage.
  - Make sure no receptacles are buried under the wallboard.
  - Drywall installers are very conscientious about exposing all the receptacles, but occasionally they will miss one.
  - The listing instructions for receptacles and switches normally allow only one wire to be terminated to each terminal screw.
  - However, many electrical boxes containing receptacles (or switches) will have many circuit conductors requiring connections to the device terminal screws.
  - The best way to make the necessary connections so that only one conductor gets connected to a terminal screw is to use a pigtail.
  - The first step in connecting a receptacle to the branch-circuit wiring is to locate the proper conductors and pull them out from the device box.
  - Once all of the circuit conductors have been properly identified, the conductors are ready to be connected to the receptacle.
  - Three wires for regular duplex receptacles.
    - Ungrounded black conductor.
      - Connected to the brass-colored terminal.
    - Grounded white conductor.
      - Connected to the silver screw terminal.
    - Bare grounding conductor.
      - Connected to the green grounding screw.

      ○ A split-wired receptacle has a switch controlling half of the receptacle and the other half is "hot" all the time.

          ■ Before installing, remove the tab connecting the terminal screws on the short slot side of the receptacle.

          ■ Do not remove the tab connecting the silver terminal screws.

- Once the particular receptacle you are installing is connected to the electrical system, secure it to the device box.

- Make sure the conductors inside the box are pushed to the back of the device box, leaving enough room to install the receptacle.

- Next, position the receptacle so the grounding slot is on the top.

      ○ Lessens the chances of a piece of metal coming in contact with both the hot and grounded conductors, causing arcing and a short circuit.

- Carefully push the receptacle into the device box, checking that the ears on the top and bottom of the receptacle yoke will rest against the wallboard when the receptacle is installed.

- Once you have determined that the receptacle ears will rest against the wallboard properly, attach the receptacle to the device box using the provided 6-32 machine screws.

- Install the receptacle cover plate.

      ○ Attach the cover to a regular receptacle with a single screw in the center of the cover plate.

      ○ For cover plates with rectangular openings, use two screws.

- Make sure the receptacle is flush with the wall and straight.

      ○ This is the part of the installation the home owner and everyone who enters the house will see.

      ○ Quality of this part of the installation will determine, to a large extent, the impression others will have concerning your work.

Objective: Select the proper switch type for a specific residential application.

4) Select the proper switch type

- Selecting the appropriate switch

      ○ The primary factor is having the proper switch, or combination of switches, control the desired luminaire (light fixture) or receptacle.

      ○ Chapter 2 presented a very detailed look at the different switch types.

          ■ Review at this time.

      ○ Chapter 13 presented several different switching circuits and how various switch types are connected into those circuits.

          ■ Review at this time.

Objective: Demonstrate an understanding of the proper installation techniques for switches.

5) Understand the proper installation techniques for switches

- Installing switches
  - ○ Very similar to installing receptacles.
  - ○ The more effort put into the rough in phase, the easier the installation process will be.
  - ○ Connecting the conductors to the switch is done basically in the same manner as connecting conductors to receptacles.
- Single-pole switches
  - ○ Connect two conductors to the switch.
    - ■ Usually black.
  - ○ The two conductors include the incoming power wire and a wire running to the light fixture or receptacle.
  - ○ Ground the switch.
  - ○ Set up the switch so it will read "OFF" when the toggle is in the down position.
  - ○ Black and white conductors
    - ■ Use a connection called a switch loop.
    - ■ Use the white wire as an ungrounded conductor and identify it as such.
    - ■ Section 200.7(C) requires that the white conductor be identified at both ends as a "hot" conductor.
    - ■ Use black electrical tape to mark the conductor.
- Three-way switches
  - ○ Connect three conductors to the switch.
  - ○ Connect the traveler conductors to the brass-colored "traveler" terminals.
  - ○ Connect the black conductor that provides power or the black conductor that takes power to the light fixture to the third dark "common" terminal.
  - ○ Section 200.7(C) requires the white conductor to be reidentified if it is used as an ungrounded conductor.
  - ○ Ground the switch.
- Four-way switches
  - ○ Four-way switches have four conductors connected to them.
  - ○ Two conductors will come from one three-wire cable and two conductors will come from a second three-wire cable.
  - ○ Connect the conductors from one cable to the two screw terminals that are the same color (traveler terminal screws).
  - ○ Connect the remaining two conductors to the two screw terminals that are a different color (traveler terminal screws).
- Once the switches are connected to the electrical system, they are ready to install in the device box.
  - ○ Very similar to the receptacle installation procedure.
  - ○ Main difference is the number of switches installed in multi-gang boxes.

- ○ In residential installations, two- and three-gang switch boxes are common.
- ○ Take care to ensure there is enough room in the device box for all conductors and switches.
  - ■ Easily done by:
    - • Tilting the switch and pushing it into the device box.
    - • Pulling it back out and aligning the mounting holes.
    - • Installing the 6-32 mounting screws and tightening them.
- • Installing cover plates on switches is done in just about the same manner as installing cover plates on receptacles.
  - ○ However, instead of one screw, switch cover plates require two screws to attach them to a switch.
  - ○ Because there are so many multi-gang switch boxes, make sure all the switches are level so the cover plate will be level when it is installed.

Objective: Demonstrate an understanding of GFCI receptacle installation.

6) Understand GFCI receptacle installation

Key term: GFCI receptacle

- • Ground fault circuit interrupter (GFCI) receptacles
  - ○ Designed to protect people from electrical shock.
  - ○ Monitors current flow on both the ungrounded and grounded circuit conductors.
  - ○ When a GFCI senses a current on the ungrounded conductor that is at least 6 mA greater than the current flowing on the grounded conductor, it trips off.
  - ○ Class A GFCI receptacle devices are designed to not trip when the current to ground is less than 4 milliamps according to UL 943, Standard for Ground Fault Circuit Interrupters.
  - ○ Section 210.8(A) lists the areas in a house that require GFCI protection.
  - ○ Factors to consider when buying a GFCI receptacle
    - ■ Lock-out feature so GFCI cannot be reset if not functioning properly.
    - ■ Make sure GFCI cannot be reset if the line and load connections are reversed.
    - ■ Indicator light to show that the GFCI is working properly or has been tripped off.
  - ○ Connect GFCI receptacles to the electrical system in much the same manner as regular duplex receptacles.
    - ■ However, on the back of the GFCI receptacle, one of the brass screw terminals and one of the silver screw terminals are marked for the "Line," or incoming power conductors.
    - ■ The other set of screw terminals are marked as "Load" terminals, or the outgoing power conductors.
  - ○ If there is only one cable in the device box, the connection for the GFCI receptacle is quite simple.
    - ■ Connect the grounding conductor to the green grounding screw.
    - ■ Connect the white grounded conductor to the silver "Line" terminal screw.
    - ■ Connect the black ungrounded conductor to the brass "Line" terminal screw.

Objective: Demonstrate an understanding of AFCI devices.

7) Understand AFCI devices

Key term: AFCI device

- Arc fault circuit interrupter (AFCI) devices
    - According to Section 210.12(A) in the 2011 *NEC®*, all 120-volt, 15- and 20-amp branch circuits supplying outlets in dwelling unit living rooms, dining rooms, family rooms, parlors, libraries, dens, bedrooms, sunrooms, recreation rooms, closets, hallways, or similar locations must be AFCI protected.
    - Resulted from Consumer Product Safety Commission studies.
        - Showed that a high percentage of the 150,000 residential electrical fires that occur annually in the United States start by electrical arcs.
    - Arcing typically occurs for two reasons:
        - Overloaded extension cord or a broken wire in an extension cord or power cord.
            - Causes the insulation to break down.
            - Creates conditions where arcing between the conductors can occur.
    - When a plug is knocked out of a receptacle when furniture or other similar objects are being moved.
        - Arcing occurs when the plug blades are pulled from the receptacle slot.
    - AFCI devices are designed to trip when they sense rapid fluctuations in the current flow that are typical of arcing conditions.
    - AFCI receptacles were not commercially available at the time this textbook was published.
        - AFCI circuit breakers are used to provide the required protection and are covered in Chapter 19.

Objective: Demonstrate an understanding of TVSS devices.

8) Understand TVSS devices

Key term: transient voltage surge suppressor (TVSS)

- Transient voltage surge suppressor (TVSS)
    - TVSS devices are also known as Surge Protection Devices (SPDs).
    - The National Electrical Code covers these devices in Article 285 and refers to TVSS devices as Type 2 and Type 3 SPDs.
    - Many of the appliances used in homes today contain sensitive electronic circuitry, circuit boards, or microprocessors.
    - These components are susceptible to transient voltages, also known as voltage surges or voltage spikes.
    - Voltage surges or spikes can destroy components or cause microprocessors to malfunction.
    - Line surges can be line-to-line, line-to-neutral, or line-to-ground.
    - Transient voltages are grouped into two categories.
        - Impulse
            - Transient voltage starts outside the residence.
            - Generally caused by power company equipment.

- Ring wave
  - Line surge begins inside the residence.
  - Caused by inductive loads such as the spark igniters on gas clothes dryers, gas ranges, or gas water heaters, electric motors, and computers.
  - TVSS devices are used to minimize voltage surges and spikes.
    - Works by clamping the high surge voltage to a level that a piece of equipment like a computer can withstand.
  - Available in several different styles.
    - One style looks very much like a GFCI receptacle and mounts in a device box.
    - Another is one unit that protects the entire house for transient voltages and is hardwired directly into the main electrical panel.
    - Most familiar type is the plug-in strip.
      - Better strips will also provide filtering of the electricity being delivered to sensitive electronic equipment.
      - Filtering will help eliminate electrical "noise" that may be present on the incoming electrical conductors.

Quizzes

SAFETY

1. Electricians should wear safety glasses and observe all applicable safety rules when installing devices. True or False?

CODE/REGULATIONS/STANDARDS

1. A grounding conductor may not be _____.
   a. bare
   b. green
   c. green with white stripes

2. A _____ receptacle is one that has internal shutters that are closed behind the receptacle slots when there is no attachment plug inserted. The shutters open when a plug is inserted, allowing the plug (and cord) to become energized.
   a. duplex
   b. single
   c. tamper-resistant

3. A device intended to provide protection from the effects of arc faults by recognizing characteristics unique to arcing and by functioning to de-energize the circuit when an arc is detected is called a(n) _____.

4. The *NEC*® defines a(n) _____ as a contact device installed at an outlet for the connection of an attachment plug.

TOOLS

1. A _____ is used to attach receptacle and switch cover plates.
   a. pair of pliers
   b. screwdriver
   c. scratch awl

2. When making a terminal loop on the end of a wire so it can be terminated to a switch or receptacle terminal screw, it is best to use _____ or a wire stripper.
   a. long-nose pliers
   b. a screwdriver
   c. a torpedo level

Chapter Quiz

1. A feed-through GFCI receptacle can provide GFCI protection for regular receptacles installed "downstream" on the same branch circuit. True or False?

2. If a white conductor in a Type NM cable is connected to a traveler terminal on a three-way switch, it does not have to be reidentified because it is not "hot" all the time. True or False?

3. A multiple receptacle has more than _____ contact device(s) on the same yoke.

4. Which of the following conductors are permitted to be used for the push-in method of attaching conductors to a device?
   a. 14 AWG solid copper
   b. 14 AWG stranded copper
   c. 14 AWG solid aluminum
   d. 14 AWG solid copper-clad aluminum

5. On a 20-ampere-rated individual branch circuit, which of the following types of receptacles may *not* be used?
   a. 20-amp single receptacle
   b. 15-amp single receptacle
   c. 15-amp duplex receptacle
   d. 20-amp duplex receptacle

6. Which of the following determine the effect electrical shock has on the body?
   a. The amount of current flowing through the body
   b. The path the current takes through the body
   c. The length of time the current flows through the body
   d. All of the above

7. A _____ device operates by monitoring the current flow on both the ungrounded "hot" conductor and the grounded conductor.

8. A GFCI device will trip when it senses a current imbalance of _____ milliamps or greater.

9. Which of the following is not considered to be a line surge?
   a. ground to line
   b. ground to neutral
   c. line to line
   d. line to neutral

10. A(n) _____ line surge is caused by inductive loads in a residence such as spark igniters on gas ranges and clothes dryers, electric motors, and computers.

    a. impulse

    b. grounding

    c. shorting

    d. ring wave

11. A grounding conductor may not be _____.

    a. bare

    b. green

    c. green with a yellow stripe

    d. green with a white stripe

12. The proper NEMA designation for a 30-amp, 125/250-volt receptacle installed for an electric clothes dryer is _____.

    a. 5-15R

    b. 6-30R

    c. 14-30R

    d. 10-30R

13. Electricians create a split-wired receptacle by removing the _____ between the brass terminal screws on the "hot" side of a duplex receptacle.

    a. screw

    b. connecting tab

    c. strap

    d. wire

14. The preferred method for connecting a circuit conductor to a device is to form a _____ in the wire, put it under a terminal screw, and tighten the screw properly.

    a. terminal loop

    b. curve

    c. circle

    d. jug handle

15. A _____ receptacle is one that has internal shutters that are closed behind the receptacle slots when there is no attachment plug inserted. The shutters open when a plug is inserted, allowing the plug (and cord) to become energized.

    a. duplex

    b. single

    c. tamper-resistant

    d. GFCI

16. If a device box has been installed too far back behind a wall, a _____ can be used to bring the edge of the box flush with the wall surface.

    a. new box

    b. wall shrinker

    c. box extender

    d. different kind of device

17. On a 15- or 20 amp, 125-volt rated receptacle, the _____ conductor is connected to the silver terminal screw.

18. If the two conductors to be connected to a single-pole switch are black and a reidentified white, the connection you are using is called a(n) _____ .

19. Explain the purpose of the metal strap that runs through or behind a receptacle or switch.

_____

_____

_____

_____

20. Explain why it is recommended to install receptacles with the grounding slot up.

_____

_____

_____

_____

Troubleshooting/Quality Exercises

DISCUSSION QUESTIONS

1. Self-grounding receptacles can be used with metal electrical boxes. Discuss how the self-grounding feature of a receptacle works and why self-grounding receptacles are used even though they cost more than a regular receptacle.

_____

_____

_____

_____

2. Crooked cover plates on switches and receptacles are not acceptable. Keeping the cover plates straight is not a difficult task. Discuss some installation techniques used by electricians to keep the cover plates straight.

_____

_____

_____

_____

HANDS-ON APPLICATIONS

1. Match the following terms to their definitions. Write the correct number for the term in the blanks.

  1. AFCI   5. single receptacle

  2. duplex receptacle   6. split-wired receptacle

  3. GFCI   7. strap

  4. receptacle   8. TVSS

_____ a. A single contact device with no other contact device on the same strap

_____ b. A device installed in an electrical box for the connection of an attachment plug

_____ c. An electrical device designed to protect sensitive electronic circuit boards from voltage surges

_____ d. A device that protects people from dangerous amounts of current

_____ e. A device intended to provide protection from the effects of arc faults

_____ f. The most common receptacle type used in residential wiring

_____ g. A duplex receptacle wired so that the top half is "hot" all the time and the bottom half is switch controlled

_____ h. The metal frame that a receptacle or switch is built around

REAL-WORLD SCENARIOS

1. Some wiring situations call for a split receptacle to be used. This wiring technique will allow the bottom half of a duplex receptacle to be controlled by a switch while the top half will be "hot" at all times. Explain how an electrician creates a split duplex receptacle.

_____

_____

_____

_____

2. Many electrical contractors will not allow their employees to use the screwless push-in terminal connections on the back of a switch or receptacle. Why?

_____

_____

_____

_____

3. When installing a duplex receptacle into a metal electrical box, an electrician strips out the 6-32 tapped device mounting holes which means the receptacle cannot be attached properly. Name the tool that the electrician could use to fix the stripped-out mounting holes.

_____

_____

_____

_____

4. Sometimes spacers are needed when a device box was installed so that the front edge of the box is not perfectly flush with the wall covering. The spacers are put around the 6-32 device mounting screws and act to hold the device out from the face of the box so that the device ends up being properly positioned with respect to the wall surface. Name some common items that an electrician might use as spacers.

_____

_____

_____

_____

5. When choosing the color for switches and receptacles, be sure to purchase all of the devices and their covers from the same manufacturer. Why?

_____

_____

_____

_____

6. What is the best way to install switches in a multi-gang device box so that the multigang cover plate will fit correctly the first time?

_____

_____

_____

_____

Chapter **19**   **Service Panel Trim-Out**

Outline

Objective: Select the proper overcurrent protection device for a specific residential branch circuit or feeder circuit.

Objective: Demonstrate an understanding of the common fuses and circuit breakers used in residential wiring.

1) Select the proper overcurrent protection device

2) Understand common fuses and circuit breakers

Key terms: AFCI circuit breaker, cartridge fuse, Edison-base plug fuse, interrupting rating, overcurrent protection device (OCPD), Type S plug fuse

- Residential overcurrent protection devices (OCPDs)
  - Chapter 2 presented a good overview of the different types of overcurrent protection devices used in residential wiring.
    - Review the material.
  - Both circuit breakers and fuses are used.
    - Circuit breakers are used more often.
  - Circuit breakers
    - Available as a single-pole device for 120-volt applications.
    - Available as a twin (dual) device for 120-volt applications when there is a limited number of spaces available in an electrical panel.
    - Available as a two-pole device for both 120/240-volt and 240-volt applications.
    - Also available as:
      - Ground fault circuit interrupter (GFCI) circuit breakers.
      - Arc fault circuit interrupter (AFCI) circuit breakers.
  - Fuses in two basic styles
    - Plug fuses
      - Available as an Edison-base model and as a Type S model.
    - Cartridge fuses
      - Available as a ferrule model or a blade-type model.
- Requirements for OCPDs
  - Article 240 of the *NEC®* provides many installation requirements for OCPDs.

○ Section 240.4 states that circuit conductors must be protected against overcurrent in accordance with their ampacity as specified in Table 310.15(B)(16).

○ Not all of the ampacities listed in Table 310.15(B)(16) correspond with a standard fuse or circuit breaker size.

○ Section 240.6(A) lists standard sizes of fuses and circuit breakers.

○ Section 240.4(B) allows the next higher standard overcurrent device rating to be used, provided all of the following conditions are met.

  ■ Conductors being protected are not part of a multi-outlet branch circuit supplying receptacles for cord-and-plug-connected portable loads.

  ■ Ampacity of the conductors does not correspond with the standard ampere rating of a fuse or a circuit breaker.

  ■ The next higher standard rating selected does not exceed 800 amperes.

○ Section 210.3 indicates that branch-circuit conductors rated 15, 20, 30, 40, and 50 amperes must be protected at their ratings.

○ Section 240.4(D) requires that overcurrent protection device not exceed:

  ■ 15 amperes for 14 AWG copper.

  ■ 20 amperes for 12 AWG copper.

  ■ 30 amperes for 10 AWG copper.

  ■ 15 amperes for 12 AWG and 25 amperes for 10 AWG aluminum and copper-clad aluminum.

○ Section 240.24 addresses the location of the overcurrent protection devices in the house.

  ■ Section 240.24(A) requires that overcurrent protection devices be readily accessible.

  ■ Section 240.24(C) requires the overcurrent protection devices to be located where they will not be exposed to physical damage.

  ■ Section 240.24(D) prohibits overcurrent protection devices from being located near easily ignitable material.

    • Clothes closets, etc.

  ■ Section 240.24(E) says that in dwelling units, overcurrent protection devices cannot be located in bathrooms.

  ■ Section 240.24(F) says that overcurrent devices cannot be located over steps in a stairway.

○ Section 110.9 states all fuses and circuit breakers intended to interrupt the circuit at fault levels must have an adequate interrupting rating wherever they are used in the electrical system.

  ■ Fuses or circuit breakers that do not have adequate interrupting ratings could rupture while attempting to clear a short circuit.

  ■ *NEC*® defines interrupting rating as the highest current at rated voltage that a device is intended to interrupt under standard test conditions.

  ■ Interrupting rating

    • For most circuit breakers it is either 5000 or 10,000 amps.

    • For fuses, it can be as high as 300,000 amps.

    • It must be at least as high as the available fault current.

  ■ The available fault current is the maximum amount of current that can flow through the circuit if a fault occurs.

- General requirements for plug fuses
  - Section 240.50 applies to both Edison-base and Type S plug fuses.
  - Limits the voltage on the circuits using this fuse type to the following:
    - Circuits not exceeding 125 volts between conductors.
      - All 120-volt residential branch circuits meet this requirement.
    - Circuits supplied by a system having a grounded neutral where the line-to-neutral voltage does not exceed 150 volts.
      - All residential 120/240-volt circuits meet this requirement.
  - Section 240.50(B) requires each fuse, fuseholder, and adapter to be marked with its ampere rating.
  - Section 240.50(C) requires plug fuses of 15-ampere and lower rating to be identified by a hexagonal configuration of the window, cap, or other prominent part to distinguish them from fuses of higher ampere ratings.
- Specific requirements for Edison-base plug fuses
  - Section 240.51 applies specifically to Edison-base plug fuses.
    - Must be classified at not over 125 volts and 30 amperes or below.
    - Can be used only for replacements in existing installations where there is no evidence of overfusing or tampering.
- Specific requirements for Type S plug fuses
  - Section 240.53 applies specifically to Type S plug fuses.
    - Classified at not over 125 volts and 0 to 15 amperes, 16 to 20 amperes, and 21 to 30 amperes.
    - Found in three sizes for residential work:
      - 15-amp rated (blue)
      - 20-amp rated (orange)
      - 30-amp rated (green)
  - Type S fuses of a higher ampere rating cannot be interchanged with a lower-ampere-rated fuse.
  - They must be designed so they cannot be used in any fuseholder other than a Type S fuseholder or a fuseholder with a Type S adapter inserted.
  - Section 240.54 applies specifically to Type S fuses, adapters, and fuseholders.
    - Adapters must fit Edison-base fuseholders.
    - Fuseholders and adapters must be designed so that either the fuseholder itself or the fuseholder with a Type S adapter inserted cannot be used for any fuse other than a Type S fuse.
    - Adapters must be designed so that once inserted in a fuseholder, they cannot be removed.
    - Fuses, fuseholders, and adapters must be designed so that tampering or shunting (bridging) would be difficult.
    - Dimensions of fuses, fuseholders, and adapters must be standardized to permit interchangeability regardless of the manufacturer.
- Specific requirements for cartridge fuses
  - Sections 240.60 and 240.61 apply specifically to cartridge fuses.
  - Use fuses and fuseholders of the 300-volt type in circuits not exceeding 300 volts between conductors.

- Use fuses rated 600 volts or less for voltages at or below their ratings.

- Fuseholders must be designed so that it will be difficult to put a fuse of any given class into a fuseholder designed for a current that is lower, or a voltage that is higher, than that of the class to which the fuse belongs.

- Fuses and fuseholders are to be classified according to their voltage and amperage ranges.

- Fuses must be plainly marked, either by printing on the fuse barrel or by a label attached to the barrel showing the following:

  - Ampere rating.

  - Voltage rating.

  - Interrupting rating where other than 10,000 amperes.

  - Current limiting, where applicable.

  - Name or trademark of the manufacturer.

- Specific requirements for circuit breakers

  - Section 240.81 states circuit breakers must clearly indicate whether they are in the open "OFF" or closed "ON" position.

  - Circuit breakers must be designed so that any fault must be cleared before the circuit breaker can be reset.

  - Even if the handle is held in the "ON" position, the circuit breaker will remain tripped as long as there is a trip-rated fault on the circuit.

  - Section 240.83 requires that circuit breakers be marked with their ampere rating in a manner that will be durable and visible after installation.

  - Section 240.83(B) says circuit breakers rated at 100 amperes or less and 600 volts or less must have the ampere rating molded, stamped, etched, or similarly marked on their handles.

  - Section 240.83(C) requires every circuit breaker having an interrupting rating other than 5000 amperes to have its interrupting rating shown on the circuit breaker.

  - Section 240.83(D) requires circuit breakers used as switches in 120-volt and 277-volt fluorescent lighting circuits to be listed and marked SWD or HID.

  - Section 240.83(E) says circuit breakers must be marked with a voltage rating not less than the nominal system voltage that is indicative of their capability to interrupt fault currents between phases or phase to ground.

- GFCI circuit breakers

  - Looks very similar to regular circuit breaker

  - Two differences

    - GFCI breaker has a white pigtail attached to it.

    - GFCI breaker has a "Push to Test" button located on the front.

- AFCI circuit breakers

  - Section 210.12 requires that AFCI protection be provided for all 120-volt, 15- and 20-ampere branch circuits supplying outlets in dwelling unit family rooms, dining rooms, living rooms, parlors, libraries, dens, bedrooms, sunrooms, recreation rooms, closets, hallways, and similar areas.

  - AFCI circuit breakers look very similar to GFCI circuit breakers.

  - The "Push-to-Test" button is typically a different color than that of a GFCI breaker.

Objective: Demonstrate an understanding of installing circuit breakers or fuses in an electrical panel.

3) Understand installing circuit breakers and fuses in a panel

Key term: stab

- Installing circuit breakers in a panel
    - Before installing circuit breakers
        - Make sure the circuit breakers being used are compatible with the panel already installed.
        - Be sure that your panel will handle the number of circuits the electrical system requires.
    - Most 100-amp-rated electrical panels are designed for 20 or 24 overcurrent devices.
    - Most 200-amp-rated electrical panels are designed for 40 or 42 circuits.
    - Electrical panels designed for over 42 circuits are allowed by the *NEC®*.
- Safety rules when installing circuit breakers
    - Always turn off electrical power at the main circuit breaker when working in an energized main breaker panel.
    - The LOAD side of the panel will be disconnected, but the LINE side will still be energized.
    - If you are working on an energized subpanel, find the circuit breaker in the service panel, turn it off, and lock it in the OFF position.
    - Test the panel you are working on with a voltage tester to verify that the electrical power is off.
        - NEVER assume the panel is de-energized.
- Circuit breaker installation
    - Circuit breakers are installed by attaching them to the bus bar assembly in the panel.
    - Bus bar assembly:
        - Connected to the incoming service entrance conductors.
        - Connected (in the case of a subpanel) to the incoming feeder conductors.
        - It distributes the electrical power to each of the circuit breakers located in the panel.
    - Circuit breakers are attached to the bus bar by contacts in the breakers snapped onto the bus bar at specific locations, commonly called stabs.
        - A single-pole circuit breaker has one stab contact.
        - A two-pole circuit breaker has two stab contacts.
    - Most branch circuits are 120-volt circuits.
        - Wired with 14 AWG or 12 AWG copper conductors.
    - Multiwire branch circuits are often installed in residential wiring.
        - A multiwire circuit consists of two ungrounded conductors that have 240 volts between them and a grounded conductor that has 120 volts between it and either of the two ungrounded conductors.
        - Section 210.4(B) states that each multiwire branch circuit must be provided with a means that will simultaneously disconnect all ungrounded conductors at the point where the multiwire branch circuit originates.
        - Section 210.4(D) states that ungrounded and grounded conductors of each multiwire branch circuit must be grouped together by wire ties or some other means in at least one location in the electrical panel.

o Many branch circuits serve appliances such as electric water heaters, air conditioners, and electric heating units.

  ▪ These loads require 240 volts to operate properly.

o 240-volt branch circuits require

  ▪ 15-amp circuit breaker when wired with 14 AWG wire.

  ▪ 20-amp circuit breaker when wired with 12 AWG wire.

  ▪ 30-amp circuit breaker when wired with 10 AWG wire.

  ▪ However, since it is a 240-volt circuit, it needs a two-pole circuit breaker.

o Can also supply 120/240 volts to appliances such as electric clothes dryers and electric ranges.

  ▪ This installation requires a two-pole circuit breaker, just like the 240-volt-only application.

Objective: Demonstrate an understanding of the common techniques for trimming out a residential electrical panel.

4) Understand the common techniques for trimming out a panel

• Review the procedures for trimming out a panel located at the end of Chapter 19.

Quizzes

SAFETY

1. Electricians should wear safety glasses and observe all applicable safety rules when trimming out a loadcenter. True or False?

2. Fuses and circuit breakers that are subject to fault currents that exceed their interrupting ratings may _____ .

  a. explode like a bomb

  b. turn red

  c. have a light come on that indicates a problem

3. When replacing fuses, never use a fuse that has a lower interrupting rating than the one that is being replaced. True or False?

CODE/REGULATIONS/STANDARDS

1. *NEC®* Article _____ provides many installation requirements for overcurrent protection devices.

  a. 230

  b. 240

  c. 250

2. *NEC®* Table _____ lists the ampacity of conductors used in residential wiring.

  a. 240.6

  b. 250.66

  c. 310.15(B)((16)

3. All plug fuses are limited to circuits where the voltage between conductors does not exceed _____ volts. (general rule)

   a. 240

   b. 125

   c. 277

4. A circuit breaker used to switch fluorescent lighting must be listed and have _____ marked on it.

   a. okay for fluorescent lighting

   b. SWT

   c. SWD or HID

5. Because all of the ampacities listed in Table 310.15(B)(16) do not correspond with a standard fuse or circuit breaker size, the *NEC*® allows the next _____ standard overcurrent device to be used up to 800 amps.

6. For circuit breakers, if the interrupting rating is different than _____ amperes, it must be marked on the breaker.

Chapter Quiz

1. A(n) _____ fuse uses different fuse bases for each size fuse and an adapter that matches each fuse size is required.

   a. cartridge

   b. Edison-base

   c. Type S

   d. stab

2. A(n) _____ fuse is enclosed in an insulating tube that confines the arc when the fuse blows and is available in either a ferrule or blade style.

   a. cartridge

   b. Edison-base

   c. Type S

   d. stab

3. In residential wiring, a 12 AWG copper conductor is protected by a _____-ampere or lower fuse or circuit breaker.

4. Which of the following will not cause an overcurrent condition?

   a. open circuit

   b. ground fault

   c. overload

   d. short circuit

5. Type S fuses are classified at _____ amperes.

   a. 0–15

   b. 16–20

   c. 21–30

   d. all of the above

6. A GFCI circuit breaker should be tested at least once a _____.

7. When installing an ungrounded conductor in a circuit breaker, approximately _____ inch(es) of insulation is stripped from the end of the conductor.

8. A circuit breaker used to switch fluorescent lighting must be listed and have _____ marked on it.

   a. okay for fluorescent lighting

   b. FL

   c. SWD only

   d. SWD or HID

9. A 240-volt electric water heater branch circuit will require a _____ circuit breaker.

   a. single-pole

   b. double-pole

   c. HACR

   d. three-pole

10. Overcurrent protection devices must be readily accessible. True or False?

11. The color coding for a 30-amp Type S plug fuse is _____.

    a. blue

    b. orange

    c. green

    d. any of the above

12. The color coding for a 15-amp Type S plug fuse adapter is _____.

    a. blue

    b. orange

    c. green

    d. none of the above

13. The highest current at rated voltage that a device is intended to interrupt under standard test conditions is called the _____ rating.

14. Because electric motors draw more current to get started than they do when running, _____ fuses are often used to protect motor circuits.

15. Because all of the ampacities listed in Table 310.15(B)(16) do not correspond with a standard fuse or circuit breaker size, the *NEC*® allows the next _____ standard overcurrent device to be used up to 800 amps.

16. If an electrician chooses to use plug fuses in a new electrical installation, they must be _____-plug fuses.

17. Fuseholders for current-limiting fuses cannot permit the insertion of fuses that are not _____.

18. The circuit breakers are attached to the bus bar by contacts in the breakers commonly called _____.

19. Explain how a 240-volt circuit is wired differently than a 120/240-volt circuit.

    _____

    _____

    _____

    _____

20. Explain what a hexagonal configuration on a plug fuse indicates.

_____

_____

_____

_____

Troubleshooting/Quality Exercises

DISCUSSION QUESTIONS

1. Explain why circuits that supply electric motor loads need to be protected by fuses or circuit breakers that have a time-delay feature.

_____

_____

_____

_____

2. Why is it so important that overcurrent protection devices have an interrupting rating high enough to handle any available fault current?

_____

_____

_____

_____

3. It is very important to install the correct Type S fuse adapter. How do you know which adapter goes with which Type S fuse?

_____

_____

_____

_____

HANDS-ON APPLICATIONS

1. Match the following terms to their definitions. Write the correct number for the term in the blanks.

   1. AFCI circuit breaker
   2. cartridge fuse
   3. Edison-base plug fuse
   4. interrupting rating

   5. OCPD
   6. stab
   7. Type S plug fuse

   _____ a. A term used to identify the location on a loadcenter's ungrounded bus bar where a circuit breaker is snapped on

   _____ b. The highest current at rated voltage that a device is intended to interrupt

_____ c. A fuse enclosed in an insulating tube and may either be a ferrule or blade type

_____ d. A circuit breaker intended to provide protection from the effects of arc faults

_____ e. A fuse type that uses the same standard screw base as an ordinary incandescent lightbulb

_____ f. A fuse type that uses different fuse bases for each fuse size

_____ g. A fuse or circuit breaker used to protect an electrical circuit from an overload, a short circuit, or a ground fault

REAL-WORLD SCENARIOS

1. An electrician decides to use plug fuses as the circuit overcurrent protection devices in a new residential wiring system. What style of plug fuse must be used?

    _____

    _____

    _____

    _____

2. What must be done if you take out too many "twistouts" from the cover of a loadcenter?

    _____

    _____

    _____

    _____

Chapter 20

Checking Out and Troubleshooting Electrical Wiring Systems

Outline

Objective: Follow a checklist to determine if the basic requirements of the *NEC*® were met in the electrical system installation.

1) Follow a checklist of the basic *NEC*® requirements

- Determining if applicable *NEC*® requirements are met

  ○ At the end of the rough-in stage and upon completion of the residential electrical system trim-out stage, review the following list to make sure you have met the basic installation requirements of the *NEC*®.

  ○ General Circuitry (*NEC*® Sections 210.11(C) and 422.12)

    ■ In addition to the branch circuits installed to supply general illumination and receptacle outlets in dwelling units, the following minimum requirements also apply:

      • Two 20-amp circuits for the kitchen receptacles.

      • One 20-amp circuit for the laundry receptacles.

      • One 20-amp circuit for the bathroom receptacles.

      • One separate, individual branch circuit for central heating equipment.

  ○ General Circuitry (*NEC*® Section 210.52(B)(3))

    ■ Supply receptacles installed in the kitchen to serve countertop surfaces must be supplied by at least two separate small-appliance branch circuits.

  ○ General Circuitry (*NEC*® Section 300.3(B))

    ■ All conductors of the same circuit, including grounding and bonding conductors, must be contained in the same raceway, cable, or trench.

  ○ General Circuitry (*NEC*® Section 408.4)

    ■ All circuits and circuit modifications must be identified as to purpose or use on a directory located on the face or inside of the electrical panel doors.

  ○ General Circuitry (*NEC*® Section 406.4)

    ■ Receptacle outlets must be of the grounding type, be effectively grounded, and be wired to have the proper polarity.

  ○ General Circuitry (*NEC*® Section 406.9(A) and (B))

    ■ All 15- and 20-amp, 125- through 250-volt receptacles installed in a wet location must have a weatherproof cover.

    ■ The receptacle, unless of the locking-type, must be a listed weather-resistant type.

- General Circuitry (*NEC*® Section 210.52)

  - Install receptacle outlets in habitable rooms so that no point measured horizontally along the floor line in any wall space is more than 6 feet (1.8 meters) from a receptacle outlet. (Receptacle must be installed in each wall space 2 feet (600 millimeters) or more in width.)

  - At kitchen countertops, install receptacle outlets so that no point along the wall line is more than 24 inches (600 millimeters) measured horizontally from a receptacle outlet in that space.

  - Install a receptacle outlet at each counter space that is 12 inches (300 millimeters) or wider, and at each island counter or peninsular space 24 inches (600 millimeters) by 12 inches (300 millimeters) or larger. (Countertop spaces separated by range tops, sinks, or refrigerators are separate spaces.)

  - Install outdoor receptacles, accessible at grade level and no more than 6.5 feet (2 meters) above grade, at the front and back of a dwelling.

  - Install at least one receptacle on each balcony, deck, or porch that is accessible from inside the dwelling unit. There is no need for a receptacle if the area is less than 20 square feet (1.86 square meters).

  - Foyers that have an area greater than 60 square feet have a receptacle(s) located in each wall space that is 2 feet (600 mm) or wider.

- The receptacles required by Section 210.52 must be tamper-resistant (Section 406.12).

  - Exceptions include receptacles more than 5½ feet (1.7 m) above the floor and receptacles located within dedicated space for appliances that in normal use are not easily moved from one place to another and that are cord-and-plug connected.

- General Circuitry (*NEC*® Section 210.12)

  - All branch circuits supplying 120-volt, 15- and 20-ampere outlets in dwelling unit family rooms, dining rooms, living rooms, parlors, libraries, dens, bedrooms, sunrooms, recreation rooms, closets, hallways, or similar areas must be protected by a listed combination-type arc-fault circuit interrupter device (AFCI).

- Required Ground Fault Circuit Interrupter (GFCI) Protection (*NEC*® Section 210.8)

  - In dwellings, provide GFCI protection for all 15- and 20-amp, 125-volt-rated receptacle outlets installed in bathrooms, garages, crawl spaces, unfinished basements, locations within 6 feet (1.8 meters) of sinks in other than kitchens, boathouses and boat hoists, outdoors, and in kitchens for all receptacles serving the countertop.

- Wiring Methods (*NEC*® Section 314.23)

  - All electrical boxes must be securely supported by the building structure.

- Wiring Methods (*NEC*® Section 314.27(C))

  - When boxes are used as the sole support for a ceiling-suspended paddle fan, they must be listed and labeled for such use.

- Wiring Methods (*NEC*® Section 334.30)

  - Type NM cable must be secured at intervals not exceeding 4½ feet (1.4 meters) and within 12 inches (300 millimeters) of each electrical box and within 8 inches (200 millimeters) of each single-gang nonmetallic electrical box.

- Wiring Methods (*NEC*® Section 314.17(C) *Exception*)

  - The outer jacket of Type NM cable must extend into a single-gang nonmetallic electrical box a minimum of ¼ inch (6 millimeters).

- Wiring Methods (*NEC*® Section 300.14)

  - Minimum length of conductors at all boxes must be at least 6 inches (150 millimeters). At least 3 inches (75 millimeters) must extend outside the box; this includes grounding conductors.

- Wiring Methods (*NEC*® Section 300.4(A))

  - Where cables are installed through bored holes in joists, rafters, or wood framing members, bore the holes so the edge of the hole is not less than 1.25 inches (8 millimeters) inch from the nearest edge of the wood member. Where this distance cannot be maintained, or where screws or nails are likely to penetrate the cable, it must be protected with a steel plate at least ¹⁄₁₆ inch (1.6 millimeters) thick and of appropriate length and width.

- Wiring Methods (*NEC*® Section 300.22(C) *Exception*)

  - Do not install Type NM cable in spaces used for environmental air. However, Type NM cable is permitted to pass through perpendicular to the long dimension of such spaces.

- Wiring Methods (*NEC*® Sections 250.134, 314.4, and 404.9(B))

  - Ground all metal electrical equipment, including boxes, cover plates, and plaster rings. Ground all switches, including dimmer switches.

- Wiring Methods (*NEC*® Sections 110.12(A) and 314.17(A))

  - Close unused openings in boxes. When openings in nonmetallic boxes are broken out and not used, the entire box must be replaced.

- Wiring Methods (*NEC*® Section 110.14(A))

  - Install only one conductor under a terminal screw unless it is identified for more than one. In boxes with more than one ground wire, splice the ground wires with an approved mechanical connector. Then using a "jumper" or "pigtail," attach to the grounding terminal screw of the device. In metal boxes, connect the equipment grounding wires to the box with a green grounding screw or other approved method.

- Wiring Methods (*NEC*® Sections 110.14(B) and 300.15)

  - Splices must be made with an approved method, like a "wirenut," and must be made in listed electrical boxes or enclosures. When splicing underground conductors, the method and items used must be identified for such use.

- Wiring Methods (*NEC*® Sections 314.25 and 410.22)

  - In a completed installation, all outlet boxes must have a cover, lamp holder, canopy for a luminaire (light fixture), and an appropriate cover plate for switches and receptacles.

- Wiring Methods (*NEC*® Section 314.29)

  - Install junction boxes so the wiring contained in them can be rendered accessible without removing any part of the building.

- Wiring Methods (*NEC*® Section 314.16)

  - The volume of electrical boxes must be sufficient for the number of conductors, devices, and cable clamps contained within the box. Nonmetallic boxes are marked with their cubic inch capacity.

- Wiring Methods (*NEC*® Sections 410.2 and 410.16)

  - Storage space, as applied to an electrical installation in a closet.

  - The volume bounded by the sides and back closet walls and planes extending from the closet floor vertically to a height of 6 feet (1.8 meters) or the highest clothes hanging rod.

  - Parallel to the walls at a horizontal distance of 24 inches (600 millimeters) from the sides and back of the closet walls, respectively.

  - The volume continuing vertically to the closet ceiling.

  - Parallel to the walls at a horizontal distance of 12 inches (300 millimeters) or the shelf width, whichever is greater.

- ○ Luminaires (lighting fixtures) installed in clothes closets must have the following minimum clearances from the defined storage area.

  - 12 inches (300 millimeters) for surface incandescent or LED fixtures with completely enclosed lamps.

  - 6 inches (150 millimeters) for recessed incandescent or LED fixtures with completely enclosed lamps or fluorescent recessed fixtures.

  - 6 inches (150 millimeters) for surface-mounted fluorescent fixtures.

  - ○ Incandescent luminaires with open or partially enclosed lamps and pendant fixtures or lamp holders are not permitted in clothes closets.

- Wiring Methods (*NEC*® Section 410.116)

  - ○ Recessed lighting fixtures installed in insulated ceilings or installed within ½ inch (13 millimeters) of combustible material must be approved for insulation contact and labeled Type IC.

- Equipment Listing and Labeling (*NEC*® Section 110.3(B))

  - ○ Install and use all electrical equipment in accordance with the listing requirements and manufacturer's instructions.

  - ○ All electrical equipment, including luminaires, devices, and appliances are listed and labeled by a nationally recognized testing laboratory (NRTL) as having been tested and found suitable for a specific purpose.

    - UL and the CSA are two of the recognized agencies.

- Service Entrances (*NEC*® Section 310.15(B)(7))

  - ○ Service entrance conductor sizes for 120/240-volt residential services must not be smaller than those given in Table 310.15(B)(7).

- Service Entrances (*NEC*® Section 110.14)

  - ○ Do not mix conductors of dissimilar metals in a terminal or splicing device unless the device is listed for the purpose.

  - ○ Use listed antioxidant compound on all aluminum conductor terminations, unless information from the device manufacturer specifically states that it is not required.

- Service Entrances (*NEC*® Section 300.7(A))

  - ○ Seal portions of raceways and sleeves subject to different temperatures (where passing from the interior to the exterior of a building) with an approved material to prevent condensation from entering the service equipment.

- Service Entrances (*NEC*® Section 230.54)

  - ○ Where exposed to weather, enclose service entrance conductors in raintight enclosures and arrange in a drip loop to drain.

- Service Entrances (*NEC*® Section 300.4(G))

  - ○ Where raceways containing ungrounded conductors 4 AWG or larger enter a cabinet, box, or electrical enclosure, protect the conductors with an insulated bushing providing a smoothly rounded insulating surface.

- Service Entrances (*NEC*® Section 230.70(A)(1))

  - ○ Install the electrical service disconnecting means at a readily accessible location either outside a house or inside at a location nearest to the point of entrance of the service entrance conductors. No excess "inside run" is allowed.

- Service Entrances (*NEC*® Sections 230.70(A)(2) and 240.24(C)(D)(E))

  o Electrical panels containing fuses or circuit breakers must be readily accessible and must not be located in bathrooms, above stairway steps, or in the vicinity of easily ignitable materials such as in clothes closets.

- Service Entrances (*NEC*® Section 110.26)

  o Provide sufficient working space around electrical equipment.

  o When the voltage to ground does not exceed 150 volts, the depth of that space in the direction of access to live parts must be a minimum of 3 feet (900 millimeters).

  o The minimum width of that space in front of electrical equipment must be the width of the equipment or 30 inches (750 millimeters), whichever is greater.

  o Workspace must be clear and extend from the floor to a height of 6.5 feet (2 meters). This space cannot be used for storage.

  o All work spaces must be provided with illumination.

- Grounding and Bonding (*NEC*® Sections 250.50, 250.52, and 250.53)

  o Connect the house electrical service to a grounding electrode system consisting of a metal underground water pipe in direct contact with earth for 10 feet (3.0 meters) or more.

  o If a metal water pipe is not available as the grounding electrode, any other electrode as specified in Section 250.52 is allowed.

  o An additional electrode must supplement the water pipe electrode. If the metal water pipe is used as part of the grounding system, place a bonding jumper around the water meter.

  o Rod, pipe, or plate electrodes are supplemented by an additional electrode providing the single rod, pipe, or plate electrode does not have a resistance to earth of 25 ohms or less.

- Grounding and Bonding (*NEC*® Sections 250.64(C) and 250.66)

  o The grounding electrode conductor must be unspliced and its size determined, using the size of the service entrance conductors, by Table 250.66.

  o The conductor that is the sole connection to a rod, pipe, or plate electrode is not required to be larger than 6 AWG copper.

- Grounding and Bonding (*NEC*® Section 250.28)

  o Install a main bonding jumper or the green bonding screw provided by the panel manufacturer in the service panel to electrically bond the grounded service conductor and the equipment grounding conductors to the service enclosure.

- Grounding and Bonding (*NEC*® Section 250.104(A)(1))

  o Bond the interior metal water piping and other metal piping that may become energized to the service equipment with a bonding jumper sized the same as the grounding electrode conductor.

- Underground Wiring (*NEC*® Section 300.5)

  o Direct buried cable or conduit or other raceways must meet the minimum cover requirements of Table 300.5.

  o Underground service laterals must have their location identified by a warning ribbon placed in the trench at least 12 inches (300 millimeters) above the underground conductors.

  o Where subject to movement, arrange direct buried cables or raceways to prevent damage to the enclosed conductors or connected equipment.

  o Install conductors emerging from underground in rigid metal conduit, intermediate metal conduit, or Schedule 80 rigid PVC conduit to provide protection from physical damage.

  o Protection must extend from 18 inches (450 millimeters) below grade or the minimum cover distance, to a height of 8 feet (2.4 meters) above finished grade, or to the point of termination above ground.

Objective: Demonstrate an understanding of how to test for current and voltage in an energized circuit.

2) Understand how to test for current and voltage

Key terms: receptacle polarity

- Determining if the electrical system is working properly

  - Once you have determined that all applicable *NEC®* installation requirements have been met, do a check of the electrical system to determine if everything is working properly.

  - Usually done after an electrical inspector has approved installation and local electric utility has connected the house service entrance to the utility electrical system.

  - Main service disconnecting means is turned on, which energizes the fuse or circuit breaker panel.

  - Each circuit is then energized one by one and each receptacle and lighting outlet on the circuit is checked for proper voltage, proper polarity of the connections, and proper switch control.

  - During the checkout, use the electrical plans as a guide for the initial installation of the electrical system.

  - If a lighting outlet or power outlet is not working, initiate a troubleshooting procedure to find the problem.

  - Once the cause of the problem is found, correct the problem and then verify that the correction has caused the circuit and its components to operate properly.

- Testing receptacle outlets

  - Check each receptacle outlet to determine if the proper voltage is available and the wiring connections have resulted in the proper receptacle polarity.

  - Proper receptacle polarity check:

    - The white grounded conductor is attached to the silver screw of the receptacle and the "hot" ungrounded circuit conductor is attached to the brass screw on the receptacle.

  - Outlets can be checked with a voltage tester, a voltmeter, or a plug tester.

- Testing lighting outlets

  - It is assumed that the luminaire has been properly wired if the lamp(s) light in the luminaire when the proper switches are activated.

  - If the lamps in the lighting fixture do not come on, troubleshoot the problem.

  - Check the circuit. If it is off, turn it on.

  - Make sure the lamps are not "burned out."

  - Check that all circuit conductor connections are tight and they all have continuity.

  - Check for damage to the circuit wiring.

  - Check if the lighting outlet is being fed from a buried box that has had wallboard applied over it and never trimmed out.

  - Make sure the luminaire is properly grounded, especially if it is a fluorescent fixture.

  - Check the switching scheme for the proper connections.

- Testing switch connections

  - When a lighting circuit is energized, a single-pole switch can be checked by simply opening and closing the switch and observing if the lighting fixture or fixtures it is controlling also turn on and off.

○ Three-way and four-way switching arrangements are not as simple to check since there are several different switching combinations to be checked out.

○ Problem most often found:

■ Traveler wires were mixed up when the switches were installed as part of the trim-out process.

Objective: Demonstrate an understanding of how to test for continuity in existing branch-circuit wiring and wiring devices.

Objective: Demonstrate an understanding of the common testing techniques to determine whether a circuit has a short circuit, ground fault, or open.

Objective: Troubleshoot common residential electrical system problems.

Objective: Demonstrate an understanding of how to use a circuit tracer.

3) Understand how to test for continuity

4) Understand common testing techniques for a shorted, ground faulted or opened circuit

5) Troubleshoot common electrical system problems

6) Understand how to use a circuit tracer

Key terms: ground fault, short circuit, circuit tracer

• Troubleshooting common electrical circuit problems

○ Example: A circuit breaker immediately trips when you turn it on to energize a circuit.

○ Example: There is no electrical power on the circuit even when a circuit breaker is turned on to energize the circuit.

○ These are common situations and require good troubleshooting techniques to locate and fix the problem.

○ The likely cause of a circuit breaker tripping (or fuse blowing) as soon as the circuit is energized is a short circuit condition.

■ Unintended low-resistance path through which current flows around, rather than along, a circuit's intended current path.

■ Occurs when two circuit conductors come in contact with each other unintentionally.

■ Because the short circuit current path has a very low resistance, the actual current flow on a 120-volt or 240-volt residential circuit becomes very high and causes a circuit breaker to trip the instant the circuit is energized.

■ Until the short circuit is fixed, the circuit breaker will not stay on.

○ Another possible cause for having a circuit breaker trip (or fuse blow) as soon as a circuit is energized is a ground fault condition.

■ Unintended low-resistance path in an electrical circuit through which some current flows to ground using a pathway other than the intended pathway.

■ Occurs when an ungrounded "hot" conductor unintentionally touches a grounded surface or grounded conductor.

- The likely cause of no electrical power in a circuit (or parts of a circuit) is an open circuit condition.

  - Results in a circuit that is energized but does not allow useful current to flow on the circuit because of a break in the current path.

- When performing troubleshooting tests with a continuity tester, be aware of the following:

  - Feed-through.

  - Occurs when a load, like a light bulb, is still connected in a branch circuit that is being tested.

  - Current from the continuity tester can flow through the load and cause a false reading.

- When testing lighting branch circuits with a continuity tester, remove all lamps from lighting fixtures.

- When using a continuity tester to test a receptacle circuit, make sure there are no portable items plugged into any of the receptacles on the circuit.

- Using a circuit tracer allows an electrician to zero in on the correct circuit breaker in a quick and easy manner.

  - Circuit tracers typically consist of a transmitter that is connected to the circuit in some way and a receiver.

  - The transmitter is designed to send a signal with a special frequency on the circuit wires.

  - The receiver is handheld and is positioned close to the circuit breakers in the electrical panel.

  - Flashing lamps on the receiver or an audible signal tells you that the correct circuit breaker has been found.

Objective: Demonstrate an understanding of how to perform a successful service call.

7) Understand how to perform a service call

- Service calls

  - A service call is required if a home owner finds something wrong with the house electrical system.

  - There are a few steps to follow for a successful service call:

    - Display a professional, courteous, and intelligent attitude when dealing with customers.

    - Do not track dirt or mud into the house.

    - Make sure your tools do not scratch walls, floors, or furniture.

    - Be prepared to show some identification.

    - Find the problem.

    - Fix the problem.

    - Explain what you found and how you fixed it to the customer.

    - Fill in the appropriate paperwork in a legible manner. Make one last inspection of the work area.

    - Clean up any mess you made.

Quizzes

SAFETY

1. Electricians should wear safety glasses and observe all applicable safety rules when checking out and troubleshooting electrical wiring systems. True or False?

CODE/REGULATIONS/STANDARDS

1. According to the *NEC®*, lighting fixtures must be installed in clothes closets. True or False?

2. In a sub-panel, a green bonding screw provided by the panel manufacturer must be installed to electrically bond the grounded feeder conductor and the equipment grounding conductors to the panel enclosure. True or False?

3. At least _____ 20-amp small-appliance branch circuit(s) is/are required to serve kitchen countertop surfaces.

   a. one

   b. two

   c. three

4. GFCI protection is not required for 125-volt, 15- or 20-amp receptacles located _____.

   a. in bathrooms

   b. in finished basements

   c. in garages

5. When holes are bored through wood framing members, they must be at least _____ inch(es) from the nearest edge of the wood member.

   a. 1

   b. 1⅛

   c. 1¼

6. The minimum size of service for a single-family dwelling unit permitted by the *NEC®* is _____ amperes.

   a. 60

   b. 75

   c. 100

7. Type NM cable must be secured no more than _____ inches from each single-gang nonmetallic device box.

8. Recessed lighting fixtures installed in insulated ceilings so that the insulation is covering the fixture must be labeled _____.

Chapter Quiz

1. A(n) _____ is a circuit that is energized but does not allow useful current to flow on the circuit because of a break in the current path.

   a. grounded circuit

   b. short circuit

   c. ground fault

   d. open circuit

2. A(n) _____ is an unintended low-resistance path through which current flows around, rather than along, a circuit's intended current path.

   a. grounded circuit

   b. open circuit

   c. short circuit

   d. grounding circuit

3. At least _____ 20-amp small-appliance branch circuits are required to serve kitchen countertop surfaces.

4. The rating of the _____ determines the rating of a branch or feeder circuit.

   a. fuse or circuit breaker

   b. conductor

   c. receptacle

   d. main panel

5. Receptacle outlets in habitable rooms must be installed so that no point measured horizontally along the floor line in any wall space is more than _____ feet from a receptacle outlet.

6. The required 15- or 20-amp, 125-volt outdoor receptacles must be installed _____.

   a. only at the front of the dwelling

   b. every 12 feet along the wall line

   c. no more than 6.5 feet above final grade

   d. on each side of the dwelling

7. All branch circuits supplying 125-volt, 15- and 20-ampere outlets in residential bedrooms, family rooms, living rooms, libraries, recreation rooms, closets, hallways, parlors, and sun rooms must be protected by a(n) _____.

   a. fuse

   b. circuit breaker

   c. GFCI device

   d. AFCI device

8. GFCI protection is not required for 125-volt, 15- or 20-amp receptacles located _____.

   a. in bathrooms

   b. in finished basements

   c. in garages

   d. outdoors

9. When holes are bored through wood framing members, they must be at least _____ inches from the nearest edge of the wood member.

10. Which of the following is not a *NEC®* requirement for minimum clearance of a lighting fixture from the storage space in clothes closets?

    a. 12 inches for surface incandescent or LED fixtures with completely enclosed lamps

    b. 6 inches for recessed incandescent or LED fixtures with completely enclosed lamps

    c. 12 inches for surface-mounted fluorescent fixtures

    d. 6 inches for recessed fluorescent fixtures

11. The house electrical system may be grounded using a metal water pipe in direct contact with the earth for at least _____ feet.

12. *NEC®* Table _____ contains the requirements for direct burial of conductors and raceways.

13. An electrician is testing a 15-amp, 125-volt rated duplex receptacle with a voltage tester. The tester indicates 120 volts between the ungrounded slot and the grounded slot. The tester also indicates 120 volts between the grounded slot and the grounding slot. The likely cause is _____.

    a. an open equipment grounding path

    b. a short circuit in the cable feeding the receptacle

    c. an open grounded circuit conductor

    d. reversed polarity at the receptacle

14. The *NEC*® requires a minimum of _____ 20-amp bathroom branch circuit(s) in a dwelling unit.

15. An electrician tests a 30-amp, 250-volt dryer receptacle with a voltage tester. The tester indicates 120 volts between the X and Y slots. The tester also indicates 240 volts between the X and W slots. The tester also indicates 0 volts between the Y and the grounding slot. The likely cause is _____.

    a. an open grounding path

    b. the wires are reversed between the W and the Y slots

    c. the wires are reversed between the X and the Y slots

    d. an open grounded conductor

16. An electrician tests a three-way switching circuit. The lighting load can be turned OFF and then ON at each of the three-way switch locations. However, when one three-way switch toggle is left in the up position, toggling the other three-way switch does not cause the lighting load to come on. The likely cause is _____.

    a. an open grounding path

    b. an open grounded path

    c. the traveler wires are incorrectly connected to the traveler terminals on the switch

    d. there is no electrical power at the switch

17. An electrician is testing a circuit where a single-pole switch controls an incandescent lighting fixture in a bedroom. When the single-pole switch is toggled ON, the lighting fixture does not come on. Which of the following items could *not* be a cause for this?

    a. There is an open grounding path.

    b. The circuit breaker protecting the circuit is off.

    c. There are no lamps installed in the lighting fixture.

    d. The connections to the lighting fixture are not making a good connection.

18. When testing a "live" 120-volt receptacle using a voltage tester, you should get a 120-volt reading when measuring from the grounded conductor connection (long slot) to the grounding conductor connection (U-shaped slot). True or false?

19. When testing a "live" 50-amp, 125/250-volt range receptacle using a voltage tester, you should get a 240-volt reading when measuring from the grounded conductor connection (W slot) to either ungrounded conductor connection (X or Y slot). True or false?

20. When the white grounded circuit conductor is connected to the silver screw on a receptacle and the "hot" ungrounded conductor is connected to the brass screw on the receptacle, you have proper receptacle _____.

    a. troubleshooting

    b. grounding

    c. polarity

    d. timing

Troubleshooting/Quality Exercises

DISCUSSION QUESTION

1. There are many times when a ground fault condition is caused by a bare grounding conductor making contact with one of the device screw terminals in the box. Describe how you might find the problem and then fix it.

_____

_____

_____

_____

HANDS-ON APPLICATIONS

1. There are a few steps that an electrician needs to follow for the successful completion of a service call. Place a check next to the steps that should be followed.

_____ Display a professional, courteous, and intelligent attitude when dealing with a customer.

_____ Do not track dirt or mud into the house.

_____ Be careful not to let your tools put scratches on walls, floors, or furniture.

_____ Ask the customer to get you a cup of coffee before you start.

_____ Introduce yourself to the home owner.

_____ Do not listen to the home owner's version of what is wrong since they typically do not know what they are talking about.

_____ Explain to the home owner what you found to be the problem and how you fixed it.

_____ Keep good paperwork and have the customer sign off on the job if possible.

_____ Make a last inspection and clean up any debris and tools that you left behind.

_____ Do not worry about cleaning up any mess you made—the customer will clean it up.

REAL-WORLD SCENARIOS

1. Sometimes an incandescent lamp gets broken off while still screwed into the socket of a lighting fixture. Two methods for getting the broken base out of the lighting fixture are presented in the *House Wiring* textbook. List the steps for each of the two methods.

_____

_____

_____

_____

Chapter 21   Green Wiring Practices

Outline

Objective: Demonstrate an understanding of how to advise a house building team about energy efficient wiring practices.

1) Understanding how to advise the building team about energy efficient wiring

Key terms: Energy Star, green, phantom load, photovoltaic

- Energy Efficiency
  - Energy efficiency is one of the most important features of a green home.
  - The electrician can influence the electrical efficiency of a home by:
    - Advising the building team on the selection of energy efficient lighting, appliances, and equipment.
    - Recommending and installing electronic control equipment that reduces the electricity used in a home
- Selecting and Installing Energy Efficient Electrical Equipment
  - Lighting
    - Fluorescent hardwired Energy Star qualified lighting fixtures
    - Light emitting diode (LED) lighting fixtures.
      - LEDs are more energy efficient than fluorescent lamps and last longer.
  - Appliances
    - Energy Star rated models
    - Reduced operation expense of energy efficient appliances offsets the upfront cost over the life of the equipment.
  - Equipment
    - There are energy efficient models available for electrical equipment that is hardwired into homes.
    - The electrician will work cooperatively with another trade to supply and install equipment.
    - Electric water heaters
      - A remotely located tank type that supplies the hot water needs of an entire house
      - A tankless type that is located near the fixture(s) it supplies
    - Exhaust Fans and Fresh Air Ventilation Fans
      - There are three exhaust fan features to consider when selecting one: electrical efficiency, noise level, and power.

- Bathroom exhaust fans
  - Special electrical controls include: timers, dehumidistats, 24-hour timers, and fan speed controls
- Kitchen exhaust fans
- Whole house fresh air ventilation equipment; Heat Recovery Ventilators (HRVs) or Energy Recovery Ventilators (ERVs)

- Selecting and Installing Electric Controls
  - Lighting Control Equipment
    - A motion detector controls lighting by sensing when someone is in a room or hallway.
    - Automatic lighting controls can reduce electricity by only turning on the lights when they are needed.
    - Other controls sense the level of light (photo cells) and can automatically turn on lights inside or outside the house when the natural light fades.
    - Timer control
      - Commonly used for exterior lighting.
  - Programmable Thermostat
    - Regulates the heating and air-conditioning systems.
    - A properly programmed thermostat can reduce the energy to heat and cool a home by up to an estimated 15%.
  - Controls for Phantom Electric Loads
    - Some appliances and electric devices consume electricity even when they are not performing their intended function.
    - Total electrical load of all these devices can add up to 5% or more of a monthly electric bill
  - Maintaining the Integrity of the Air Barrier and Insulation
    - Avoid holes in the air barrier
  - A list of common penetrations made by the electrician through the air barriers of a house:
    - Cables to exterior lights
    - Electrical boxes for exterior lights
    - Recessed light fixtures
    - Ceiling light fixture electrical boxes
    - Cables or conduits run into an attic, basement, or crawl space
    - Service entrance cable
    - Exterior receptacle outlets
    - Interior receptacle and switch boxes on exterior walls
    - Wiring routed from an interior partition wall into an exterior wall
  - During the planning stage the electrician should discuss with the building team the steps for air sealing penetrations.
  - Electrical boxes that penetrate the air barrier must be special airtight models.
  - On-site Electric Power Generation

- Photovoltaic (PV) systems (also called solar electric systems), wind turbine generators, and micro-hydro electric generators.

- The electrician advises a building team considering an on-site power generation system of the options:

  - Type of system (solar, wind, or water)

  - Capacity of the system

  - Whether the system is tied to the electric grid or independent

- PV systems can consist of the following:

  - Separate solar panels mounted to a roof (the most common method) or on a ground level frame

  - Solar panels integrated into roof shingles, tiles, or metal roof panels

  - Not all houses or building sites are suited to PV systems.

  - The house should be in a sunny region, the roof of the house should face south (roof-mounted PV system), no trees or structures should shade the PV panels during the day.

- There are many residential sized wind turbine systems available now.

  - Wind turbines require a minimum constant wind speed in order to work.

- Residential micro-hydro electric systems

Objective: Demonstrate an understanding of how to advise the building team about durability and water management when installing the electrical system.

2) Understanding how to advise the building team about durability and water management

- Durability and Water Management

  - Seal Roofing and Siding Penetrations

  - Some common electric penetrations that need to be flashed and sealed on a house:

    - Service entrance conductors

    - Overhead service mast

    - Exterior receptacles

    - Air conditioner / heat pump disconnect

    - Exterior-mounted lighting

  - Wall flashing and sealing

    - Pre-made wall flashings for electrical boxes

    - Custom flashing

  - Roof Flashing

Objective: Demonstrate an understanding of how to advise the building team about selecting green products whenever they are available.

3) Understanding how to advise the building team about selecting green products

- Green Product Selection
  - o An electrician can recommend and install:
    - Products and materials manufactured with low environmental impact.
    - Products and materials made with recycled content.
    - Materials made locally reduce energy to transport.
    - Materials that are more environmentally friendly.
  - o Select electric products and materials manufactured with low environmental impact
  - o Select products and materials made with recycled content
  - o Select locally made materials and fixtures
  - o Try to limit the distance from manufacturing plant to the jobsite to 500 miles or less.
  - o Select and install environmentally friendly products

Objective: Demonstrate an understanding of how to advise the building team about reducing material use and waste when installing the house electrical system.

4) Understanding how to advise the building team about reducing material use and waste

- Reduce Material Use and Recycle Waste
  - o Every green home project should have a material recycling and waste management plan.
  - o The electrician can incorporate waste from rough-in and trim-out wiring phases into the jobsite recycling system.
    - Design a material-efficient branch-circuit layout
    - Use leftover materials
    - Save and sort lengths of different size cable and either use the cable on the current job or save them for the next job.
    - Cable cut-offs 6 inches or longer can still be used for things like pigtails, short box to box runs, switch legs, and other short runs.
  - o Integrate electrical waste into the jobsite-recycling program
    - Copper and aluminum cable are valuable and can be sold to a recycling company.
    - Steel can be recycled.
    - Cardboard, paper, and plastic packaging materials
      - Account for the largest percentage of waste on a typical new home building project.

Objective: Demonstrate an understanding of how to advise the building team about what electrical system items to include in a home owner education and reference manual.

5) Understanding how to advise the building team about electrical system items to be included in a home owner education and reference manual

- Home Owner Education and Reference Manual
  - ○ A reference manual of the entire electrical system should be compiled for home owners.
  - ○ Printed documentation from all the electrical fixtures and equipment and information about the electrical system should be saved and organized for the home owners.
  - ○ Include the following in the reference manual:
    - Locations of the main electrical panel, sub-panels, and disconnects.
    - Low voltage transformer locations for door chimes, lighting, alarm, and telecommunications equipment.
    - System service checklist.
    - Photos and drawings of the rough wiring within walls, floors, and ceilings.
    - How to operate AFCI circuit breakers.
    - How to operate GFCI circuit breakers and receptacles.
    - How to operate lighting, ventilation, HVAC, and other electrical control equipment.
    - How to reset a circuit breaker.
    - How to turn off the main breaker and other disconnects.
  - ○ Save and include manufacturers' printed installation instructions, information and warrantee forms for the fixtures, appliances, and equipment in the reference manual.
  - ○ Teach the owners about practices that reduce electricity use such as:
    - Turn off lights and electrical devices when not in use.
    - Adjust automatic lighting controls with each season change.
    - Replace light bulbs with compact florescent or LEDs.
    - Use fans or open windows instead of running air-conditioning.
    - Use a microwave to heat food instead of an electric range or cooktop.
    - Use a clothes line instead of a clothes dryer.

Quizzes

SAFETY

1. Always wear PPE when installing energy efficient electrical materials in a green home. True or False?

CODE/REGULATIONS/STANDARDS

1. _____ is the international standard for energy efficient consumer products.
   a. The National Electrical Code
   b. Energy Star
   c. The National Electrical Manufacturer's Association

Chapter Quiz

1. Green building is the process of designing and building a home that minimizes its impact on the environment both during construction and over its useful life. True or False?

2. A load that consumes electricity even when the load is not performing its intended function is called a ghost load. True or False?

3. Attenuation is the internationally recognized unit of loudness and is measured in decibels (dB). True or False?

4. Fluorescent lamps only use one-quarter of the electricity that incandescent lamps use and they last up to ten times longer. True or False?

5. Energy efficient appliances may cost _____ less efficient models; but the reduced operation expense more than offsets the upfront cost over the life of the equipment.
   a. less than
   b. more than
   c. the same as
   d. none of the above

6. Electric water heaters supply domestic hot water and come in a remotely located tank type that supplies the hot water needs of an entire house or a _____ type that is located near the fixture(s) it supplies.
   a. tankless
   b. round
   c. square
   d. seamless

7. Electric heat pump water heaters are more than _____ as efficient as electric resistance water heaters.
   a. two times
   b. three times
   c. four times
   d. five times

8. Without mechanical ventilation, the _____ level inside the house will rise to an unhealthy level and the durability of the structure may suffer as well.
   a. light
   b. sound
   c. humidity
   d. smoke

9. There are three features of exhaust fans to consider when selecting one: electrical efficiency, noise level, and _____.
   a. color
   b. mounting options
   c. location
   d. power

10. Fan power is measured in _____.
    a. foot pounds
    b. cubic feet per minute
    c. square feet per minute
    d. none of the above

11. One way to eliminate lights being left on when either they are not needed or when no one is in a room is to install a _____ to turn lights on and off automatically.
    a. motion detector
    b. dimmer switch
    c. photo cell
    d. double-pole switch

12. A _____ can be set to lower the temperature inside the house at night while home owners are sleeping and when they are away during the day at work or school.
    a. motion sensor
    b. occupancy sensor
    c. programmable thermostat
    d. dimmer switch

13. The _____ of a house blocks air from moving between the inside of the house and the outside and holds the conditioned (heated or cooled) air in the house.
    a. floor
    b. basement
    c. roof
    d. air barrier

14. Ideally, the roof of the house should face _____ so a roof-mounted PV system receives the best exposure.
    a. north
    b. south
    c. east
    d. west

15. _____ requires a minimum constant wind speed in order to work so a lot of home sites are not suitable for installation.

16. A home's biggest enemy is _____.

17. Steel, copper, and _____ are three of the most easily recycled materials.

18. A _____ minimizes the amount of waste that ends up in landfills and saves money in trash hauling and dumping fees.

19. A _____ of the entire electrical system should be compiled so owners can refresh their memories and so the information is carried forward to future owners.

20. It is estimated that a programmable thermostat can reduce the energy to heat and cool a home by up to _____ %.

Troubleshooting/Quality Exercises

DISCUSSION QUESTION

1. Discuss why some home designers and building contractors do not want an electrician to install lighting fixture outlet boxes in the ceiling.

_____

_____

_____

HANDS-ON APPLICATIONS

1. Match the following terms to their definitions. Write the correct number for the term in the blanks.

   1. dioxins
   2. Energy Star
   3. flashing
   4. green

   5. micro-hydro electric system
   6. phantom load
   7. photovoltaic
   8. sones

   _____ a. sheets of metal or other material used to weatherproof locations on exterior surfaces, like a roof, where penetrations have been made

   _____ b. an alternative energy system that uses water pressure to move a turbine which then drives a generator to produce electricity

   _____ c. the name used to describe the effect of sunlight striking a specially prepared surface and producing electricity

   _____ d. the process of designing and building a home that minimizes its impact on the environment both during construction and over its useful life

   _____ e. a family of chlorinated chemicals; toxic under certain exposure conditions; emitted when combustion of carbon compounds is inefficient

   _____ f. an internationally recognized unit of loudness, measured in decibels (dB)

   _____ g. a load that consumes electricity even when the load is not performing its intended function

   _____ h. an international standard for energy efficient consumer products

REAL-WORLD SCENARIOS

1. To save energy in a home, electricians can build a charging station so that items like cell phone chargers do not stay on for an extended period of time and consume electrical energy. Explain how a charging station could be built.

_____

_____

_____

_____

Chapter 22 Alternative Energy System Installation

Outline

Objective: Demonstrate an understanding of the different types of photovoltaic systems used in residential wiring.

1) Understand the different types of photovoltaic systems

Key term: hybrid system

- Introduction to Photovoltaic (PV) Systems
    - PV system advantages and disadvantages
        - Advantages of PV systems
            - Produce electricity at the fixed price of the original installation for the life of a PV system
            - Reliable
            - Durable
            - Low maintenance costs
            - No fuel costs and no need to burn fuels like oil and gas
            - Very little sound pollution
            - Modularity; parts can be added easily to a system to increase their electrical output
            - PV systems are very safe.
            - Allow for energy independence.
        - Disadvantages of PV systems
            - PV systems are expensive.
            - Solar radiation amounts vary across the country.
            - Energy storage; batteries cost a lot of money and they have to be replaced every few years.
            - Many people have not been educated about PV systems and are nervous about installing technology they do not understand.
- Types of PV systems
    - Day use only systems
        - Simplest and least expensive
        - Modules wired directly to DC loads
        - No electrical storage capabilities

- Examples are remote water pumping; fans, blowers, or circulators for solar water heating or ventilation systems.
- Not used in homes because electrical loads in homes must also be powered at night
- Direct current system with storage batteries
  - Loads can be powered day or night.
  - Can supply the extra surge current needed for starting large electric motors.
  - A charge controller is needed.
  - Not used in very many homes because almost all home electrical loads use alternating current (AC) and not direct current (DC)
- Direct current systems powering alternating current loads
  - Must use an inverter to convert the DC electricity to AC electricity
  - Commonly used in residential applications.
  - Newer type of PV system used to supply AC loads is available where individual micro-inverters are installed at each PV panel so that each panel produces AC electricity
- Hybrid system
  - Incorporates a gas or diesel powered generator along with a PV system
  - Generator can provide the extra power needed and charge the system batteries
  - Many hybrid systems also include a small wind turbine
- Stand-alone or grid-tied
  - Stand-alone system—no connection to the local electric utility's grid system; can be installed if the location of a home is far away from a main road and the utility grid
  - Utility Grid Interconnected System, often called a Utility-Connected, Grid-Tie, Intertie, or Line-Tie system—connected to the utility grid; does not need battery storage and will automatically shut down if the utility grid goes down
    - Home owner "sells" excess energy to the utility company
- PV grid-tie systems and net metering
  - Most locations in the United States offer some type of net metering agreement for home owners who have installed a grid-tie PV system

Objective: Demonstrate an understanding of the components that make up a PV system installation.

Objective: List the system components that make up a typical stand-alone PV system.

Objective: List the system components that make up a typical interactive (grid-tie) PV system.

2) Understand the components that make up a PV system installation

3) List the system components for a stand-alone PV system

4) List the system components for a grid-tie PV system

Key terms: array, charge controller, inverter, module, panel, solar cell

- PV System Components
  - PV modules
    - PV cell—converts sunlight into DC electricity

- PV modules or "panels" are assemblies of PV cells wired in series, parallel, or series/parallel to produce a desired voltage and current.
- The PV reaction
  - PV module types and characteristics
    - The wattage rating of a module is equal to its output voltage multiplied by its operating current.
    - "I-V curve"
    - Maximum Power Point (MPP) indicates the maximum output of the module and is the result of the maximum voltage (Vmp) multiplied by the maximum current (Imp).
    - Open Circuit Voltage (Voc) is the maximum voltage when no current is being drawn from the module.
    - Short Circuit Current (Isc) is the maximum current output of a module under conditions of a circuit with no resistance (short circuit).
    - *NEC*® Section 690.51
  - PV module mounting
    - Pole-mount systems
    - Ground-mount systems
    - Roof-mount systems
  - PV modules and diodes
    - Blocking diodes
    - Bypass diodes
  - Batteries
    - Liquid vented style
    - Sealed style
    - Gel cell
    - Absorbed glass mat (AGM)
      - Alkaline, nickel-cadmium, or nickel-iron
    - Hydrogen gas is very explosive! Make sure to properly vent the area where lead-acid batteries used in a PV system are present.
  - There are several things to consider when installing batteries as part of a PV system:
    - Batteries should be placed in a sturdy enclosure
    - Place batteries as near as is safely possible to the electrical equipment and loads to minimize wire sizes and length of runs
    - Batteries need to be configured to obtain the desired voltage and amp-hours
  - Charge controllers
    - Prevents the batteries from being overcharged by the arrays and over discharged by the electrical load.
    - Section 690.8(A)(1)

○ Inverters

   ■ Grid-tie inverters

   ■ Stand-alone inverters

      • Some inverters may have both capabilities built-in for future utility connection.

      • Also classified for the type of AC waveform they produce

         ○ Square wave

         ○ Modified square wave

         ○ Pure sine wave

   • IEEE Standard 1547 and UL 1741

Objective: Demonstrate an understanding of how a typical PV system is installed.

Objective: List several *NEC*® requirements that pertain to PV system installation.

5) Understand how a PV system is installed

6) List several *NEC*® requirements that pertain to PV system installation

   • PV System Wiring and the *NEC*®

      ○ PV System Conductors

         ■ Copper wire should be used when wiring PV systems

         ■ Aluminum wire can be used where allowed but most PV system installers still use copper wire

      ○ Section 690.31(F)

      ○ PV system installers must use conductors with insulation suitable for where the conductor is used.

         ■ Wet locations; types used must have a "W" in its letter designation

         ■ Table 310.104(A)

         ■ Color coding is used to identify the grounded, grounding, and ungrounded conductors

      ○ PV system conductors are installed as a cable assembly or in a conduit system.

      ○ Section 690.31(B)

      ○ MC (multi-contact) connections

      ○ *NEC*® Chapter 9 Informational Annex C and Tables 4 and 5

   • Conductor sizing

      ○ Installer must consider ampacity and voltage drop

         ■ Ampacity refers to the current-carrying ability of a conductor.

         ■ Tables 310.15(B)(16) and 310.15(B)(17)

         ■ Section 310.15(B)(3)(c) and Table 310.15(B)(3)(c)

      ○ Sizing the conductors based on ampacity

         ■ Section 690.8

      ○ PV circuit type locations in a typical stand-alone system

      ○ PV circuit type are locations in a typical utility interactive (grid-tie) system

- Finding the maximum circuit current on the conductors
  - Section 690.8(A)
  - Section 690.8(B)
  - Section 690.8(B)(1)(a)
  - Table 310.15(B)(16)
  - Table 310.15(B)(3)(c)
  - Table 310.15(B)(2)(a)
- Battery circuit sizing
- Inverter output circuit sizing
- Inverter input circuit sizing for stand-alone systems
  - Section 690.8(A)(4)
- Conductor sizing based on voltage drop
  - Common formula used to calculate voltage drop:

$$VD = \frac{K \times I \times L \times 2}{cm}$$

  - Formula for the minimum conductor size that would need to be installed to allow a certain voltage drop can be calculated:

$$cm = \frac{K \times I \times L \times 2}{VD}$$

- Overcurrent protection
  - Two types of overcurrent devices used in PV systems are fuses and circuit breakers.
  - Overcurrent protection placement
    - Section 240.20
    - Section 690.9(A)
  - Overcurrent protection sizing
    - Section 240.6(A) and (B)
    - Section 240.4(D)
    - Section 690.8(B)(1)
- Disconnects
  - Section 690.13
  - Section 690.15
  - Circuit breakers used as disconnects
  - Fuses used as disconnects
  - Section 690.14(C)
  - Section 690.17
  - Section 250.118(10)
  - Section 690.31(E)

○ Grounding

  ■ There are two types of grounding in a PV system: system grounding and equipment grounding.

    • Section 250.4

○ Equipment grounding

  ■ Section 690.43

○ Module and rack grounding

  ■ Section 690.43(C), (D), (E), and (F)

  ■ Section 250.110

○ System grounding

  ■ Section 690.41

  ■ Section 690.42

  ■ Section 240.21

○ Ground fault protection

  ■ Section 690.5

○ Sizing the equipment grounding conductor

  ■ Section 690.45(A)

  ■ Table 250.122

  ■ Sections 690.46 and 250.120(C)

○ Sizing the grounding electrode conductor

  ■ Section 250.66 for the AC side of the system

  ■ Section 250.166 for the DC side of the system

○ Grounding electrodes

  ■ Section 250.52

  ■ Section 690.47(C)

○ Surge suppression

○ Plaque requirements

  ■ Section 690.56(A) and (B)

○ NEC® requirements specific to stand-alone systems

  ■ Section 690.10(A)

  ■ Section 690.10(C)

  ■ Section 690.10(E)

  ■ Section 408.36(D)

  ■ Section 690.11

○ NEC® requirements specific to grid-tie systems

  ■ Section 690.4(D)

  ■ Section 690.60

- Section 690.64
- Section 705.12
- 705.12(D),
- 705.12 (D) (1), (2), (4), (5), (6)
- Section 690.14(D)

Objective: Demonstrate an understanding of small wind turbine system installation.

Objective: List the components that make up a small wind turbine system.

Objective: List several *NEC*® requirements that pertain to a small wind turbine system installation.

7) Understand small wind turbine system installation

8) List the components of a small wind turbine system

9) List *NEC*® requirements that pertain to small wind turbine system installation

Key terms: tower, wind turbine, wind turbine system, nacelle, guy, anemometer

- A wind turbine system includes the wind machine, tower, and all associated equipment.
  - ○ Can be interactive with other electrical power production sources or may be stand-alone systems
- Introduction to wind power
  - ○ Wind power is quite reliable.
  - ○ Three main components to a wind turbine are the rotor, the alternator, and the tower.
  - ○ The nacelle contains the alternator and other parts of a small wind turbine.
  - ○ The size of the blades and rotor really matters
    - Larger blade areas allow more wind power to exert a force on the blades that can turn larger generators.
    - Small wind turbines can come in micro, mini, and household sizes.
- Small wind applications
  - ○ Supplement existing electrical energy requirements
  - ○ Energy independence
  - ○ Two types of small wind turbines suitable for producing utility compatible power
    - Turbines that produce an AC voltage with a proper 60 Hz waveform that can be directly connected to the utility grid
    - Turbines that produce a voltage and/or frequency that cannot be directly connected to the grid
- Measuring the wind
  - ○ Measuring the wind at that site is an important part of determining the suitability of a particular site.
    - 1 knot = 1.15 mph
    - 1 m/s = 2.24 mph
    - 1 km/h = .621 mph

- ○ Wind speed and power vary with the height above the ground.
  - There is a lot of published wind data available. It is usually gathered near population centers. The main sources in the United States are the National Weather Service (NWS) and the Federal Aviation Administration (FAA). The wind speed data that these organizations have compiled are available to you at the National Climatic Data Center (NCDC) in Ashville, North Carolina (http://lwf.ncdc.noaa.gov/oa/ncdc.html). The National Renewable Energy Lab (NREL) website also has wind speed data available (www.nrel.gov/wind/).
- ○ When surveying the proposed site for a small wind turbine system try to determine:
  - The expected maximum instantaneous wind speed (wind gust)
  - The average annual wind speed
  - The distribution of wind speeds over a significant length of time
- ○ Using an anemometer, mast, and a recorder will allow you to get an accurate measurement of wind speeds over a period of time
- Small wind turbine system towers
  - ○ Choose a tower that is strong enough to withstand the strongest wind gusts that could be encountered.
    - Using taller towers; for small wind turbines today's minimum recommended tower height is 80 feet.
    - Never install a wind turbine on a tower that is not at least 20–30 feet above any buildings or trees in the vicinity.
- Standards in the United States require withstanding a 120-mph wind with no damage.
- Two common tower types, a freestanding tower and a guyed tower, require a deep concrete foundation to properly support them.
  - ○ Freestanding towers can be broken down into two types: the lattice type and tubular type.
  - ○ Some towers are hinged at the bottom and can be tipped up and into place
  - ○ Guyed towers are the most common for small wind turbine systems
- Small wind turbine installation
  - ○ All small wind turbine system wiring must be done according to the *NEC®* and any local rules.
    - *NEC®* Article 694
    - Sections 314.16 or 314.28
    - Section 694.20
    - Section 694.22
  - ○ The disconnecting means must consist of manually operated switches or circuit breakers and be compliant with all of the following requirements:
    - Located where readily accessible
    - Externally operable without exposing the operator to contact with live parts
    - Able to plainly indicate whether in the open or closed position
    - Have an interrupting rating sufficient for the circuit voltage and the current that is available at the line terminals of the equipment
    - Suitable for the environment if located outside
    - A plaque must be installed in accordance with Section 705.10
    - Section 694.24

- ○ Conductor sizing
  - ▪ Section 694.12
- ○ Wiring method
  - ▪ Section 694.30(A), (B), and (C)
- ○ Informational Annex C is used when conductors are all the same size and have the same type of insulation.
  - ▪ When using different combinations of conductor sizes and/or the conductor insulations are not the same, use the information in Tables 4 and 5 in Chapter 9 of the *NEC®*.
- ○ Grounding
  - ▪ Section 694.40(A), (B), and (C)
  - ▪ Auxiliary electrodes are permitted to be installed in accordance with Section 250.54.
- ○ Wind turbine systems are grounded to limit voltage surges from nearby lightning strikes.
- ○ Some additional items to remember when installing wind turbine wiring are:
  - ▪ Properly mount all equipment so it is secure.
  - ▪ Be sure to seal all holes or knock-outs that are open.
  - ▪ Make sure there is enough clearance (working space) in front of the equipment.
  - ▪ All terminals, connectors, and conductors must be compatible and if using aluminum wire make sure it is prepared properly.
- Small wind system installation safety
  - ○ Mechanical safety hazards
    - ▪ The main rotor is the most serious mechanical safety risk.
    - ▪ Never approach the turbine when it is operating; always shut down the turbine by waiting until the turbine is stationary on a windless day.
    - ▪ Follow the manufacturer's recommendation on how to stop the blades from turning for their specific small wind turbine.
    - ▪ If a grid-tie inverter or batteries are part of the system follow the same safety procedures as those covered in the PV system part of this chapter.
  - ○ Tower safety
    - ▪ Only people who are directly involved with the tower installation should be allowed in the work area.
    - ▪ All persons on the tower or in the vicinity must wear an OSHA approved hard hat.
    - ▪ All tower work should be done under the supervision of a trained professional.
    - ▪ Towers should never be erected near utility power lines.
    - ▪ Workers should always use a full body harness and be tied off when climbing or working on a tower.
    - ▪ Tool belts should always be used to properly hold the tools that you will be using.
    - ▪ Never carry tools or parts in your hands when climbing a tower. Use a bucket and a hoist line to raise and lower tools and parts.
    - ▪ Never stand or work directly below someone else who is working on the tower.
    - ▪ Never work on a tower alone. Always work with at least one other person.

- Never climb a tower unless the turbine is furled and the alternator is locked out by short circuiting the output conductors in the disconnect switch located at the base of the tower.

- Never work on a tower during thunderstorms, high winds, tower icing, or severe weather of any kind.

o Electrical safety hazards

- Be sure to install the proper size wiring that will bring the electrical power from the turbine to the house electrical system.

- Be sure to properly size and install overcurrent protection devices.

- Always use insulated electrical tools when working on the battery's electrical connections.

- A properly sized fuse or circuit breaker should be used in the cables connected to the battery.

- Do not attempt to move heavy batteries by yourself.

- Always keep these batteries in an upright position and do not allow the electrolyte to come into contact with your skin or face.

- Always follow the manufacturer's safety instructions when handling lead-acid batteries.

Quizzes

SAFETY

1. Always wear PPE when installing a solar PV system. True or False?

2. There is no need for an electrician to wear PPE when installing a small wind turbine system because the small size means there are no hazards to worry about. True or False?

3. Wearing a fall protection harness and being tied off while working on a roof is the best way to protect you from falling off the roof and seriously hurting yourself. True or False?

4. The hydrogen gas produced around lead-acid batteries can cause explosions and fire if the batteries are not properly vented and a spark of some kind causes the hydrogen to ignite. True or False?

5. The main rotor of a small wind turbine is the most serious mechanical safety risk. True or False?

CODE/REGULATIONS/STANDARDS

1. Article _____ in the 2011 *NEC*® applies specifically to solar photovoltaic (PV) systems.
   a. 380
   b. 690
   c. 820

2. Article _____ in the 2011 *NEC*® applies specifically to small wind turbine systems.
   a. 380
   b. 430
   c. 694

Chapter Quiz

1. A typical PV solar cell is about 4 inches across and produces about 1 watt of power in full sunlight at about _____ volts DC.
   a. .3
   b. .5
   c. 1
   d. 2

2. A _____ is a configuration of PV cells laminated between a clear outer superstrate (glazing) and an encapsulating inner substrate.
   a. solar cell
   b. module
   c. charge controller
   d. inverter

3. Solar panels are connected together to form a PV _____.
   a. module
   b. inverter
   c. charge controller
   d. array

4. A(n) _____ regulates the battery voltage and makes sure that the PV system batteries (if used) are charged properly.
   a. array
   b. charge controller
   c. inverter
   d. module

5. PV modules are connected to a(n) _____ that "converts" the DC electricity produced by most solar arrays to the AC electricity commonly used in a house.
   a. array
   b. module
   c. inverter
   d. charge controller

6. A DC system with storage batteries is a PV system where loads can be powered _____.
   a. day only
   b. night only
   c. day or night
   d. none of the above

7. Electricity is the flow of _____ through a circuit.
   a. neutrons
   b. charges
   c. electrons
   d. protons

8. One amp of current flowing for one hour is equal to one _____.

   a. volt

   b. watt

   c. ohm

   d. amp-hour

9. The rate at which an electrical load uses electrical energy or the rate at which electrical energy is produced is called _____.

   a. voltage

   b. amperage

   c. resistance

   d. power

10. 1000 watt-hours equal 1 _____.

    a. kVA

    b. kWh

    c. mA

    d. kV

11. Voltage sources connected in series result in total voltage _____ but the _____ flow remains the same.

    a. increases, current

    b. decreases, current

    c. increases, voltage

    d. decreases, voltage

12. Solar _____ is a measure of how much solar power is striking a specific location.

    a. insolation

    b. irradiance

    c. power

    d. sunlight

13. Solar _____ is a measure of solar irradiance over a period of time, typically over the period of a single day.

    a. insolation

    b. power

    c. sunlight

    d. current

14. The northern hemisphere is tilted toward the sun from _____ so there is more available solar energy in summer than winter.

    a. January through May

    b. June through August

    c. September through December

    d. August through December

15. Shading greatly affects a PV array's performance and as a general rule a PV array should be free of shade from _____.

    a. 6:00 a.m. to 9:00 p.m.

    b. 10:00 a.m. to 2:00 p.m.

    c. 9:00 a.m. to 3:00 p.m.

    d. 5:00 a.m. to 5:00 p.m.

16. A wind turbine system includes the wind machine, _____, and all associated equipment.

    a. tower

    b. base

    c. fence

    d. road

17. Small wind turbine electric systems consist of one or more wind electric generators with individual systems up to and including _____ kW.

    a. 20

    b. 50

    c. 100

    d. 600

18. The _____ is the enclosure housing the alternator and other parts of a wind turbine.

    a. box

    b. rotor

    c. nacelle

    d. compartment

19. Using a(n) _____, mast, and a recorder will allow you to get an accurate measurement of wind speeds over a period of time.

    a. ammeter

    b. anemometer

    c. voltage tester

    d. ground resistance meter

20. The maximum current output of a PV module under conditions of a circuit with no resistance is called the _____.

    a. short circuit current

    b. open circuit voltage

    c. maximum power point

    d. none of the above

Troubleshooting/Quality Exercises

DISCUSSION QUESTIONS

1. The amount of available sunlight varies because of atmospheric attenuation (loss). Discuss some of the reasons for atmospheric loss.

_____

_____

_____

_____

2. True south is not indicated on a compass. Discuss why a compass will point to a direction that is not true south.

_____

_____

_____

_____

HANDS-ON APPLICATIONS

1. Match the following terms to their definitions. Write the correct number for the term in the blanks.

1. amorphous	14. maximum power point
2. anemometer	15. micro-hydro
3. array	16. module
4. azimuth	17. nacelle
5. charge controller	18. open circuit voltage
6. diode	19. panel
7. fuel cell	20. short circuit current
8. guy	21. solar cell
9. hybrid system	22. thermals
10. insolation	23. tower
11. irradiance	24. wind turbine
12. inverter	25. wind turbine system
13. magnetic declination	

_____ a. a system comprised of multiple power sources; the power sources may include photovoltaic, wind, micro-hydro generators, or engine-driven generators

_____ b. a cable that mechanically supports a wind turbine tower

_____ c. a small wind electric generating system

_____ d. the term used for the measure of solar radiation striking the earth's surface at a particular time and place

_____ e. an electrochemical system that consumes a fuel like natural gas or LP gas to produce electricity

_____ f. a mechanical device that converts wind energy to electrical energy

_____ g. a measure of how much solar power is striking a specific location

_____ h. a semiconductor device that allows current to pass through in only one direction

_____ i. a pole or other structure that supports a wind turbine

_____ j. a device that changes DC input to an AC output

_____ k. equipment that controls DC voltage or DC current, or both, used to charge a battery

_____ l. rising currents of warm air that go up and over land during sunny daylight hours

_____ m. the deviation of magnetic south from true south

_____ n. the sun's apparent location in the sky east or west of true south

_____ o. the basic photovoltaic device that generates electricity when exposed to light

_____ p. indicates the maximum output of the module and is the result of the maximum voltage (Vmp) multiplied by the maximum current (Imp)

_____ q. a mechanically integrated assembly of modules or panels with a support structure and foundation, tracker, and other components, as required, to form a DC power-producing unit

_____ r. the maximum current output of a module under conditions of a circuit with no resistance (short circuit)

_____ s. a small hydroelectric alternative energy system that uses water flow to turn a generator that produces electricity; the water flow can come from a stream or a small reservoir

_____ t. a device used to measure wind speed

_____ u. a collection of modules mechanically fastened together, wired, and designed to provide a field-installable unit

_____ v. a complete, environmentally protected unit consisting of solar cells, optics, and other components, exclusive of tracker, designed to generate DC power when exposed to sunlight

_____ w. having no definite form or distinct shape

_____ x. the maximum voltage when no current is being drawn from the module

REAL-WORLD SCENARIOS

1. There are several advantages to installing a solar PV system in a house. Name what is considered to be the biggest advantage.

_____

_____

_____

_____

2. There are a few disadvantages to installing a PV system in a home. Name the three major disadvantages.

_____

_____

_____

_____

PART

TWO

Lab Manual

Chapter 1

Residential Workplace Safety

Lab 1.1: Demonstrate an Understanding of Both General Safety and Electrical Safety by Scoring 100% on a Comprehensive Safety Exercise

Name: _____ Date: _____ Score: _____

Introduction

Safety should be the main concern of every worker. Too often, failure on the part of workers to follow recommended safe practices results not only in serious injury to themselves and fellow workers, but also in costly damage to equipment and property. The electrical trades, perhaps more than most other occupations, require constant awareness of the hazards associated with the occupation. The difference between life and death is a very fine line. There is no room for mistakes or mental lapses. Trial-and-error practices are not acceptable! Electricity plays a big part in our lives and serves us well. Being able to control electricity allows us to make it do the things we want it to do. Control comes with a good understanding of how electricity works and an appreciation of the hazards and consequences involved when this control is not present. A good residential electrician will, in addition to being proficient in the technology of the trade, possess and display respect for the hazards associated with the occupation. Residential electrical workers must realize from the beginning of their training that if they do not observe safe practices when installing, maintaining, and troubleshooting an electrical system, there will be a good chance that they could be injured on the job. Both general and electrical safety is serious business. In this lab exercise, you will show how much you know about safety by answering several questions concerning both general and electrical safety. Because there is no room for error when it comes to safety, you are required to answer 100% of the questions correctly.

Materials and Equipment

House Wiring 3e textbook, pencil

Procedure

1. Use Chapter 1 of the *House Wiring* 3e textbook to answer each of the items in the Review section.

2. For the multiple choice items, indicate your choice by circling the letter next to the correct answer.

3. For the true or false items, in the space provided write in a **T** to indicate a true statement or an **F** to indicate a false statement.

4. For the short answer items, write clear and complete statements.

Review

MULTIPLE CHOICE

1. In a work situation where a residential electrician's hands get wet while operating a faulty electric drill, which of the following would be true?
   a. Body resistance increases and any shock would be mild.
   b. Body resistance remains the same and there is no danger as long as rubber boots are worn.
   c. Body resistance is substantially decreased and severe shock could occur.
   d. There is no danger as long as a three-prong plug is used.

2. The amount of current flow though a human body that it takes to cause ventricular fibrillation is:
   a. 5 milliamperes
   b. 100 milliamperes
   c. 1000 to 4300 milliamperes
   d. 10,000 milliamperes

3. Which of the following is **not** an insulator?
   a. Porcelain
   b. Copper
   c. Plastic
   d. Rubber

4. Who publishes the *National Electrical Code® (NEC®)*?
   a. OSHA
   b. The National Electrical Contractors Association
   c. The National Fire Protection Association
   d. The Department of Labor

5. The unit of measure for electrical force is the:
   a. Ampere
   b. Volt
   c. Ohm
   d. Watt

6. Energized electrical equipment is dangerous. How does OSHA suggest that a worker deal with this potential hazard?
   a. Work only on equipment that is marked or tagged "dead."
   b. Inspect and test all equipment, assuming all to be energized until proven different.
   c. Work only on equipment that the foreman says is "dead."
   d. Ask someone nearby what the condition is.

7. Which of the following is considered to be the most important safeguard against serious or fatal accidents on the residential jobsite?
   a. Knowledge of both general safety and electrical safety
   b. Experience
   c. Common sense
   d. Bravery

8. Which of the following types of extinguishers should **not** be used on electrical fires? (Hint: There are two.)
   a. Carbon dioxide
   b. Foam
   c. Dry chemical
   d. Pressurized water

9. When rescuing a coworker who is being shocked from an energized electrical conductor, all of the following should be followed **except:**
   a. Shut off the electrical supply (if possible).
   b. Use a non-conductor of electricity to remove the victim from the source.
   c. Have someone call for help.
   d. If you are wearing rubber soled boots it is safe to pull the victim away with bare hands.

10. Fires that occur in or near electrical equipment such as motors, switchboards, and electrical wiring are classified as:
    a. Class A fires
    b. Class B fires
    c. Class C fires
    d. Class D fires

TRUE/FALSE

11. _____ If the circuit voltage stays the same, the higher the resistance in a circuit the lower the amount of current that will be flowing in the circuit.

12. _____ The resistance of the human body is fixed, regardless of the conditions.

13. _____ OSHA requires voluntary compliance, and each employer may comply if they want to.

14. _____ The purpose of NFPA 70E is to provide a practical and safe working area for employees relative to the hazards arising from the use of electricity.

15. _____ Electrical shock at lower voltages of 120V or 240V cannot cause death.

16. _____ Water is not a conductor of electricity, because only solid materials conduct electricity.

17. _____ The third prong (grounding prong) on a three-prong plug is optional, and may be removed.

18. _____ Safety eye glasses or goggles should be worn only when working on energized circuits or equipment.

19. _____ The *National Electrical Code® (NEC®)* gives minimum safety standards for electrical work and is not a "how-to" manual.

20. _____ Electricians should wear metal helmets on jobsites to avoid injury from falling objects.

SHORT ANSWER

21. List five (5) safety rules that should be followed when using ladders.

   1. _____

   2. _____

   3. _____

   4. _____

   5. _____

22. List five (5) types of personal protective equipment (PPE) and when they must be worn.

   1. _____
      _____

   2. _____
      _____

   3. _____
      _____

   4. _____
      _____

   5. _____
      _____

23. List five (5) electrical safety rules that a residential electrician should follow.

   1. _____
      _____

   2. _____
      _____

   3. _____
      _____

   4. _____
      _____

   5. _____
      _____

24. List the three (3) components of the fire triangle.

   1. _____

   2. _____

   3. _____

25. General safety in residential electrical work includes how an electrician should behave on the job. List two (2) behavior rules to follow while on the jobsite.

   1. _____

   2. _____

Lab 1.2: Find Information in the National Electrical Code®

Name: _____ Date: _____ Score: _____

Introduction

The *National Electrical Code®* (*NEC®*) is the guide for safe practices and procedures in the electrical field and is published by the National Fire Protection Association. Every three years the *NEC®* is brought up to date to reflect the latest changes and trends in the electrical industry. It contains specific rules to help safeguard people and property from the hazards arising from the use of electricity. Its content should be very familiar to residential electricians because all electrical work done in a dwelling must conform to the *NEC®*. Residential electricians often have to find information in the *NEC®*. If they are not familiar with the *NEC®*, a procedure for finding the information they want needs to be used. In this lab exercise use the procedure outlined below to find the answers to the items in the Review section.

Materials and Equipment

2011 National Electrical Code®, pencil

Procedure

1. **Determine the main topic area for the information you want.**

   This is a very important first step. If you are not able to identify the proper **main topic area,** the information you eventually locate in the *Code* probably will not be the information you want. Identifying the proper **main topic area** gets easier with practice.

2. **Locate the main topic area in the Index.**

   The Index lists **main topic areas** in bold print and in alphabetical order to help you find them more easily. If you are unable to find the **main topic area** in the Index, try alternative wording. For example, the topic area "electric motors" is not listed in the Index but the topic area "motors" is.

3. **Determine the appropriate sub-topic.**

   In most cases, the Index lists several **sub-topics** under the main topic area heading. The sub-topics are listed in alphabetical order. Identifying the proper sub-topic gets easier with practice. **Note:** If the **main topic area** you have chosen does not have any sub-topics listed under it, skip this step and proceed to step 4.

4. **Determine which Article, Part, Section, or Table is referenced for the topic area.**

   If more than one reference is given, choose the one that will most likely have information pertaining to your application. If you find that the reference you chose does not result in the information you are looking for, simply try the next reference. Continue to try the references until you find the proper information.

5. **In the Table of Contents, find the page number of the Article that contains the reference.**

   This step is optional if you have a good feel for where an Article is located in the *Code*. If this is the case, you can simply open up to the Article and skip to step 6. Also, you can look at the top left and right corners in the *NEC®* and see the Section number that begins the left page and ends the right page. By using this information, you can quickly scan to a specific area in the *Code* that was referenced in the Index. Another method that can be used to quickly locate an Article is to use *Code* tabs. Tabs are commercially available from several companies and when installed on your *NEC®*, make finding the major Articles and Tables in the *Code* much easier.

6. **Turn to the Article and scan through until you find the referenced Part, Section, or Table.**

   As you become more experienced with using the *Code,* you may be able to simply jump to the Part, Section, or Table in the Article instead of scanning through the Article from the beginning.

7. **Read all of the material in the referenced area until you find the information that applies to your specific application.**

   Now that you have located the area in the *Code* that was referenced in the Index, you must be able to determine what material in this area applies to your particular application. Once again, the more experience you have finding information in the *Code* will make this step much easier.

Review

Note: The main topic area and the sub-topic have been included to help you find the answers. Some main topic areas do not have sub-topics. Use this information and the procedure outlined above to find the *NEC®* section that applies to the topic and then choose the best answer to the question.

1. A receptacle in a dwelling unit laundry area must be installed within _____ feet of the laundry equipment location.

   a. 2   b. 4   c. 6   d. 12

Index main topic area:	Laundry
Index sub-topic:	Outlets, dwelling
NEC® reference:	
Answer:	

2. A lighting outlet controlled by a wall switch must be installed in each habitable room and bathroom of a dwelling unit.

   a. True   b. False

Index main topic area:	Lighting outlets
Index sub-topic:	None
NEC® reference:	
Answer:	

3. When flexible cord is used to supply a 120-volt room air conditioner, the length of the cord must not exceed _____ feet.

   a. 3   b. 5   c. 6   d. 10

Index main topic area:	Cords
Index sub-topic:	Flexible, Air conditioner
NEC® reference:	
Answer:	

4. A recessed incandescent luminaire (lighting fixture) installed in a clothes closet must have at least _____ of clear space from the storage area.

   a. 3 inches   b. 6 inches   c. 12 inches   d. 18 inches

Index main topic area:	Clothes closets
Index sub-topic:	Luminaires
NEC® reference:	
Answer:	

5. A continuous load is defined as a load where the maximum current is expected to continue for _____ hours or more.

   a. 2   b. 3   c. 4   d. 5

   Index main topic area: Continuous load

   Index sub-topic: Definition

   NEC® reference:

   Answer:

6. How many small-appliance branch circuits at a minimum must be installed in each dwelling unit?

   a. 2   b. 3   c. 4   d. 5

   Index main topic area: Branch circuits

   Index sub-topic: Small appliance

   NEC® reference:

   Answer:

7. Which of the following locations in a house require ground fault circuit interrupter (GFCI) protection for 15- and 20-amp, 125-volt-rated receptacles?

   a. Kitchen   b. Bathroom   c. Garage   d. All of the above

   Index main topic area: Ground fault circuit interrupters

   Index sub-topic: Receptacles

   NEC® reference:

   Answer:

8. How often does a length of nonmetallic sheathed cable need to be supported?

   a. 2 feet   b. 3 feet   c. 4½ feet   d. None of the above

   Index main topic area: Nonmetallic sheathed cable

   Index sub-topic: Supports

   NEC® reference:

   Answer:

9. What is the minimum rating of the service entrance disconnecting means for a one-family dwelling?

   a. 60 amps   b. 100 amps   c. 150 amps   d. 200 amps

   Index main topic area: Service equipment

   Index sub-topic: Disconnecting means, rating

   NEC® reference:

   Answer:

10. Receptacle faceplates may be installed so that a small crack not more than ⅛ inch wide can be seen after the installation.

    a. True   b. False

    Index main topic area: Receptacles, cord connectors, and attachment plugs

    Index sub-topic: Faceplates

    NEC® reference:

    Answer:

Lab 1.3: Use a Material Safety Data Sheet (MSDS)

Name: _____ **Date:** _____ **Score:** _____

Introduction

A material safety data sheet (MSDS) must be made available to a worker using any hazardous material. MSDSs are designed to inform the electrician of a material's physical properties, as well as the effects on health that make the material dangerous to handle. The proper type and style of protective equipment needed when using the material is given and if you are exposed to the material or its hazards, the proper first aid treatment is given. Other information, such as the proper storage methods, how to safely handle spills of the material, and how to properly dispose of the material, is given in the MSDS. Each company is required by law to develop a hazard communications (HazCom) program that must contain, at a minimum, warning labels on containers of hazardous material, employee training on the safe use and handling of hazardous material, and MSDSs. The law requires the employer to keep the MSDS up to date and to keep them in a readily accessible location so that employees can access them quickly when they are needed. It is very important for residential electricians to be able to use an MSDS. In this lab exercise, you will use the information found in a typical MSDS to answer a few questions about a product often used in the electrical trades.

Materials and Equipment

A sample MSDS for an item typically used in residential wiring, a pencil

Procedure

1. Using the sample MSDS shown below, answer the questions in the Review section of this lab exercise. The sample MSDS is for a commonly used antioxidant that is used to keep aluminum conductors from oxidizing.

MATERIAL SAFETY DATA SHEET
GENERAL INFORMATION

PRODUCT NAME OR NUMBER (as it appears on label)	CATALOG NUMBER
Noalox® Anti-Oxidant	All "30" Series
MANUFACTURER'S NAME	EMERGENCY TELEPHONE NO.
IDEAL INDUSTRIES, INC.	**(815) 895-5181**

ADDRESS (Number, Street, City, State, Zip Code)
Becker Place, Sycamore, IL 60178

HAZARDOUS MATERIAL DESCRIPTION, PROPER SHIPPING NAME, HAZARD CLASS, HAZARD CLASS NO., HAZARD ID NO. (49 CFR 172.101)
None

CHEMICAL DESCRIPTION	FORMULA
Petroleum-Based Mixture	Proprietary

SECTION I - INGREDIENTS

CAS REGISTRY NO.	%W	CHEMICAL NAME(S)*	Listed as a carcinogen in NTP, I ARC or OSHA 1910(z) (specify)
9003-29-6	<80	Polybutene	No
7440-66-6	20	Zinc Dust	No
7631-86-9	<5	Silicon Dioxide	No

SECTION II - PHYSICAL DATA

BOILING POINT	SPECIFIC GRAVITY (H₂O=1)	PERCENT VOLATILE BY VOLUME (%)
>500°F °C.	1.04	N.E.
SOLUBILITY IN WATER	pH =	PERCENT SOLID BY
Moderate	6.5 - 8.0	WEIGHT (%) 100
APPEARANCE AND ODOR		IS MATERIAL: LIQUID SOLID GEL
Gray solid paste, mild odor		GAS PASTE

SECTION III - FIRE AND EXPLOSION HAZARD DATA

FLASH POINT 310 F	method used C.O.C	FLAMMABLE LIMITS	LEL	UEL
			N.E.	N.E.

EXTINGUISHING MEDIA
Use dry chemical, carbon dioxide or foam.

SPECIAL FIRE FIGHTING PROCEDURES
Self-contained respiratory protection should be provided for fire fighters. Keep fire exposed containers cool with water.

UNUSUAL FIRE AND EXPLOSION HAZARDS
Water or foam may cause a frothing reaction. (Water reacts with zinc dust).

* None of the chemical raw materials contained in this formulation are considered hazardous under the Federal Hazards Communication Standard 29 C. F. R 1910.1200

SECTION IV - HEALTH HAZARD INFORMATION

EFFECTS OF OVEREXPOSURE - Conditions to Avoid	None normally expected. Upon prolonged contact, may cause temporary eye discomfort.
THRESHOLD LIMIT VALUE	Zinc dust or silicon dioxide as dust: 10mg/m.
PRIMARY ROUTES OF ENTRY Inhalation ☐ Skin Contact ☒ Other (specify)	
EMERGENCY FIRST AID PROCEDURES SKIN CONTACT:	Wash with soap and water for 15 minutes
EYE CONTACT:	Flush with water for 15 minutes
INGESTION:	Induce vomiting and consult physician or local poison control center.

SECTION V - REACTIVITY DATA

STABILITY	UNSTABLE		CONDITIONS TO AVOID
	STABLE	X	Avoid conditions of moisture or high humidity.
INCOMPATIBILITY (materials to avoid)			Avoid strong oxidizers, strong acids and water.
HAZARDOUS DECOMPOSTION PRODUCTS:			Excessive heat and burning may release oxides of carbon.
HAZARDOUS POLYMERIZATION	MAY OCCUR		CONDITIONS TO AVOID None
	WILL NOT OCCUR	X	

SECTION VI - SPILL AND LEAK PROCEDURES

STEPS TO BE TAKEN IF MATERIAL IS RELEASED OR SPILLED	Wipe up, shovel or vacuum spilled material. Clean up spills immediately.
Use absorbent media.	
WASTE DISPOSAL METHOD	Comply with Federal, state and local regulations for solid landfill.
CERCLA (Superfund) REPORTABLE QUANTITY (in lbs)	None Required
RCRA HAZARDOUS WASTE NO. (40CFR 261.33)	None Required
VOLATILE ORGANIC COMPOUND (VOC) (as packaged, minus water)	120 g/l, calculated

[a] Theoretical _____ lb/gal N/A	[a] Analytical _____ lb/gal N/A

SECTION VII - PERSONAL PROTECTION INFORMATION

RESPIRATORY PROTECTION (specify type)		If TLV exceeded, use NIOSH respirator		
VENTILATION	LOCAL EXHAUST (Specify Rate)	Necessary above TLV	SPECIAL	None
	MECHANICAL (General) (Specify Rate) Recommended in closed areas		OTHER	None
PROTECTIVE GLOVES (specify type) None normally needed -Neoprene if necessary		EYE PROTECTION (specify type) Safety glasses or splash goggles.		
OTHER PROTECTIVE EQUIPMENT		Eye fountain in work area is recommended.		

SECTION VIII - SPECIAL PRECAUTIONS

PRECAUTIONS TO BE TAKEN IN HANDLING AND STORING	Store in dry conditions at temperatures between 40 - 120 F.
OTHER PRECAUTIONS	Keep away from children, infants and pets.

SECTION IX - ADDITIONAL INFORMATION

This product contains the following materials that are subject to the reporting requirements of Section 313 of EPCRA:

CAS # 7440-66-6, Zinc Dust, 20%

N/A = Not Applicable, N.E. = None Established

THIS MATERIAL SAFETY DATA SHEET PREPARED BY:	
NAME James R. MacMurdo	SIGNATURE
TITLE Director, Corporate Quality Assurance	*James R. MacMurdo*
DATE 03/10/2006	

Review

1. What is the product name in this MSDS?

   _____

2. Who manufactures this product?

   _____

3. List the three chemicals found in this product.

   _____

4. Describe the appearance and color of this product.

   _____

5. What type of extinguisher would be used on this product should it catch on fire?

   _____

6. If somebody ingested this product by mistake, what should be done?

   _____

7. Is this product considered to be stable or non-stable as far as reactivity?

   _____

8. What steps should be taken if this product is spilled?

   _____

9. Name the type of eye protection that should be used when using this product.

   _____

10. How should this product be handled and stored?

    _____

Chapter 2

Hardware and Materials Used in Residential Wiring

Lab 2.1: Identify the Parts of a Typical Metal Device Box

Name: _____ **Date:** _____ **Score:** _____

Introduction

Metal device boxes are used widely in residential wiring. The standard metal device box has a 3 × 2-inch opening and can vary in depth from 1½ inch to 3½ inches. The most common metal device box used is the 3 × 2 × 3½-inch box because it can contain several wires and a device. A metal device box has many features for the electrician to use. One often-used feature involves taking the sides off the box and ganging the boxes together to make a box that can accommodate more than one device.

Boxes with mounting ears are typically called "old work" boxes because the ears allow them to be installed in an existing wall. Boxes without the mounting ears are referred to as "new work" boxes and are the kind used when installing an electrical system in a new home. The device mounting holes are designed to accommodate the screws of a switch or receptacle and are used to secure the device to the box. The holes are tapped for a 6-32 screw size. In the back of the box you will find a tapped hole that accommodates a 10-32 grounding screw. This screw will be hex-headed and green in color, and its purpose is to attach the circuit grounding conductor to the metal box. Knockouts (KO) are included on the sides, top, and bottom of the box. When a KO is removed an opening exists for a connector to be used to secure a wiring method to the box. The pryouts (PO) located at the rear top and bottom of the box are pried off and provide an opening through which a cable can be secured to the box. Most of these boxes come with internal cable clamps that are tightened onto the cable after it has been inserted through the pryout opening. The cable clamps secure the cable to the box. In this lab exercise, you will identify the various parts of a metal device box.

Materials and Equipment

Pencil

Procedure

1. Identify the parts of the metal device box shown in Figure 2.1-1.

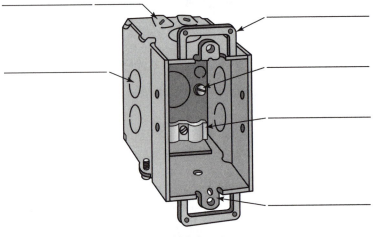

FIGURE 2.1-1

Lab 2.2: Identify Different Types of Wire Connectors

Name: _____ Date: _____ Score: _____

Introduction

Conductors are terminated to electrical equipment and each other with a variety of connector types, terminals, and lugs. A wirenut is the most common wire connector used by electricians to splice together two or more conductors. In this lab exercise, you will identify common wire connectors and indicate the wire sizes that can be used with them.

Materials and Equipment

Pencil

Procedure

1. Identify the wire connectors shown in Figure 2.2-1 and list the wire sizes that each connector type may be used with.

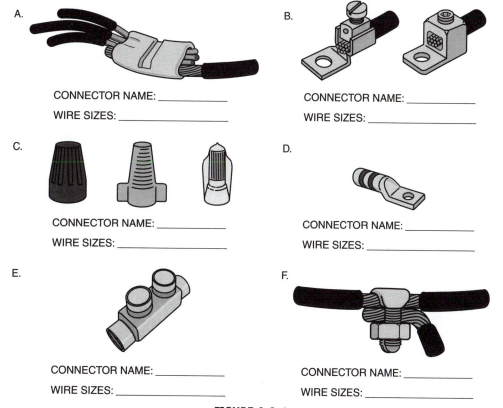

A.

CONNECTOR NAME: _____

WIRE SIZES: _____

B.

CONNECTOR NAME: _____

WIRE SIZES: _____

C.

CONNECTOR NAME: _____

WIRE SIZES: _____

D.

CONNECTOR NAME: _____

WIRE SIZES: _____

E.

CONNECTOR NAME: _____

WIRE SIZES: _____

F.

CONNECTOR NAME: _____

WIRE SIZES: _____

FIGURE 2.2-1

Lab 2.3: Identify the Parts on a Duplex Receptacle

Name: _____ Date: _____ Score: _____

Introduction

Receptacles are probably the most recognizable parts of a residential electrical system. They provide ready access to the electrical system and are defined as a contact device installed at the outlet for the connection of an attachment plug. Even though many people, including electricians, refer to a receptacle as an outlet, it is the wrong term to use. An outlet is the point on the wiring system at which current is taken to supply equipment. A receptacle is the device that allows us to access current from the wiring system and deliver it through a cord and attachment plug to a piece of equipment. A single receptacle is a single contact device with no other contact device on the same yoke. The most common type of receptacle used in residential wiring is a duplex receptacle rated for 15 amperes and 125 volts. It consists of two single receptacles on the same mounting strap. The short contact slot on the receptacle receives the ungrounded conductor from the attached cord. The long contact slot receives the grounded conductor from the attached cord. There is also a U-shaped grounding slot that receives the grounding conductor. Silver screws are located on the side with the long contact slot and are used to terminate the white, grounded circuit conductor. Brass or bronze colored screws are located on the same side as the short contact slot and are used to terminate the ungrounded circuit conductor. A green screw is located on the duplex receptacle for terminating the circuit bare or green grounding conductor. In this lab exercise, you will identify the various parts of a duplex receptacle.

Materials and Equipment

Pencil

Procedure

1. Identify the parts of the duplex receptacle shown in Figure 2.3-1.

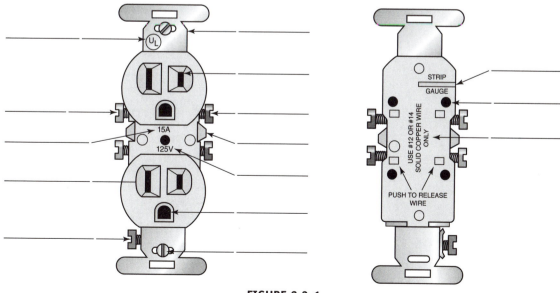

FIGURE 2.3-1

Lab 2.4: Identify the Parts on Single-Pole, Double-Pole, Three-Way, and Four-Way Switches

Name: _____ **Date:** _____ **Score:** _____

Introduction

Devices called switches are used to control the various lighting outlets installed in residential wiring. Article 404 in the *National Electrical Code®* provides the installation requirements for switches. This device type can be called many different names such as "toggle switch," "snap switch," or "light switch" but in this section we will be referring to these devices simply as "switches." Single-pole, double-pole, three-way, and four-way switches are the switch types commonly used in residential wiring. In this lab exercise, you will identify the various parts of single-pole, double-pole, three-way, and four-way switches.

Materials and Equipment

Pencil

Procedure

1. Identify the parts of the single-pole switch shown in Figure 2.4-1.

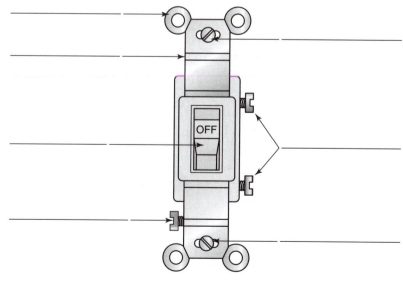

FIGURE 2.4-1

2. Identify the parts of the double-pole switch shown in Figure 2.4-2.

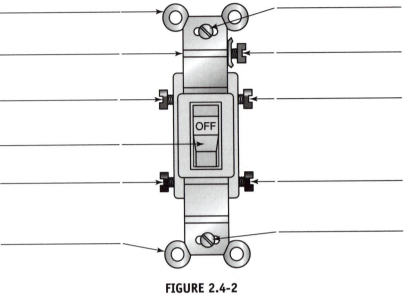

FIGURE 2.4-2

3. Identify the parts of the three-way switch shown in Figure 2.4-3.

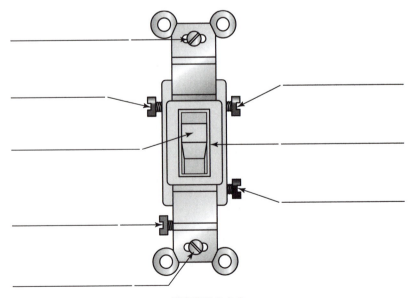

FIGURE 2.4-3

4. Identify the parts of the four-way switch shown in Figure 2.4-4.

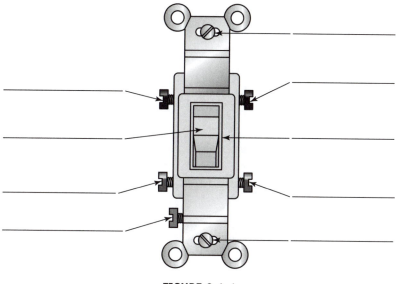

FIGURE 2.4-4

Chapter 3 Tools Used in Residential Wiring

Lab 3.1: Identify Several Guidelines for the Care and Safe Use of Electrical Hand Tools

Name: _____ Date: _____ Score: _____

Introduction

Using hand tools to install a residential electrical system can be hazardous. Injuries to yourself and others can result if you are not careful. Because of this, it is very important for electricians to be aware of some common guidelines for the care and safe use of electrical hand tools. In this lab exercise, you will list nine rules to follow for the care and safe use of electrical hand tools.

Materials and Equipment

House Wiring 3e textbook, pencil

Procedure

1. List the nine points covered in Chapter 3 of the *House Wiring* 3e textbook that should be considered whenever you use hand tools.

   1. _____
   _____
   _____

   2. _____
   _____
   _____

   3. _____
   _____
   _____

4. _____

_____

_____

5. _____

_____

_____

6. _____

_____

_____

7. _____

_____

_____

8. _____

_____

_____

9. _____

_____

_____

    1. _____

    _____

    _____

    2. _____

    _____

    _____

    3. _____

    _____

    _____

4. _____

_____

_____

5. _____

_____

_____

6. _____

_____

_____

7. _____

_____

_____

8. _____

_____

_____

9. _____

_____

_____

Lab 3.2: Using Lineman Pliers

Name: _____ Date: _____ Score: _____

Introduction

Side-cutting pliers, commonly called lineman pliers, are used extensively by electricians to cut larger conductors and cables to specific lengths. It is important to use pliers large enough for the job. Typically, the handles should be around 9 inches long so that a minimum of hand pressure is required to cut the conductor or cable. In most cases, one hand is all that is needed to make the cut; however, at times when the conductor or cable is fairly large, two hands may be required to provide enough pressure to make the cut. In this lab exercise, you will use lineman pliers to cut nonmetallic sheathed cable to specific lengths. Before you proceed, review the procedure for using lineman pliers to cut cable in Chapter 3 of the *House Wiring* 3e textbook.

Materials and Equipment

House Wiring 3e textbook, safety glasses, lineman pliers, a tape measure, a pencil or marker, Type NM cable, tool catalogs

Procedure

1. Put on safety glasses and observe all safety rules.

2. Ask your instructor for the size of the Type NM cable used in this lab exercise.
   Type NM cable size: _____

3. Using lineman pliers and a tape measure, measure out and then cut off a 2-foot length of Type NM cable.

4. Using a tape measure and a pencil or marker, measure and mark the 2-foot length of Type NM cable at 3-inch intervals.

5. Open the lineman pliers and put the cable into the cutting jaws at the first 3-inch mark. Have the cable positioned in the cutting jaws as close to the joint as possible.

6. Squeeze the lineman pliers' handles together and cut off the 3-inch piece of cable. One hand should be used to make the cut.

7. Repeat the process until all of the 2-foot length is used.

8. Complete the Review of this lab.

9. Show all completed work to the instructor. There should be eight 3-inch pieces of cable to show the instructor.

10. Clean up the work area and return all tools and materials to their proper locations.

11. Get the instructor to sign off upon satisfactory completion of this lab exercise.

Review

1. List at least three safety items to consider when cutting cable with lineman pliers.

   _____

   _____

   _____

2. List at least three other electrical tools that could also be used to cut Type NM cable.

   _____

   _____

   _____

3. Using a tool catalog as a reference, describe at least two types of lineman (side-cutting) pliers used by electricians. Include the manufacturer's name and catalog numbers of the models that you describe.

   _____

   _____

   _____

   _____

Lab 3.3: Using a Wire Stripper

Name: _____ **Date:** _____ **Score:** _____

Introduction

Wire stripping tools are used extensively by electricians to remove the insulation from conductors. It is important to use a wire stripper that is right for the job. Conductors can be as small as 18 AWG and as large as 2000 kcmil. One wire stripper will not work for all conductor sizes. Also, remember that there is a difference in stripping hole sizes when dealing with stranded wire versus solid wire. In this lab exercise, you will use wire strippers to strip the insulation from the ends of wires. Before you proceed, review the procedure for using a wire stripper in Chapter 3 of the *House Wiring* 3e textbook.

Materials and Equipment

House Wiring 3e textbook, safety glasses, lineman pliers, T-strippers, long-nose pliers, a tape measure, a pencil or marker, Type NM cable, tool catalogs

Procedure

1. Put on safety glasses and observe all safety rules.

2. Ask your instructor for the size of the Type NM cable used in this lab exercise.

   Type NM cable size: _____

3. Using lineman pliers and a tape measure, measure out and then cut off a 2-foot length of Type NM cable.

4. Using a tape measure and a pencil or marker, measure and mark the 2-foot length of Type NM cable at 8-inch intervals.

5. Open the lineman pliers and put the cable into the cutting jaws at the first 8-inch mark. Have the cable positioned in the cutting jaws as close to the joint as possible.

6. Squeeze the lineman pliers' handles together and cut off the 8-inch piece of cable. One hand should be used to make the cut.

7. Repeat the process until all of the 2-foot length is used. There should now be three 8-inch pieces of cable.

8. Now, using long-nose pliers, pull the individual conductors out of the 8-inch lengths of cable. You will have three white insulated conductors, three black insulated conductors, and three bare conductors. Discard the bare conductors.

   Note: This is a common way for electricians get what are called "pigtails" or "jumpers" that are often used when installing receptacles and switches.

9. Using the wire strippers, strip approximately ¾ inch of conductor insulation from both ends of the 8-inch lengths.

   - Insert the conductor in the proper stripping slot. Refer to photo A in the procedure for using a wire stripper in Chapter 3 of the *House Wiring* 3e textbook.

   - Close the jaws until you feel that you have reached the conductor and then open the jaws slightly.

- With an even pressure, pull back the strippers and remove the conductor insulation. Refer to photo B in the procedure for using a wire stripper in Chapter 3 of the *House Wiring* 3e textbook.

- Check the conductor for a ring or nick. If a nick occurs, restrip until the insulation is stripped without conductor damage. Refer to photo C in the procedure for using a wire stripper in Chapter 3 of the *House Wiring* 3e textbook.

10. Now, use the long-nose pliers or the special hole in the T-stripper to form a terminal loop on each end of each of the six lengths of wire. Review the procedure for using terminal loops to connect conductors to the terminal screws of a receptacle or switch in Chapter 18 of the *House Wiring* 3e textbook.

11. Complete the Review of this lab exercise.

12. Show all completed work to the instructor. There should be six 8-inch pieces of wire to show the instructor: three white insulated and three black insulated conductors. Each end should be stripped and have a terminal loop.

13. Clean up the work area and return all tools and materials to their proper locations.

14. Get the instructor to sign off upon satisfactory completion of this lab exercise.

Review

1. List at least three safety items to consider when stripping conductor insulation with wire strippers.

_____

_____

_____

2. List at least three other electrical tools that could also be used to strip insulation from conductors.

_____

_____

_____

3. Using a tool catalog as a reference, describe at least two types of wire strippers used by electricians. Include the manufacturer's name and catalog numbers of the models that you describe.

_____

_____

_____

_____

Lab 3.4: Using an Electrician's Knife

Name: _____ **Date:** _____ **Score:** _____

Introduction

An electrician's knife is one of the most important tools in a tool pouch. Knives are used for stripping insulation from larger conductors, cutting off the outside sheathing of cords and cables, and even cutting open cardboard boxes containing various pieces of electrical equipment. It is also one of the more dangerous tools that an electrician will use. Care must be taken when cutting with a knife. It is a good idea to always have the knife be as sharp as possible—more injuries come from dull knives than from sharp ones. In this lab exercise, you will use an electrician's knife to strip the insulation from a large conductor size. Before you proceed, review the procedure for using a knife to strip insulation from large conductors in Chapter 3 of the *House Wiring* 3e textbook.

Materials and Equipment

House Wiring 3e textbook, safety glasses, gloves, lineman pliers, long-nose pliers, a tape measure, a pencil or marker, electrician's knife, 2 AWG insulated aluminum conductor, tool catalogs

Procedure

1. Put on safety glasses and observe all safety rules. Gloves should be used when using a knife in this lab exercise.

2. Using lineman pliers and a tape measure, measure and then cut off a 1-foot length of 2 AWG aluminum conductor.

3. Using the tape measure and a pencil or marker, measure and make a mark at approximately 1½ inches from each end of the 2 AWG aluminum conductor.

4. Place the blade of the electrician's knife on the insulation at the 1½ inch mark and carefully cut around the conductor to a depth that is just shy of touching the actual conductor. Do this on each end. Refer to photo A in the procedure for using a knife to strip insulation from large conductors in Chapter 3 of the *House Wiring* 3e textbook.

5. Now, position the knife blade almost parallel to the conductor insulation at the cut done in the previous step. Refer to photo B in the procedure for using a knife to strip insulation from large conductors in Chapter 3 of the *House Wiring* 3e textbook.

6. Using a pushing motion, cut the insulation to a depth that is as close as possible to the actual conductor. Refer to photo C in the procedure for using a knife to strip insulation from large conductors in Chapter 3 of the *House Wiring* 3e textbook.

7. Continue cutting to the end of the conductor. Be careful to not nick the conductor in this process.

8. Using your fingers (or your long-nose pliers), peel off the remaining insulation. Refer to photo D in the procedure for using a knife to strip insulation from large conductors in Chapter 3 of the *House Wiring* 3e textbook.

9. Now strip the other end of the conductor by following steps 4 through 8.

10. Complete the Review of this lab exercise.

11. Show all completed work to the instructor. There should be a 1-foot length of 2 AWG aluminum insulated conductor with both ends stripped back to a distance of 1½ inches.

12. Clean up the work area and return all tools and materials to their proper locations.

13. Get the instructor to sign off upon satisfactory completion of this lab exercise.

Review

1. List at least three safety items to consider when stripping conductor insulation with an electrician's knife.

   _____

   _____

   _____

2. List at least three other electrical tools that could also be used to strip insulation from larger conductors.

   _____

   _____

   _____

3. Using a tool catalog as a reference, describe at least two types of knives used by electricians. Include the manufacturer's name and catalog numbers of the models that you describe.

   _____

   _____

   _____

   _____

Lab 3.5: Using an Electrician's Knife and Rotary Stripping Tool to Strip Off the Outer Sheathing of Various Sizes of Type NM, Type UF, and Type MC Cables

Name: _____ **Date:** _____ **Score:** _____

Introduction

The most common type of cable used in residential wiring is nonmetallic sheathed cable, or as it is called in the field, Romex™. It is defined as a factory assembly of two or more insulated conductors having an outer sheath of moisture-resistant, flame- retardant, nonmetallic material. It is very important for an electrician to be able to recognize the various types and sizes of Romex™ that are used and to be able to properly prepare the cable for use. Article 334 in the *NEC*® covers nonmetallic sheathed cable. This Article should be studied carefully by a student electrician.

Type UF cable is covered in Article 340 of the *NEC*®. It is used for underground installations of branch circuits and feeder circuits. It can also be used in interior installations but must be installed following the installation requirements for nonmetallic sheathed cable. The burial depth for Type UF cable is found in Table 300.5 of the *NEC*®. Additionally, Type UF can be used for outside residential installations, but only if the cable is listed and marked as sunlight-resistant. The outside sheathing of a Type UF cable is much more difficult to remove than the outside sheathing on a nonmetallic sheathed cable. In a Type UF cable, the outer sheathing is molded around each of the circuit conductors. This manufacturing process helps make the cable suitable for underground installation but does make the cable rather difficult to strip.

Article 330 covers Type MC and defines it as a factory assembly of one or more insulated circuit conductors enclosed in an armor of interlocking metal tape or a smooth or corrugated metallic sheath. There is no limit to the number of conductors found in Type MC. A green insulated grounding conductor is typically included in the cable assembly. It is used in house wiring when a nonmetallic sheathed cable cannot be used but a cable wiring method is desired.

In this lab exercise, you will strip off the outer sheathing of Type NM cable, Type UF cable, and Type MC cable using an electrician's knife for the nonmetallic sheathed cables and a rotary armored cable cutter for the metal sheathed cable. Before you proceed, review the procedures for using a knife to strip sheathing from a Type NM cable and for using a rotary armored cable cutter in Chapter 3 of the *House Wiring* 3e textbook. Also, review the procedure for stripping the outside sheathing of Type UF cable in Chapter 15 of the *House Wiring* 3e textbook.

Materials and Equipment

House Wiring 3e textbook, safety glasses; gloves; electrician's tool kit; rotary armored cable cutter; 2-foot lengths of 14/2, 12/2, and 10/2 Type NM cable; 2-foot lengths of 14/3, 12/3, and 10/3 Type NM cable; 2-foot lengths of 14/2 and 14/3 Type UF cable; 2-foot lengths of 14/2 and 14/3 Type MC cable

Procedure

1. Put on safety glasses and observe all safety rules. Gloves should be used when using a knife in this lab exercise.

2. Using lineman pliers and a tape measure, measure and cut a 2-foot length of each of the following cable types:

   - 14/2, 12/2, and 10/2 Type NM
   - 14/3, 12/3, and 10/3 Type NM
   - 14/2 and 14/3 Type UF
   - 14/2 and 14/3 Type MC

3. After observing an instructor demonstration of how Type NM cable should be stripped using a knife, remove the outside sheathing on the Type NM cables.

   - Using the electrician's knife, carefully strip the outside sheathing of each piece of Type NM cable so that there is 8 inches of free conductor on each end. Be careful not to cut through the conductor insulation. A cable ripper may be used to make the initial cut on the outside sheathing of two wire cables **ONLY**.

   - Trim off any paper or nylon wrappings used with the Type NM cable. The electrician's knife may be used for this step. Diagonal cutting pliers also will work very well.

4. After observing an instructor demonstration of how Type UF cable should be stripped using a knife, remove the outside sheathing on the Type UF cables.

   - At the end of the cable you wish to strip, use a knife to cut a 1-inch slit in the cable above the bare grounding conductor.

   - Using your fingers, open up the cable sheathing and grab the end of the bare grounding conductor with your long-nose pliers.

   - While holding on to the cable end with one hand, pull the grounding conductor back about 8 inches. This action will slit the cable sheathing as you are pulling the grounding conductor.

   - Now, using your long-nose pliers, grab the end of the other conductors in the cable and pull them back to the same point as the grounding conductor.

   - Cut off the cable sheathing with a knife or your diagonal cutting pliers.

5. After observing an instructor demonstration of how Type MC cable should be stripped using a rotary stripping tool, remove the outside sheathing on the Type MC cables.

   - Using a rotary stripping tool, strip the outer armor sheathing from each end, making sure to leave at least 8 inches of free conductor.

   - Make sure to take off any plastic wrapping left on the conductors.

6. Using a wire stripper, remove ¾ inch from the ends of each of the cable conductors.

7. Complete the Review of this lab exercise.

8. Show all completed work to the instructor. There should be 2-foot lengths of 14/2, 12/2, and 10/2 Type NM cable; 14/3, 12/3, and 10/3 Type NM cable; 14/2 and 14/3 Type UF cable; and 14/2 and 14/3 Type MC cable. Each end of these cables will be stripped back 8 inches and each of the cable conductors will have ¾ inch of insulation removed.

9. Clean up the work area and return all tools and materials to their proper locations.

10. Get the instructor to sign off upon satisfactory completion of this lab exercise.

Review

1. List and describe the three types of Type NM cable that is listed by Underwriters Laboratories. (Hint: Refer to Chapter 2 in the *House Wiring* 3e textbook.)

   _____

   _____

   _____

2. Name the outside sheathing colors that are used to identify the following Type NM cable sizes:

   14 AWG _____

   12 AWG _____

   10 AWG _____

3. Name the minimum size and the maximum size of Type NM cable, Type MC cable, and Type UF cable.

	Min. Size	Max. Size
Type NM		
Type MC		
Type UF		

Lab 3.6: Using a Screwdriver and Appropriate Fasteners to Install Electrical Boxes

Name: _____ Date: _____ Score: _____

Introduction

Cable connectors are used to clamp cables securely to each outlet box. Many boxes have built-in cable clamps and do not require separate clamps. Unless the listing by Underwriters Laboratories of a specific cable connector indicates that the connector has been tested for use with more than one cable, the rule is **one cable, one connector.** Wirenuts are types of solderless connectors used to connect two or more wires together. They are used extensively in electrical work because they are easy to use, relatively inexpensive, and they work well. In this lab exercise you will be installing surface-mounted electrical boxes and connecting Type NM cable to them with cable connectors. You will also be splicing the conductors in the boxes together using wirenuts. See Figure 3.6-1 to see how this lab project should look. Before you proceed, review the procedure for using a screwdriver in Chapter 3 of the *House Wiring* 3e textbook.

Materials and Equipment

House Wiring 3e textbook, safety glasses, gloves, electrician's tool kit, 6 feet of 14/2 Type NM cable, two 4 × 1½-inch square box with ½-inch KOs, five ½-inch Type NM cable clamps, four 10 × ¾-inch pan head sheet metal screws, two grounding screws, two 4-inch square flat blank covers, six wirenuts

Procedure

1. Put on safety glasses and observe all safety rules. Gloves should be used when using a knife in this lab exercise.

2. Using a screwdriver, mount the 4-inch square boxes on the plywood mockup wall at 48 inches to center from the floor. They should be 12 inches apart, measured edge to edge. Secure them to the wall using two sheet metal screws for each box.

3. Remove the appropriate box knockouts, install cable connectors, and then install a Type NM cable between the two boxes. The cable ends need to be stripped so that there are 6 to 8 inches of free conductor in the boxes.

4. Remove the appropriate box knockouts, install cable connectors, and then install a 12-inch length of Type NM into the top of the right-hand box *and* the top of the left-hand box. There should be approximately 4 to 6 inches of cable outside the box. The cable ends need to be stripped so that there are 6 to 8 inches of free conductor in the boxes.

5. Remove the appropriate box knockout, install a cable connector, and then install a 12-inch length of Type NM cable into the bottom of the left-hand box. There should be approximately 4 to 6 inches of cable outside the box. The cable end needs to be stripped so that there are 6 to 8 inches of free conductor in the box.

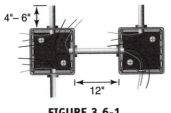

FIGURE 3.6-1

6. Ground the boxes:

   - Install a green grounding screw in each box.

   - Wrap one of the bare grounding conductors around the grounding screw as shown in Figure 10-10 in the *House Wiring* 3e textbook.

   - Tighten the screw.

   - Use a wirenut and connect the other bare grounding conductors in the box to the wire that was placed under the screw. Refer to the procedure for connecting wires together using a wirenut in Chapter 2 of the *House Wiring* 3e textbook.

7. Place all of the grounding wires as far as you can into the back of the box.

8. Strip approximately ¾-inch from the ends of the white conductors and splice all of them together using a wirenut. Refer to the procedure for connecting wires together using a wirenut in Chapter 2 of the *House Wiring* 3e textbook.

9. Strip approximately ¾ inch from the ends of the black conductors and splice all of them together using a wirenut. Refer to the procedure for connecting wires together using a wirenut in Chapter 2 of the *House Wiring* 3e textbook.

10. Position all of the conductors into the boxes and attach 4-inch-square flat blank covers.

11. Complete the Review of this lab exercise.

12. Show all completed work to the instructor.

13. Clean up the work area and return all tools and materials to their proper locations.

14. Get the instructor to sign off upon satisfactory completion of this lab exercise.

Review

1. If a wire connector has a marking of AL/CU on it, what types of conductor can be used with that connector?

   _____

   _____

2. Describe the proper way to install a wirenut.

   _____

   _____

   _____

   _____

   _____

3. Why is it a good idea to use at least two screws to secure each electrical box?

_____

_____

4. Why does the *National Electrical Code®* require at least 6 inches of free conductor be left in each electrical box?

_____

_____

_____

Lab 3.7: Stripping Large Electrical Conductors with an Electrician's Knife

Name: _____ **Date:** _____ **Score:** _____

Introduction

A residential electrician will often work with larger types of electrical wire such as 2 AWG aluminum or 4/0 copper. It is important to know how to properly strip the insulation on these conductors and connect them together. Wirenuts used on wire sizes such as 14, 12, and 10 AWG are not available in sizes large enough to handle large wire sizes. Connectors such as split-bolt, insulated Unitap type, or compression connectors will have to be used. Also, if aluminum wire is used and the connector is not made for aluminum wire, make sure to properly prepare the wire for use with the connector so that the rapid oxidation of aluminum does not adversely affect the connection. In this lab exercise, you will strip the insulation from large electrical conductors and then connect them together using a split-bolt connector or an insulated unitap type connector. Before you proceed, review the procedure for using a knife to strip insulation from large conductors in Chapter 3 of the *House Wiring* 3e textbook.

Materials and Equipment

House Wiring 3e textbook, safety glasses, gloves, electrician's tool kit, Allen wrench set, 18-inch length of insulated 2 AWG aluminum wire, one split-bolt connector or one insulated unitap type connector that can be used with 2 AWG aluminum, vinyl electrical tape, rubber splicing tape, tool catalogs

Procedure

1. Put on safety glasses and observe all safety rules. Gloves should be used when using a knife in this lab.

2. Ask your instructor for the type of connector used in this lab exercise.

   Split-bolt connector: _____

   Insulated unitap type connector: _____

3. Using lineman pliers and a tape measure, measure and cut an 18-inch length of 2 AWG aluminum insulated wire.

4. Using lineman pliers and a tape measure, measure and cut the 18-inch length of 2 AWG aluminum insulated wire exactly in half. You should now have two 9-inch lengths.

5. Using the electrician's knife, strip back 1 inch from **each** end of the 9-inch lengths.

6. If you are using a split-bolt connector, follow Steps 7 – 13. If you are using an insulated unitap type of connector, skip to Step 14.

7. Using the electrician's knife, strip off 1½ inches **from the center** of one 9-inch length of 2 AWG aluminum wire.

8. Connect one length of wire to the other using a split-bolt connector. Refer to the bottom illustration of Figure 2-16 in Chapter 2 of the *House Wiring* 3e textbook for the correct way to place the conductors in the split-bolt connector.

9. Tighten the split-bolt connector using two adjustable wrenches or an adjustable wrench and a pair of pump pliers.

10. Show this work to the instructor for approval before you tape up the connection.

11. Once the connection has been approved, tape up the connection. Use rubber splicing tape first and vinyl electrical tape last. **Note:** Make sure that all exposed surfaces are covered adequately. Do not use too much tape when covering the connection.

12. Complete the Review of this lab exercise.

13. Show all completed work to the instructor. There should be two lengths of 2 AWG aluminum insulated wire spliced together using a split-bolt connector and properly taped.

14. Remove all of the caps from the holes in the insulated unitap type of connector.

15. Insert the stripped end of each of the 9-inch lengths of 2 AWG aluminum wire into separate holes.

16. Using the proper size Allen wrench, tighten the set screws to securely fasten the conductors in the connector. Replace the caps on all exposed holes.

17. Show all completed work to the instructor. There should be two lengths of 2 AWG aluminum insulated wire spliced together using an insulated unitap type of connector.

18. Clean up the work area and return all tools and materials to their proper locations.

19. Get the instructor to sign off upon satisfactory completion of this lab exercise.

Review

1. Describe the suggested preparation procedure for aluminum conductors. Hint: Refer to the Procedures section in Chapter 2 of the *House Wiring* 3e textbook.

   _____

   _____

   _____

   _____

   _____

   _____

2. Name and describe two other tools that can be used to strip larger sizes of conductors. Tool catalogs may have to be consulted.

   _____

   _____

   _____

   _____

3. What is the largest size conductor allowed to be used with a wirenut? A wirenut manufacturer's catalog may have to be consulted.

   _____

4. What does the *National Electrical Code*® say about the insulation (electrical tape) used to re-insulate spliced conductors? Hint: *NEC*® Section 110.14(B).

   _____

   _____

   _____

Lab 3.8: Use Various Electrical Tools to Cut, Strip, and Install an Attachment Plug to a Length of Flexible Cord

Name: _____ Date: _____ Score: _____

Introduction

An electrician must often build cord sets and power cord pigtails for various wiring installations in the field using flexible cords. Attachment plugs, also called cord caps, are used to allow the cord to be "plugged in" to a receptacle. Article 400 of the *NEC*® covers flexible cords and cables and Article 406 covers attachment plugs. In this lab exercise, you will use various electrical hand tools to cut and strip flexible cord. You will then install an attachment plug on one end of the cord so that it can be used as a power cord in future wiring lab exercises.

Materials and Equipment

House Wiring 3e textbook, safety glasses, gloves, electrician's tool kit, continuity tester, 4-foot length of flexible cord, one NEMA 5-15P attachment plug

Procedure

1. Put on safety glasses and observe all safety rules. Gloves should be used when using a knife in this lab.

2. Using lineman pliers or diagonal cutting pliers and a tape measure, measure and cut a 4-foot long length of flexible cord. **Note:** Your instructor will tell you what size and type of cord to use.

   Flexible cord size and type: _____

3. After observing an instructor demonstration of how flexible cord should be stripped, use an electrician's knife and strip 1 inch of the black outside sheathing from one end of the 4-foot cord length and 8 inches of outside sheathing from the other end.

4. Attach a three-prong grounding type attachment plug (NEMA 5-15P) to the end of the 4-foot cord length with 1 inch of conductor showing. Check the attachment plug for a strip gauge that will tell you how much insulation to keep and how much insulation to strip. This information can also be found on the package that the attachment plug came in. The end with 8 inches of free conductor is left untouched in this lab exercise. It will be installed in electrical boxes during future labs to energize the circuits.

5. Using a continuity tester, test the cord for proper connections. Be sure to test all three of the conductors. The black insulated conductor must be attached to the brass-colored ungrounded termination point. The white insulated conductor must be attached to the silver-colored grounded termination point. The green insulated conductor must be attached to the green grounding termination point. See Chapter 4 in the *House Wiring* 3e textbook for information on how to use a continuity tester.

6. Complete the Review of this lab exercise.

7. Show all completed work to the instructor. You should have a 4-foot length of flexible cord with one end stripped back 8 inches and an attachment plug on the other end.

8. Clean up the work area and return all tools and materials to their proper locations.

9. Get the instructor to sign off upon satisfactory completion of this lab exercise.

Review

1. Using Table 400.4 in the *NEC®*, list the insulation type, wire size(s), outer covering, and uses of Type SJ flexible cord.

   Insulation type: _____

   Wire sizes: _____

   Outer covering: _____

   Uses: _____

2. Give the allowable ampacity for 16/3, 14/3, and 12/3 SJ cord. This information can be found in Table 400.5(A)(1) of the *NEC®*. Subheading B will apply since this cord type will have only two conductors that are current-carrying.

   16/3 ampacity: _____

   14/3 ampacity: _____

   12/3 ampacity: _____

3. List three safety items that should be considered when doing this lab exercise.

   _____

   _____

   _____

Lab 3.9: Set Up and Use a Pistol-Grip Drill

Name: _____ **Date:** _____ **Score:** _____

Introduction

A pistol grip type of electric drill is very popular because it is small, relatively lightweight, and easy to use. This style gets its name from the fact that it looks like, and is held in your hand like, a pistol. They are available in three common chuck sizes: ¼-inch, ⅜-inch, and ½-inch. A chuck is the part of the drill that holds the drill bit securely in place. Most drills have chucks that are tightened and loosened with a chuck key or wrench. Some newer models use a keyless chuck. The pistol grip drill can be used with a wide variety of drill bits. Drill bits are the tools that are attached to the drill and actually do the hole boring. Bits used with a pistol grip drill should be designed for use at higher speeds. A twist bit is designed to drill wood or plastic at high speed and metal at a lower speed. A flat-bladed spade bit, sometimes called a speed-bore bit, is used to drill holes in wood at high speed. A masonry bit is used to drill holes in concrete, brick, and other masonry surfaces. An auger bit is used for drilling wood framing members at a relatively slow speed. In this lab exercise you will be using a cordless pistol grip drill to drill a hole with a bit type specified by your instructor. Before you proceed, review the procedure for using a cordless pistol grip power drill in Chapter 3 of the *House Wiring* 3e textbook.

Materials and Equipment

House Wiring 3e textbook, safety glasses, electrician's tool kit, a cordless pistol grip drill, a drill bit specified by the instructor, a material to be drilled as specified by the instructor

Procedure

1. Put on safety glasses and observe all safety rules.

2. Ask your instructor for the drill bit type to be used in this lab exercise.
   Drill bit type: _____

3. Ask your instructor what material is to be drilled.
   Material to be drilled: _____

4. Place the drill bit in the drill chuck in the following manner:

   • Open the chuck by turning it with your hand in a counterclockwise direction until it can accommodate the shank of the drill bit.

   • Insert the drill bit shank into the chuck and tighten the chuck by hand as much as you can.

   • Using the chuck key, tighten the bit securely in the chuck. If a chuck has more than one tightening hole, use the key to tighten at each hole. If the drill has a keyless chuck, tighten the bit securely in the chuck by hand. Refer to photo A in the procedure for using a cordless pistol grip power drill in Chapter 3 of the *House Wiring* 3e textbook.

   • Remove the chuck key if one is used.

5. Now make a mark at the location you want to drill and using an awl for wood and plastic or a center punch for metal, make a small indent exactly where you want to drill.

6. If necessary, securely clamp the material that is being drilled.

7. Firmly grip the drill and place the tip of the bit at the indent.

8. Hold the drill perpendicular to the work so the bit will not make a hole at an angle. Start the drill slowly at first so the bit will not stray from the spot where you want the hole. Squeeze the trigger harder and speed the drill up while applying moderate pressure to the drill. Refer to photo B in the procedure for using a cordless pistol grip power drill in Chapter 3 of the *House Wiring* 3e textbook.

9. Continue to drill through the material until you sense that the bit is about to go through the material. Then reduce the amount of pressure you are exerting on the drill and let the drill bit complete drilling the hole.

10. While the drill is still turning the bit in a clockwise direction, slowly pull the drill and bit out and then release the trigger.

11. Take the bit out of the chuck and place everything in its proper storage area.

12. Complete the Review of this lab exercise.

13. Show all completed work to the instructor.

14. Clean up the work area and return all tools and materials to their proper locations.

15. Get the instructor to sign off upon satisfactory completion of this lab exercise.

Review

1. List at least three safety items to consider when using a pistol grip drill.

_____

_____

_____

2. Describe the part of this lab that you found to be the hardest to do. How do you think this part of the lab could be made easier?

_____

_____

_____

_____

_____

Lab 3.10: Using a Manual Knockout Set

Name: _____ **Date:** _____ **Score:** _____

Introduction

One of the tasks that an electrician has to perform is to make holes in metal enclosures to allow for the connection of electrical cable or conduit to the enclosure. This job usually requires the electrician to drill a hole through the enclosure at the spot where the hole needs to be made. The drilled hole should be just large enough for the threaded stud of the knockout set to fit easily through. The knockout set is then used to make a hole that will match the trade size of the conduit or cable connector that is going to be used. Typically a manual knockout set is used to make trade size holes from ½ inch up to and including 1¼ inches. For these sizes, and for sizes up to 6 inches, there are also hydraulic knockout sets. The electrician must be familiar with both the manual and the hydraulic types of knockout tools. In this lab exercise, you will make a knockout in an enclosure using the manual knockout punch. Before you proceed, review the procedure for using a manual knockout punch to cut a hole in a metal box in Chapter 3 of the *House Wiring* 3e textbook.

Materials and Equipment

House Wiring 3e textbook, safety glasses, electrician's tool kit, a center punch, 10-inch or larger adjustable wrench, corded or cordless pistol grip drill, extension cord if needed, metal drill bit index, manual knockout set, metal enclosure as specified by the instructor, a vise or C-clamps to hold the metal enclosure

Procedure

1. Put on safety glasses and observe all safety rules.

2. Ask your instructor for the size of the hole to be knocked out in this lab exercise.
   Knockout size: _____

3. Ask your instructor which metal enclosure is to be used in this lab exercise.
   Metal enclosure used: _____

4. Set up the pistol grip drill for use. Remember to do a complete check of the power tool before using it. Choose a metal drill bit that is slightly larger than the threaded stud of the knockout tool and install the bit into the drill chuck. Make sure it is tightened securely in the chuck using the chuck key. **Note:** You may want to drill a pilot hole with a small drill bit and then use the larger bit to enlarge the hole to the size required for easy insertion of the knockout set's threaded stud.

5. Drill a hole slightly larger than the knockout threaded stud in the center of the space you are going to punch. A center punch can be used to make an indentation for your drill to start in. This will reduce the tendency of the drill bit to move off line when you start to drill. Refer to photo A in the procedure for using a manual knockout punch to cut a hole in a metal box in Chapter 3 of the *House Wiring* 3e textbook. CAUTION: Hold the drill firmly while drilling. A loose grip could cause an accident. Remember that the drill bit will be hot so use caution when handling it until it cools.

6. Insert the knockout punch threaded stud through the drilled hole and put the cutting die on the threaded stud. Make sure the cutting die is aligned so the cutting edge is toward the metal enclosure. Refer to photo B in the procedure for using a manual knockout punch to cut a hole in a metal box in Chapter 3 of the *House Wiring* 3e textbook.

7. Tighten the drive nut with the adjustable wrench. Refer to photo C in the procedure for using a manual knockout punch to cut a hole in a metal box in Chapter 3 of the *House Wiring* 3e textbook.

8. Continue to tighten the drive nut until the cutting die is pulled all the way through the metal. Remove the knockout punch when the cutting die is finally pulled through.

9. Remove the cutting die from the threaded stud and shake out the punched metal. Refer to photo D in the procedure for using a manual knockout punch to cut a hole in a metal box in Chapter 3 of the *House Wiring* 3e textbook.

10. Replace the cutting die onto the threaded stud and place the knockout punch in its proper storage area.

11. Complete the Review of this lab exercise.

12. Show all completed work to the instructor. You should have the punched-out metal from a knockout hole to show the instructor.

13. Clean up the work area and return all tools and materials to their proper locations.

14. Get the instructor to sign off upon satisfactory completion of this lab exercise.

Review

1. List at least three safety items to consider when using a knockout set.

   _____

   _____

   _____

2. What part of this lab did you find to be the hardest to do? How might this part be made easier?

   _____

   _____

   _____

   _____

   _____

Lab 3.11: Set Up and Use a Hacksaw

Name: _____   Date: _____   Score: _____

Introduction

An electrician must know how to set up a hacksaw and use it to cut various pieces of electrical material like conduit to specific lengths. It is important to know what kind of hacksaw blade is best for cutting a certain type of material. Electricians will supply their own hacksaws. However, in most electrical businesses, the company will supply the electrician with hacksaw blades. In this lab exercise, electrical metallic tubing (EMT) is the material that you will be cutting. You will set up and use a hacksaw to cut a length of EMT conduit into several smaller lengths. Electricians most often use a hacksaw with a 24-teeth-per-inch blade to cut EMT. Before you proceed, review the procedure for setting up and using a hacksaw in Chapter 3 of the *House Wiring* 3e textbook.

Materials and Equipment

House Wiring 3e textbook, safety glasses, gloves, electrician's tool kit, tape measure, pencil, hacksaw frame, 24-teeth-per-inch hacksaw blade, metal file or some other tool used to ream the ends of EMT, 5-foot length of ½-inch EMT conduit, a pipe vise to hold the conduit while it is being cut

Procedure

1. Put on safety glasses and observe all safety rules. It is a good idea to wear gloves when cutting material with a hacksaw.

2. Insert the blade in the frame of the hacksaw. Be sure the teeth angles are pointed toward the front of the saw. There is usually an arrow that indicates the proper direction for the blade to be installed. Refer to photo A in the procedure for setting up and using a hacksaw in Chapter 3 of the *House Wiring* 3e textbook. **Note:** The hacksaw you are using may already have a blade. However, go ahead and take off the blade and then reinstall it so you can see how the blade is installed on the hacksaw frame.

3. Using the tape measure, mark the 5-foot length of conduit at 1-foot intervals. **Note:** You may have to cut a 10-foot (or some other) length of conduit to the 5-foot size.

4. Secure the 5-foot length of EMT conduit in the pipe vise for cutting.

5. Set the blade of the hacksaw on the conduit at the point to be cut as indicated by your marks at the 1-foot intervals.

6. Push gently forward until the cut is started. Do not exert too much pressure on the saw. Refer to photo B in the procedure for setting up and using a hacksaw in Chapter 3 of the *House Wiring* 3e textbook.

7. Make reciprocal strokes until the cut is finished. Remember that the cutting stroke is the forward stroke. Your cut should be straight and relatively smooth. **Note:** Excessive speed while cutting can ruin blades.

8. Repeat steps 5 through 7 until there are five 1-foot lengths that have been cut.

9. According to the *National Electrical Code®*, the ends of the cut conduit must all be reamed. Use a file or some other reaming tool to ream the ends.

10. Complete the Review of this lab exercise.

11. Show all completed work to the instructor. You should have five 1-foot lengths with each end properly reamed.

12. Clean up the work area and return all tools and materials to their proper locations.

13. Get the instructor to sign off upon satisfactory completion of this lab exercise.

Review

1. List at least three safety items to consider when using a hacksaw to cut conduit.

   _____

   _____

   _____

2. Name the specific section of the *NEC*® that requires reaming of the ends of field-cut EMT.

   _____

3. Pipe vises are often not available out in the field for use by an electrician when EMT conduit needs to be cut. Describe at least two methods that an electrician could use to hold the EMT while it is being field cut with a hacksaw.

   _____

   _____

   _____

   _____

   _____

   _____

Lab 3.12: Set Up and Use a Right-Angle Drill

Name: _____ Date: _____ Score: _____

Introduction

In residential wiring, electricians need to drill holes for the installation of cables through wood framing members. The tight space between wall studs can make drilling the required holes with a pistol grip drill a slow and awkward process. To allow easier drilling of wood framing members in tight spaces, the right-angle drill was introduced by the Milwaukee Electric Tool Company in 1949. The head of the drill is at a right angle (90°) in relation the rest of the drill, which allows the drill body to be located away from the material being drilled. Right-angle drills are usually used with a ½-inch chuck and work well with auger bits, Forstner bits, or hole saws to drill holes in wood framing members. See Figure 3-50 in the textbook for examples of drilling bits used with a right-angle drill. Auger bits and hole saws are the bits of choice for making the larger holes because they work best at the lower speeds at which ½-inch drills operate. Care must be taken by the electrician when using this type of drill. They are very powerful and can cause serious injury if they are not used properly. In this lab exercise, you will use a right-angle drill to drill a hole in a wooden framing member with an auger bit and cut a hole in a wooden framing member with a hole saw. Before you proceed, review the procedure for drilling a hole in a wooden framing member with an auger bit and a corded right-angle drill in Chapter 3 of the *House Wiring* 3e textbook. Also, before you proceed, review the procedure for cutting a hole in a wooden framing member with a hole saw and a corded right-angle drill in Chapter 3 of the *House Wiring* 3e textbook.

Materials and Equipment

House Wiring 3e textbook, safety glasses, electrician's tool kit, a corded or cordless ½ inch right-angle drill, an auger bit size specified by the instructor, a hole saw size specified by the instructor, an extension cord if needed, a length of wooden framing lumber specified by the instructor, tool catalogs

Procedure

1. Put on safety glasses and observe all safety rules.

2. Ask your instructor for the size of the auger bit used in this lab exercise.
   Auger bit size: _____

3. Ask your instructor for the size of the hole saw used in this lab exercise.
   Hole saw size: _____

4. Ask your instructor which length of wooden framing lumber is to be used in this lab exercise.
   Wooden framing member used: _____

5. Make a mark on the wooden framing lumber at the location where you want to bore a hole with the auger bit. **Note:** Check with the instructor if you are not sure which wooden framing member to use.

6. Set up the right-angle drill with the auger bit. Make sure that the bit is secure in the drill chuck by tightening the chuck evenly with the chuck key. Remember that the cord must be unplugged while working on the drill if using a corded drill. Refer to photo A in the procedure for drilling a hole in a wooden framing member with an auger bit and a corded right-angle drill in Chapter 3 of the *House Wiring* 3e textbook.

7. Plug the drill in if using a corded drill and place the tip of the bit at the spot that was marked in step 5.

8. Keeping an even force on the drill, start the drill and allow the auger bit to feed itself completely through the wooden framing lumber. Make sure the drill is turning in a clockwise direction. Refer to photo B in the procedure for drilling a hole in a wooden framing member with an auger bit and a corded right-angle drill in Chapter 3 of the *House Wiring* 3e textbook.

9. Once the auger bit has gone through the wood, stop the drill; reverse it; and while pulling back with an even pressure, back out the bit.

10. Unplug the cord if using a corded drill. Take the auger bit out of the drill chuck and put it in its proper storage location.

11. Now, assemble the hole saw and arbor for the size hole you wish to cut. Set up the hole saw so the pilot bit on the hole saw extends past the end of the hole saw by approximately 1 inch. Refer to photo A in the procedure for cutting a hole in a wooden framing member with a hole saw and a corded right-angle drill in Chapter 3 of the *House Wiring* 3e textbook.

12. Make sure the drill is unplugged if using a corded drill and secure the hole saw in the drill chuck by tightening the chuck evenly with the chuck key. Refer to photo B in the procedure for cutting a hole in a wooden framing member with a hole saw and a corded right-angle drill in Chapter 3 of the *House Wiring* 3e textbook.

13. Make a mark on the wooden framing lumber at the location where you want to cut a hole with the hole saw. **Note:** Check with the instructor if you are not sure which wooden framing member to use.

14. Plug the drill in if using a corded drill and place the tip of the hole saw pilot bit at the marked spot.

15. Keeping an even force on the drill, start the drill and allow the hole saw to feed itself completely through the wooden framing lumber. Make sure the drill is turning in a clockwise direction. Refer to photo C in the procedure for cutting a hole in a wooden framing member with a hole saw and a corded right-angle drill in Chapter 3 of the *House Wiring* 3e textbook.

16. Once the hole saw has gone through the wood, stop the drill; reverse it; and pulling back with an even pressure, back out the hole saw.

17. Unplug the drill if using a corded drill and remove the wood from the hole saw by using a screwdriver or some other narrow tool to dislodge the wood. Refer to photo D in the procedure for cutting a hole in a wooden framing member with a hole saw and a corded right-angle drill in Chapter 3 of the *House Wiring* 3e textbook.

18. Disassemble the hole saw and put it back in the proper storage location.

19. Complete the Review of this lab exercise.

20. Show all completed work to the instructor. You should have two holes to show the instructor.

21. Clean up the work area and return all tools and materials to their proper locations.

22. Get the instructor to sign off upon satisfactory completion of this lab exercise.

Review

1. List at least three safety items to consider when using a ½-inch right-angle drill with either an auger bit or a hole saw.

   _____

   _____

   _____

2. Using tool catalogs as a reference, describe at least two types of power drills that can be used to drill holes with auger bits or cut holes with a hole saw. Include the manufacturer's name and catalog numbers of the models that you describe.

_____

_____

_____

_____

Lab 3.13: Set Up and Use a Hammer Drill

Name: _____ Date: _____ Score: _____

Introduction

Hammer drills are used to drill holes in masonry or concrete walls and floors. When installing anchors to hold electrical equipment on a masonry wall, hammer drills are used to drill the anchor holes. Hammer drills are used with special masonry bits to bore holes in masonry. The bits used are of the percussion carbide-tip type. While the drill is turning the masonry bit, it is also moving the bit in a reciprocating, or hammering motion. Some hammer drills can switch back and forth from a drill-only mode to a hammer-drill mode. With this feature you actually have two drill styles in one package. The drill-only mode turns off the hammering action so you can drill wood, metal, or plastic. The hammer-drill mode restores the hammering action and allows you to drill holes in masonry. Pistol-style hammer drills come with ⅜-inch or ½-inch chucks. Most hammer drill mechanisms are designed so the more pressure you exert on the drill, the more hammering action you get. You will know that the right amount of pressure is being used when the hammering is even and smooth. In this lab exercise, you will use a hammer drill to drill a hole in masonry and then install a plastic anchor. Before you proceed, review the procedure for drilling a hole in masonry with a corded hammer drill in Chapter 3 of the *House Wiring* 3e textbook.

Materials and Equipment

House Wiring 3e textbook, safety glasses, electrician's tool kit, a corded or cordless hammer drill, masonry bit, a masonry block, a plastic anchor as specified by the instructor, tool catalogs

Procedure

1. Put on safety glasses and observe all safety rules.

2. Ask your instructor for the size of the masonry bit used in this lab exercise.
   Masonry bit size: _____

3. Ask your instructor which masonry block or concrete wall area is to be used in this lab exercise.
   Masonry block or concrete wall location: _____

4. Select the proper size carbide-tip masonry drill bit for the hole you are boring and secure it into the chuck of the hammer drill. Make sure the drill is unplugged if using a corded drill. Refer to photo A in the procedure for drilling a hole in masonry with a corded hammer drill in Chapter 3 of the *House Wiring* 3e textbook.

5. Check to be sure the hammer drill is in the hammer-drill mode. If it is not, place the drill in hammer-drill mode.

6. Make a mark at the location where you want to drill the hole on the concrete block or concrete wall.

7. Plug in the hammer drill if using a corded drill and while gripping the drill firmly, place the masonry drill bit at the mark.

8. Apply some pressure to the drill and slowly squeeze the trigger to increase the speed. Refer to photo B in the procedure for drilling a hole in masonry with a corded hammer drill in Chapter 3 of the *House Wiring* 3e textbook.

9. Continue to apply pressure until the hammering is smooth and even and then continue drilling out the hole to the desired depth. **Note:** Do not get the masonry bit too hot as it will drastically cut the life of the bit.

10. Discontinue putting pressure on the drill and the hammering will stop. While the drill is still turning, slowly pull the bit from the hole. This will help clean the hole of masonry debris. Refer to photo C in the procedure for drilling a hole in masonry with a corded hammer drill in Chapter 3 of the *House Wiring* 3e textbook.

11. Unplug the drill if using a corded drill, remove the bit, and put everything away.

12. Install a plastic anchor in the masonry. Refer to the procedure for installing plastic anchors in Chapter 2 of the *House Wiring* 3e textbook.

13. Complete the Review of this lab exercise.

14. Show all completed work to the instructor.

15. Clean up the work area and return all tools and materials to their proper locations.

16. Get the instructor to sign off upon satisfactory completion of this lab exercise.

Review

1. List at least three safety items to consider when using a hammer drill.

   _____

   _____

   _____

2. Using tool catalogs as a reference, describe at least two types of hammer drills that can be used to drill holes in masonry. Include the manufacturer's name and catalog numbers of the models that you describe.

   _____

   _____

   _____

   _____

Lab 3.14: Set Up and Use a Reciprocating Saw

Name: _____ **Date:** _____ **Score:** _____

Introduction

A reciprocating saw has a blade that moves back and forth to make a cut rather than using a rotating blade. It can make a straight or curved cut in many materials, including wood and metal. When cutting wood, a wood blade must be used; when cutting metal, a metal blade must be used. Electricians usually refer to this saw type as a Sawzall®. The name Sawzall® is a trademark of the Milwaukee Electric Tool Corporation, Inc. Reciprocating saws are usually found with a cord and plug, but like drills and circular saws, manufacturers are now producing well-engineered cordless models. Residential electricians use the reciprocating saw much more often than a circular saw. It can be used to cut heavy lumber and to cut off conduit, wood, and fasteners flush to the surface. It works well when doing new or remodel work for such things as cutting box holes in plywood, or fine-tuning framing members so an electrical box can be properly mounted. It is great for sawing in tight locations where other saw types cannot reach. In this lab exercise, you will use a reciprocating saw to cut a length of wood framing lumber into several smaller lengths. Before you proceed, review the procedure for using a corded reciprocating saw in Chapter 3 of the *House Wiring* 3e textbook.

Materials and Equipment

House Wiring 3e textbook, safety glasses, electrician's tool kit, a corded or cordless reciprocating saw, tape measure, pencil, wood cutting blades for the reciprocating saw, a length of 2 × 4-inch lumber, a vise or C-clamps to hold the wood in place, tool catalogs

Procedure

1. Put on safety glasses and observe all safety rules.

2. Set up the saw with a cutting blade suitable for cutting wood. Make sure that the blade is secure in the saw head by tightening the set screw with the Allen wrench provided with the saw. **Note:** Some newer reciprocating saws have a collar that turns to secure the blade. Remember that if the saw is corded, the cord must be unplugged while working on the saw. Refer to photo A in the procedure for using a corded reciprocating saw in Chapter 3 of the *House Wiring* 3e textbook.

3. Mark the location where you want to cut on the work material. For this lab exercise use a tape measure and mark the 2 × 4 lumber at three 6-inch intervals from the end. The total length of the 2 × 4 lumber used will be 18 inches.

4. Secure the work with a vise or C-clamps to reduce vibration while cutting.

5. Plug in the saw if it is corded and set the blade on the work at the point to be cut as indicated by your first mark. Always try to position the tool so that you are cutting in a downward direction and away from your body. Refer to photo B in the procedure for using a corded reciprocating saw in Chapter 3 of the *House Wiring* 3e textbook.

6. Pull gently back on the trigger switch to start the saw and begin the cut. Do not exert too much pressure on the saw. It will cut through very easily by allowing the weight of the tool itself to provide the necessary pressure. Refer to photo C in the procedure for using a corded reciprocating saw in Chapter 3 of the *House Wiring* 3e textbook.

7. Let the saw make reciprocal strokes until the cut is finished. Your cut should be straight and relatively smooth.

8. Unplug the tool if using a corded saw, remove the blade, and store the tool in an appropriate location.

9. Complete the Review of this lab exercise.

10. Show all completed work to the instructor. You should have a three 6-inch pieces of 2 × 4 lumber to show the instructor.

11. Clean up the work area and return all tools and materials to their proper locations.

12. Get the instructor to sign off upon satisfactory completion of this lab exercise.

Review

1. List at least three safety items to consider when using a reciprocating saw.

_____

_____

_____

2. Using tool catalogs as a reference, describe at least two types of reciprocating saws that can be used to cut various materials. Include the manufacturer's name and catalog numbers of the models that you describe.

_____

_____

_____

_____

Lab 4.1: Using a Voltage Tester

Name: _____ **Date:** _____ **Score:** _____

Introduction

Electricians must be able to use a variety of measuring instruments. One of the most common testing instruments used is a voltage tester. Many electricians use a solenoid type of voltage tester, commonly called a Wiggy®. This style of voltage tester is used in residential electrical work to test for voltage. A voltage tester will give an electrician an indication of the existing voltage but it will not indicate an exact amount. Voltmeters are used when an exact amount of voltage needs to be determined. Some Wiggy®-type voltage testers can also be used to test for continuity. In this lab exercise, you will be using a voltage tester to measure common voltages present in residential applications. Before proceeding with this lab exercise, review the procedure for using a voltage tester in Chapter 4 of the *House Wiring* 3e textbook. Also, read any meter instructions that come with the meter you are using.

Materials and Equipment

House Wiring 3e textbook, safety glasses, solenoid-type voltage tester (Wiggy®), an energized 120-volt duplex receptacle, an energized 120/240-volt receptacle

Procedure

1. Put on safety glasses and observe all safety rules.

2. Plug in the voltage tester test leads if they are not already connected to the tester.

3. Have your instructor indicate the location of the energized 120-volt duplex receptacle that will be used for this lab exercise. Then, using the voltage tester, carefully insert the lead probes into the grounded (long slot) and ungrounded (short slot) slots of the duplex receptacle. Refer to Illustration A in the procedure for testing a 120-volt receptacle with a voltage tester in Chapter 20 of the *House Wiring* 3e textbook. Read the indicated voltage and record it here.

   Note: If the receptacle you are testing is a tamper-resistant type you will need to insert the lead probes into the receptacle slots at the same time for the shutters to open.

   Recorded Voltage = _____ volts

4. Carefully connect the lead probes across an ungrounded (short slot) and the grounding slot (U-shaped slot). Refer to Illustration B in the procedure for testing a 120-volt receptacle with a voltage tester in Chapter 20 of the *House Wiring* 3e textbook. Read the indicated voltage and record it here.

   Note: If the receptacle you are testing is a tamper-resistant type you will need to insert the lead probes into the receptacle slots at the same time for the shutters to open. Then slowly pull out the lead probe from the grounded slot (long slot) and insert it into the grounding slot (U-shaped slot).

   Recorded Voltage = _____ volts

5. Have your instructor indicate the location of an energized 120/240-volt receptacle that will be used for this lab exercise. Then, using the voltage tester, carefully connect the lead probes across two ungrounded slots of the 240-volt receptacle. Refer to Illustration A in the procedure for testing a 120/240-volt range and dryer receptacle with a voltage tester in Chapter 20 of the *House Wiring* 3e textbook. Read the indicated voltage and record it here.

   Recorded Voltage = _____ volts

6. Carefully connect the lead probes across an ungrounded slot and the grounded neutral slot. Refer to Illustration B in the procedure for testing a 120/240-volt range and dryer receptacle with a voltage tester in Chapter 20 of the *House Wiring* 3e textbook. Read the indicated voltage and record it here.

   Recorded Voltage = _____ volts

7. Carefully connect the lead probes across an ungrounded slot and the grounding slot. Refer to Illustration D in the procedure for testing a 120/240-volt range and dryer receptacle with a voltage tester in Chapter 20 of the *House Wiring* 3e textbook. Read the indicated voltage and record it here.

   Recorded Voltage = _____ volts

8. Complete the Review for this lab exercise.

9. Show all completed work to the instructor. The work will consist of the readings that you took and recorded.

10. Clean up the work area and return all meters and materials to their proper locations.

11. Get the instructor to sign off upon satisfactory completion of this lab exercise.

Review

1. List at least three safety items to consider when using a voltage tester on energized circuits.

   _____

   _____

   _____

2. List three uses for a voltage tester.

   _____

   _____

   _____

Lab 4.2: Using a Digital Multimeter

Name: _____ **Date:** _____ **Score:** _____

Introduction

One of the more useful instruments that a residential electrician will use is a multimeter. This is a meter that could be used to measure voltage, amperage, and resistance. It can also be used to test for continuity. Electricians will use this meter most often in residential applications to measure voltage and test for continuity. In this lab exercise, you will use a digital multimeter to measure common voltages present in residential applications and use the continuity feature of the multimeter to determine if a fuse is good or bad. Before proceeding with this lab exercise, review the procedure for using a digital multimeter in Chapter 4 of the *House Wiring* 3e textbook. Also, read any meter instructions that come with the meter you are using.

Materials and Equipment

House Wiring 3e textbook, safety glasses, digital multimeter, five different sizes and types of good and blown fuses, an energized 120-volt duplex receptacle, an energized 120/240-volt receptacle

Procedure

1. Put on safety glasses and observe all safety rules.

2. Plug in the test leads as follows: black to the common jack and red to the volt/ohm jack.

3. Set the meter's selector switch to the AC voltage icon. **Caution:** If more than one AC voltage setting is available on the meter you are using, choose the setting that is over 240 volts.

4. Have your instructor indicate the location of an energized 120-volt duplex receptacle that will be used for this lab exercise. Then, using the multimeter, carefully connect the lead probes across the grounded (long slot) and ungrounded (short slot) slots of the duplex receptacle. Read the indicated voltage and record it here.

   Note: If the receptacle you are testing is a tamper-resistant type you will need to insert the lead probes into the receptacle slots at the same time for the shutters to open.

   Recorded Voltage = _____ volts

5. Carefully connect the lead probes across an ungrounded (short slot) and the grounding slot (U-shaped slot). Read the indicated voltage and record it here.

   Note: If the receptacle you are testing is a tamper-resistant type you will need to insert the lead probes into the receptacle slots at the same time for the shutters to open. Then slowly pull out the lead probe from the grounded slot (long slot) and insert it into the grounding slot (U-shaped slot).

   Recorded Voltage = _____ volts

6. Have your instructor indicate the location of an energized 120/240-volt receptacle that will be used for this lab exercise. Then, using the multimeter, carefully connect the lead probes across two ungrounded slots of the 240-volt receptacle. Read the indicated voltage and record it here.

   Recorded Voltage = _____ volts

7. Carefully connect the lead probes across an ungrounded slot and the grounded neutral slot. Read the indicated voltage and record it here.

   Recorded Voltage = _____ volts

8. Carefully connect the lead probes across an ungrounded slot and the grounding slot. Read the indicated voltage and record it here.

   Recorded Voltage = _____ volts

9. Change the selector switch on the multimeter to the continuity icon.

10. Determine which five fuses the instructor wants you to test. Then, test each of the fuses to find out whether they are good or have been blown. Do this by placing the leads across each of the fuse ends. The multimeter will give off a high-pitched sound or have some type of visual indicator to show continuity and that the fuse is good. No sound or no visual indication tells you that the fuse is bad. Indicate whether each fuse is good or blown by checking the proper blank.

    Fuse #1   Good _____   Blown _____

    Fuse #2   Good _____   Blown _____

    Fuse #3   Good _____   Blown _____

    Fuse #4   Good _____   Blown _____

    Fuse #5   Good _____   Blown _____

11. Turn off the meter.

12. Complete the Review for this lab exercise.

13. Show all completed work to the instructor. The work will consist of the readings that you took and recorded.

14. Clean up the work area and return all meters and materials to their proper locations.

15. Get the instructor to sign off upon satisfactory completion of this lab exercise.

Review

1. List at least three safety items to consider when using a digital multimeter on energized circuits.

   _____

   _____

   _____

2. List at least two other electrical instruments that could also be used to indicate that a receptacle or circuit wire is energized.

   _____

   _____

   _____

Lab 4.3: Using a Clamp-On Ammeter

Name: _____ Date: _____ Score: _____

Introduction

Another very useful instrument that a residential electrician will use is a clamp-on ammeter. This meter is used in residential work to measure the amount of current flowing in a conductor. In this lab exercise, you will be using a clamp-on ammeter to measure current flowing to a load. Before proceeding with this lab exercise, review the procedure for using a clamp-on ammeter in Chapter 4 of the *House Wiring* 3e textbook. Also, read any meter instructions that come with the meter you are using.

Materials and Equipment

House Wiring 3e textbook, safety glasses, clamp-on ammeter, an electrical load that can be energized

Procedure

1. Put on safety glasses and observe all safety rules.

2. Have your instructor indicate the location of an electrical load that can be energized.

3. Energize the load by activating the switch that controls the load.

4. Turn on the clamp-on ammeter and set the selector switch to the highest current value indicated.

5. Open the clamping mechanism and clamp the ammeter around a circuit conductor that is carrying the load current. Refer to Illustration A in the procedure for using a clamp-on ammeter in Chapter 4 of the *House Wiring* 3e textbook.

6. Read the indicated current drawn by the load and record it here.

   Clamp-on Ammeter Recorded Amperage = _____ amps

7. Turn off the meter and de-energize the load.

8. Complete the Review for this lab exercise.

9. Show all completed work to the instructor. The work will consist of the reading that you took and recorded.

10. Clean up the work area and return all meters and materials to their proper locations.

11. Get the instructor to sign off upon satisfactory completion of this lab exercise.

Review

1. List three uses for an ammeter.

   _____

   _____

   _____

2. List three important rules to follow in the care and maintenance of any meter.

_____

_____

_____

3. List at least three safety rules to follow when using a clamp-on ammeter.

_____

_____

_____

Chapter 5

Understanding Residential Building Plans

Lab 5.1: Identify Common Architectural Electrical Symbols

Name: _____ Date: _____ Score: _____

Introduction

Electrical symbols used on an architectural building electrical plan show the location and type of electrical item required. It is very important for an electrician to know and understand the meaning of the various electrical symbols so that the proper equipment is installed in the proper location. The symbols, used in conjunction with the specifications, will also allow an electrician to make an accurate estimate of the materials, cost, and time needed to do the job. In this lab exercise, you will draw many of the common electrical symbols found on residential electrical plans.

Materials and Equipment

House Wiring 3e textbook, pencil, sheet of paper

Procedure

1. Draw the architectural electrical symbol used to represent each of the electrical items listed in the Review section of this lab exercise.

2. On your first attempt, do not use the *House Wiring* 3e textbook as a reference.

3. On your second attempt, use the *House Wiring* 3e textbook to correct any that were drawn incorrectly.

4. On your third attempt, cover the drawn symbol with a piece of paper and redraw all of the symbols without using the *House Wiring* 3e textbook.

5. Compare your work with the correct symbols as shown in Chapter 5 of the *House Wiring* 3e textbook and redraw any symbols that you drew incorrectly.

6. Show the instructor your completed work.

7. Get the instructor to sign off upon satisfactory completion of this lab exercise.

Review

SYMBOLS	
Ceiling Lighting Outlet Incandescent	Wall Lamp Holder w/Pull Switch
Recessed Ceiling Lighting Outlet Incandescent	Junction Box Ceiling
Recessed Fluorescent Ceiling Outlet	Single Receptacle
Duplex Receptacle	Duplex Receptacle Split-Circuit
Weatherproof Receptacle Outlet	Special Purpose Outlet
Range Outlet	Clothes Dryer Outlet
Clock Outlet	Multioutlet Assembly
Floor Outlet	Single-Pole Switch
Three Way Switch	Double-Pole Switch
Four Way Switch	Door Switch
Dimmer Switch	Switch w/ Pilot Lamp
Three Wire Homerun	Two Wire Cable
Circuit Breaker	Television Outlet
Chime	Telephone Outlet

Lab 5.2: Identify the Structural Parts of a House

Name: _____ Date: _____ Score: _____

Introduction

Residential electricians need to become familiar with the basic structural framework of a house. During the rough-in stage of the residential wiring system installation, electrical boxes will need to be mounted on building framing members and wiring methods will have to be installed on or through building framing members. Knowing how a house is constructed will allow an electrician to install the wiring system in a safe and efficient manner. In this lab exercise, you will identify several structural parts of a house.

Materials and Equipment

House Wiring 3e textbook, pencil

Procedure

1. Using the information found in Figure 5-22 of the House Wiring 3e textbook, label the structural members in Figure 5.2-1 (balloon framing) in the Review section of this lab exercise.

2. Using the information found in Figure 5-21 of the House Wiring 3e textbook, label the structural members in Figure 5.2-2 (platform framing) in the Review section of this lab exercise.

3. Show the instructor the completed work.

4. Get the instructor to sign off upon satisfactory completion of this lab exercise.

Review

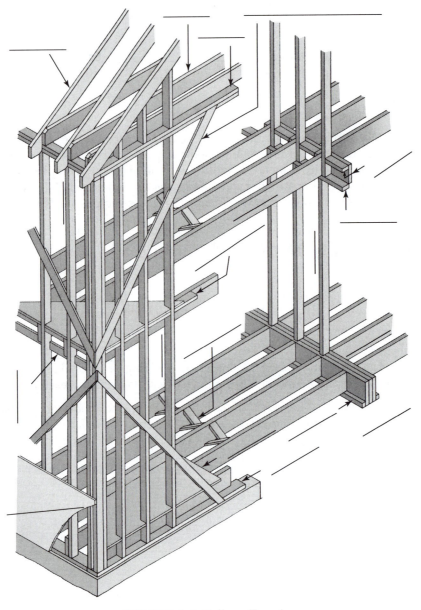

FIGURE 5.2-1 Balloon Framing

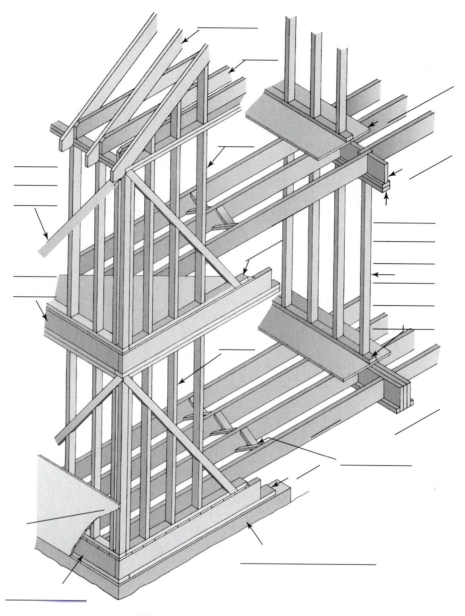

FIGURE 5.2-2 Platform Framing

Chapter 6

Determining Branch Circuit, Feeder Circuit, and Service Entrance Requirements

Lab 6.1: Calculate the Minimum Number of General Lighting Circuits

Name: _____ Date: _____ Score: _____

Introduction

A residential general lighting circuit is a branch circuit that has both lighting and receptacle loads connected to it. A good example of this circuit type would be a bedroom branch circuit that has several receptacles and a ceiling-mounted lighting fixture connected to it. This type of circuit makes up the majority of the branch circuits found in residential wiring. To determine the minimum number of general lighting circuits required in a house, a calculation of the habitable floor area is required. The calculated floor area is then multiplied by the unit load per square foot for general lighting to get the total general lighting load in volt-amperes. Section 220.12 states that a unit load of not less than that specified in Table 220.12 for occupancies listed in the table will be the minimum lighting load. The unit load for a dwelling unit, according to Table 220.12, is 3 volt-amperes per square foot. The floor area for each floor is to be computed from the *outside* dimensions of the dwelling unit. The computed floor area does not include open porches, garages, or unused or unfinished spaces not adaptable for future use. Examples of unused or unfinished spaces for dwelling units are attics, basements, or crawl spaces. A finished basement used for something such as a recreation room must be included in the floor area calculation. To determine the minimum number of general lighting circuits required in a house, the total general lighting load in volt-amperes is divided by 120 volts (the voltage of residential general lighting circuits). This will give you the total general lighting load in amperes. Then, all you have to do is divide the total general lighting load in amperes by the size of the fuse or circuit breaker that is providing the overcurrent protection for the general lighting circuits. In this lab exercise, you will determine the minimum number of 15- and 20-amp general lighting branch circuits in houses with different square foot areas.

The following formulas can be used to determine the minimum number of general lighting circuits in a dwelling:

$$\frac{3\ VA \times \text{Calculated Square Foot Area}}{120\ \text{volts}} = \text{Total General Lighting Load in Amps}$$

$$\frac{\text{Total General Lighting Load in Amps}}{15\ \text{amps}} = \text{Min. \# of 15 amp General Lighting Circuits}$$

$$\frac{\text{Total General Lighting Load in Amps}}{20\ \text{amps}} = \text{Min. \# of 20 amp General Lighting Circuits}$$

Materials and Equipment

House Wiring 3e textbook, 2011 *National Electrical Code®*, pencil, calculator

Procedure

1. Using the formulas presented in the Introduction of this lab exercise and the information found in Chapter 6 of the *House Wiring* 3e textbook, calculate the minimum number of general lighting circuits required in the residential applications shown in the Review section of this lab exercise.

2. Show the instructor your completed work.

3. Get the instructor to sign off upon satisfactory completion of this lab exercise.

Review

1. A 1200-square foot house

   Minimum # of 15 amp general lighting circuits: _____

   Minimum # of 20 amp general lighting circuits: _____

2. A 1500-square foot house

   Minimum # of 15 amp general lighting circuits: _____

   Minimum # of 20 amp general lighting circuits: _____

3. A 2000-square foot house

   Minimum # of 15 amp general lighting circuits: _____

   Minimum # of 20 amp general lighting circuits: _____

4. A 3000 square foot house

   Minimum # of 15 amp general lighting circuits: _____

   Minimum # of 20 amp general lighting circuits: _____

5. A 4500-square foot house

   Minimum # of 15 amp general lighting circuits: _____

   Minimum # of 20 amp general lighting circuits: _____

Lab 6.2: Calculate Common Residential Electric Cooking Loads, Branch-Circuit Conductor Sizes, and Circuit Breaker Sizes

Name: _____ Date: _____ Score: _____

Introduction

Individual branch circuit loads for household electric cooking appliances are based on the information found in Section 220.55 of the *NEC®* for electric ranges and other cooking appliances. Section 220.55 states that the demand load for household electric ranges, wall-mounted ovens, counter-mounted cooking units, and other household cooking appliances individually rated in excess of 1¾ kW is permitted to be computed in accordance with Table 220.55. The alternative is to use the nameplate ratings for household cooking equipment, but electricians usually do not do this because Table 220.55 and the accompanying Notes apply a demand factor, which reduces the load from the nameplate rating. Kilovolt-amperes (kVA) is considered equivalent to kilowatts (kW) for loads computed under this section. For household electric ranges and other cooking appliances, the size of the branch-circuit conductors must be determined by the rating of the range. Note that Section 210.19(A)(3) does not permit the branch-circuit rating (fuse or circuit breaker) of a circuit supplying household ranges with a nameplate rating of 8¾ kW or more to be less than 40 amperes. In this lab exercise, you will calculate electric cooking loads, the minimum wire sizes that can supply the load, and the maximum circuit breaker size that protects the branch-circuit wiring.

Materials and Equipment

House Wiring 3e textbook, 2011 *National Electrical Code®*, pencil, calculator

Procedure

1. Using the information presented in the Introduction of this lab exercise and the information found in Chapter 6 of the *House Wiring* 3e textbook, calculate the load, minimum branch-circuit copper conductor size, and maximum circuit breaker size of each of the residential electric cooking appliance applications shown in the Review section of this lab exercise.

2. Show the instructor your completed work.

3. Get the instructor to sign off upon satisfactory completion of this lab exercise.

Review

1. An electric range with a nameplate rating of 11 kW.

   Computed load: _____ kW

   Minimum conductor size for the branch circuit: _____ CU

   Maximum circuit breaker size used for the branch circuit: _____ amp

2. An electric range with a nameplate rating of 12 kW.

    Computed load: _____ kW

    Minimum conductor size for the branch circuit: _____ CU

    Maximum circuit breaker size used for the branch circuit: _____ amp

3. An electric range with a nameplate rating of 15 kW.

    Computed load: _____ kW

    Minimum conductor size for the branch circuit: _____ CU

    Maximum circuit breaker size used for the branch circuit: _____ amp

4. A counter-mounted cooktop unit with a nameplate rating of 6 kW.

    Computed load: _____ kW

    Minimum conductor size for the branch circuit: _____ CU

    Maximum circuit breaker size used for the branch circuit: _____ amp

5. A counter-mounted cooktop unit with a nameplate rating of 4 kW and a built-in oven with a nameplate rating of 8 kW. Both appliances will be connected to the same branch circuit and be located in the same room.

    Computed load: _____ kW

    Minimum conductor size for the branch circuit: _____ CU

    Maximum circuit breaker size used for the branch circuit: _____ amp

Lab 6.3: Determine the Ampacity of a Conductor

Name: _____ **Date:** _____ **Score:** _____

Introduction

Ampacity is defined as the current, in amperes, that a conductor can carry continuously under the conditions of use without exceeding its temperature rating. Section 210.19(A) requires branch-circuit conductors to have an ampacity at least equal to the maximum electrical load to be served. This rule makes a lot of sense when you think about it. For example, if you are wiring a branch circuit that will supply electrical power to a load that will draw 16 amperes, you need to use a wire size for the branch circuit that has an ampacity of at least 16 amps. Otherwise, the conductor will heat up and will cause the insulation on the conductor to melt. Electricians use Table 310.15(B)(16) to determine the ampacity of a conductor for use in residential wiring. **Caution:** Next to the 14 AWG, 12 AWG, and 10 AWG wire sizes in Table 310.15(B)(16), you will see a double asterisk (**). At the bottom of Table 310.15(B)(16) the double asterisk tells you to refer to Section 240.4(D). This section states that the overcurrent protection must not exceed 15 amperes for 14 AWG, 20 amperes for 12 AWG, and 30 amperes for 10 AWG copper; or 15 amperes for 12 AWG and 25 amperes for 10 AWG aluminum and copper-clad aluminum. So, no matter what the actual ampacity is for 14, 12, and 10 AWG conductors according to Table 310.15(B)(16), the circuit breaker or fuse size can never be larger than 15 amp for 14 AWG, 20 amp for 12 AWG, and 30 amp for 10 AWG copper conductors for residential branch-circuit applications.

The ampacity of a conductor as determined in Table 310.15(B)(16) and used in residential applications may have to be modified slightly. As the conditions of use change for a conductor, the ampacity is required to be adjusted so that at no time could the temperature of the conductor exceed the temperature of the insulation on the conductor. The following adjustments may have to be made for certain residential situations:

- The ampacities given in Table 310.15(B)(16) are based upon an ambient temperature of no more than 30°C (86°F). When the ambient temperature that a conductor is exposed to exceeds 30°C (86°F), the ampacities from Table 310.15(B)(16) must be multiplied by the appropriate correction factor shown in Table 310.15(B)(2)(a).

- The ampacities listed in Table 310.15(B)(16) are also based on not having any more than three current-carrying conductors in a raceway or cable. Section 310.15(B)(3)(a) states that where the number of current-carrying conductors in a raceway or cable exceeds three, or where single conductors or multiconductor cables are stacked or bundled longer than 24 inches (600 millimeters) without maintaining spacing and are not installed in raceways, the allowable ampacity of each conductor must be reduced by the adjustment factor listed in Table 310.15(B)(3)(a). This is called "derating" and results in a lower conductor ampacity when there are more than three current-carrying conductors in a cable or in a raceway. Where more than two Type NM cables containing two or more current-carrying conductors are installed in contact with thermal insulation without maintaining spacing between the cables, the allowable ampacity of each conductor must be adjusted according to Table 310.15(B)(3)(a).

- Section 110.14(C) requires an electrician to take a look at the temperature ratings of the equipment terminals that the electrical conductors will be connected to. The temperature rating (60°C, 75°C, or 90°C) associated with the ampacity of a conductor from Table 310.15(B)(16) must be selected so the temperature produced by that conductor cannot exceed the lowest temperature rating of any connected termination, conductor, or device. If the temperature produced by a current-carrying conductor at a termination does exceed the termination temperature rating, the termination will burn up. **Caution:** Do not forget that any electrical circuit has a beginning and an end. You must always know what the termination temperature rating is at the panel and what the temperature rating is at the equipment (like switches and receptacles) on the circuit. A typical residential circuit may have 60°C at one end and a 75°C rating at the other. Maximum conductor ampacity in this case would have to be based on the 60°C column in Table 310.15(B)(16).

In this lab exercise, you will be calculating the maximum ampacity of conductors used in various residential applications.

Materials and Equipment

House Wiring 3e textbook, 2011 *National Electrical Code®*, pencil, calculator

Procedure

1. Using the information presented in the Introduction of this lab exercise and the information found in Chapter 6 of the *House Wiring* 3e textbook, calculate the ampacity of the conductors used in the residential applications shown in the Review section of this lab exercise.

2. Show the instructor your completed work.

3. Get the instructor to sign off upon satisfactory completion of this lab exercise.

Review

1. A 14/2 Type NM copper used for a general lighting branch circuit.

   Ampacity: _____   Maximum Fuse or Circuit Breaker Size: _____

2. A 12/2 Type NM copper used for a small-appliance branch circuit.

   Ampacity: _____   Maximum Fuse or Circuit Breaker Size: _____

3. A 10/3 Type NM copper used for an electric dryer branch circuit.

   Ampacity: _____   Maximum Fuse or Circuit Breaker Size: _____

4. Three 8 AWG copper conductors with THHN insulation installed in a garage in electrical metallic tubing (EMT). The ambient temperature in the garage can reach as high as 110°F. All termination points for this circuit are rated 60°C.

   Ampacity: _____   Maximum Fuse or Circuit Breaker Size: _____

5. Four 4 AWG copper conductors with THWN insulation installed in EMT in a basement. All termination points are rated at 60°C.

   Ampacity: _____   Maximum Fuse or Circuit Breaker Size: _____

6. A 1 AWG aluminum Type USE cable installed underground from a house to a detached garage to feed a 120/240-volt subpanel. All termination points are rated at 75°C.

   Ampacity: _____   Maximum Fuse or Circuit Breaker Size: _____

7. An 8/3 Type UF copper cable installed underground from a house to a detached garage to feed a 120/240-volt subpanel. All termination points are rated at 60°C.

   Ampacity: _____   Maximum Fuse or Circuit Breaker Size: _____

8. A 12/2 Type NM copper cable installed in an attic where the ambient temperature will be as high as 120°F.

   Ampacity: _____   Maximum Fuse or Circuit Breaker Size: _____

9. An electrician installs five 3 AWG THWN copper current-carrying conductors in a raceway located in a garage. The garage area is subject to an ambient temperature of 100°F. All terminations are rated at 75°C.

   Ampacity: _____   Maximum Fuse or Circuit Breaker Size: _____

Lab 6.4: Calculate the Size of a Residential Service Entrance Using the Standard Method

Name: _____ Date: _____ Score: _____

Introduction

For dwelling units, wire sizes as listed in Table 310.15(B)(7) are permitted as 120/240-volt, three-wire, single-phase service-entrance conductors, service lateral conductors, and feeder conductors that serve as the main power feeder to a dwelling unit and are installed in raceway or cable. The wire sizes allowed in Table 310.15(B)(7) are smaller than the wire sizes given in Table 310.15(B)(16) for a particular ampacity. This reduction in wire size reflects the fact that residential service entrances are not typically loaded very heavily and the heavier loads that are on the service are not operated for long time periods. Also, the total residential electrical load is not usually energized at the same time. To be able to use this table for feeder sizing, the feeder between the main disconnect and the lighting and appliance branch-circuit panel must carry the total residential load. An example of a feeder carrying the total load would be when a service entrance is installed on the side of an attached garage. The main service disconnect equipment would be located in the garage. A feeder can be taken from the garage and used to feed a loadcenter in a basement under the main part of the house. In this case, the feeder will carry the total residential load and could be sized according to Table 310.15(B)(7). The service entrance grounded conductor, or service neutral, is permitted to be smaller than the ungrounded conductors, as long as it is never sized smaller than the grounding electrode conductor. Two methods for calculating the minimum size service entrance for dwellings are covered in Article 220. Article 220, Part III covers the standard method. Article 220, Part IV covers the optional method. In this lab exercise, you will calculate the minimum size service entrance for a house by using the standard method.

Materials and Equipment

House Wiring 3e textbook, 2011 *National Electrical Code®*, pencil, calculator

Procedure

1. Using the information presented in the Introduction of this lab exercise and the information found in Chapter 6 of the *House Wiring* 3e textbook, calculate the size of the residential service entrance shown in the Review section of this lab exercise.

2. Show the instructor your completed work.

3. Get the instructor to sign off upon satisfactory completion of this lab exercise.

Review

A 1850-square-foot dwelling unit, 120/240-volt, three-wire, single-phase service with the following electrical equipment:

ITEM	RATING
Electric range	12 kW @ 120/240 volts
Electric water heater	3 kW @ 240 volts
Electric clothes dryer	4.5 kW @ 120/240 volts
Garbage disposal	6 amps @ 120 volts
Dishwasher	1.2 kW @ 120 volts

Minimum Service Entrance Size: _____ AMPERE

Size Ungrounded Conductors: _____ SEU AL

_____ SEU CU

Minimum Size of the Service Neutral: _____ SEU AL

_____ SEU CU

Minimum Size of the Grounding
Electrode Conductor: _____ CU

Steps

1. Calculate the square foot area by using the outside dimensions of the dwelling. Do not include open porches, garages, and floor spaces unless they are adaptable for future living space. Include all floors of the dwelling. Refer to 220.12.

   _____ square feet

2. Multiply the square foot area by the unit load per square foot as found in Table 220.12. Use 3 VA per square foot for dwelling units. Refer to 220.12 and Table 220.12.

   _____ square feet × 3 VA/square foot = _____ VA

3. Add the load allowance for the required two small-appliance branch circuits and the one laundry circuit. These circuits are computed at 1500 VA each. Minimum total used is 4500 VA. Be sure to add in 1500 VA each for any additional small appliance or laundry circuits. Refer to 220.52.

   _____ VA + 4500 VA + _____ VA (additional circuits) = _____ VA

4. Apply the demand to the total of steps 1 thru 3. According to Table 220.42 the first 3000 VA is taken at 100%; 3001 to 120,000 VA is taken at 35%; and anything over 120,000 VA is taken at 25%. Refer to Table 220.42.

   If step 3 is **over** 120,000 VA, use:

   3000 VA + [(_____ − 120,000 VA) × 25%] = (117,000 VA 3 35%) = _____ VA

   If step 3 is under 120,000 VA, use:

   3000 VA + [(_____ VA − 3000 VA) × 35%] = _____ VA

5. Add in the loading for electric ranges, counter-mounted cooking units, or wall-mounted ovens. Use Table 220.55 and the Notes for cooking equipment. Refer to Table 220.55 and the Notes.

   _____ VA + _____ VA (range, etc.) = _____ VA

6. Add in the electric dryer load, using Table 220.54. Remember to use 5000 VA or the nameplate rating of the dryer if it is greater than 5 kW. Refer to 220.54 and Table 220.54.

   _____ VA + _____ VA (dryer) = _____ VA

7. Add the loading for heating (H) or the air-conditioning (AC), whichever is greater. Remember, both of these loads will not be operating at the same time. Electric heating and air-conditioning are taken at 100% of their nameplate ratings. Refer to 220.51, 220.60.

   _____ VA + _____ VA (H or AC) = _____ VA

8. Now add the loading for all other fixed appliances such as dishwashers, disposals, trash compactors, water heaters, etc. Section 220.53 allows you to apply an additional demand of 75% to the fixed appliance load total if there are four or more of the appliances. If there are three or less then take their total at 100%. Refer to 220.53.

Appliances	VA (or Watts)

   Total VA _____ × 1.00 = _____ VA (if 3 or fewer fixed appliances)

   Total VA _____ × 0.75 = _____ VA (if 4 or more fixed appliances)

   _____ VA (from step 7) + _____ Total VA (Fixed Appliances) = _____ VA

9. According to Section 220.50, which refers to Section 430.24, you must add an additional 25% of the largest motor load. Use the largest motor load included in the service entrance calculation. The largest motor load is based on the highest rated full-load current for the motor listed. Refer to 220.50 and 430.24.

   _____ VA + _____ VA (25% of the largest motor load) = _____ VA

10. The sum of steps 1 thru 9 is called the Total Computed Load. Divide this figure by 240 volts to find the minimum ampacity required for the ungrounded service entrance conductors. Size them according to Table 310.15(B)(7). If the total computed load in this step is less than 100 amps, Sections 230.42(B) and 230.79(C) require both the service entrance ungrounded conductors and the service disconnecting means to be rated not less than 100 amps. Refer to Table 310.15(B)(7) and Sections 230.42(B) and 230.79(C).

   _____ VA (Total Computed Load)/240 volts = _____ Amps

   Minimum Service Rating: _____ ampere based on Table 310.15(B)(7)

   Minimum Ungrounded Conductor Size: _____ Copper _____ Aluminum

11. Calculate the minimum neutral ampacity by referring to Section 220.61. Add the volt-amp load for the 120-volt general lighting, small appliance, and laundry circuits from step 4. Add the cooking and the drying loads and multiply the total by 70%. Now add in the loads of all appliances that operate on 120 volts. (Keep in mind that loads operating on two-wire 240 volts do not utilize a grounded neutral conductor.) Remember to apply the additional demand of 75% to the total neutral loading of the appliances if there are four or more of them. Refer to 220.61, 220.53, and 430.24.

Step 4 120-volt Circuits Load   = _____ VA

     Cooking Load from step 5 _____ VA × 0.70 = _____ VA

     Drying Load from step 6 _____ VA × 0.70 = _____ VA

     120-volt appliances from step 8

Appliances	VA

         Total VA _____ × 0.75 = _____ VA (if 4 or more)

         Total VA _____ × 1.00 = _____ VA (if 3 or fewer)

Largest 120-volt Motor Load _____ VA × 0.25 = _____ VA

         Total _____ VA (Neutral)

12. Divide the total found in step 11 by 240 volts to find the minimum ampacity of the service neutral.

     _____ VA (step 11)/240 volts = _____ ampacity of neutral

13. According to Section 310.15(B)(7), the neutral may be smaller than the ungrounded service entrance conductors, but in no case can it be smaller than the grounding electrode conductor per Section 250.24(C)(1). With this in mind, size the neutral conductor, based upon the ampacity found in step 12, by using Table 310.15(B)(7) or Table 310.15(B)(16) (if the neutral load is less than 100 amps).

     Minimum Neutral Size: _____ Copper _____ Aluminum   Table 310.15(B)(7)

     Minimum Neutral Size: _____ Copper _____ Aluminum   Table 310.15(B)(16) (75°C)

14. Calculate the minimum size copper grounding electrode conductor by using the largest size ungrounded service entrance conductor and Table 250.66. Refer to Table 250.66.

     Minimum Grounding Electrode Conductor: _____ AWG Copper

Lab 6.5: Calculate the Size of a Residential Service Entrance Using the Optional Method

Name: _____ Date: _____ Score: _____

Introduction

For dwelling units, wire sizes as listed in Table 310.15(B)(7) are permitted as 120/240-volt, three-wire, single-phase service-entrance conductors, service lateral conductors, and feeder conductors that serve as the main power feeder to a dwelling unit and are installed in raceway or cable. The wire sizes allowed in Table 310.15(B)(7) are smaller than the wire sizes given in Table 310.15(B)(16) for a particular ampacity. This reduction in wire size reflects the fact that residential service entrances are not typically loaded very heavily and the heavier loads that are on the service are not operated for long time periods. Also, the total residential electrical load is not usually energized at the same time. To be able to use this table for feeder sizing, the feeder between the main disconnect and the lighting and appliance branch-circuit panelboard must carry the total residential load. An example of a feeder carrying the total load would be when a service entrance is installed on the side of an attached garage. The main service disconnect equipment would be located in the garage. A feeder can be taken from the garage and used to feed a loadcenter in a basement under the main part of the house. In this case, the feeder will carry the total residential load and could be sized according to Table 310.15(B)(7). The service entrance grounded conductor, or service neutral, is permitted to be smaller than the ungrounded conductors, as long as it is never sized smaller than the grounding electrode conductor. Two methods for calculating the minimum size service entrance for dwellings are covered in Article 220. Article 220, Part III covers the standard method. Article 220, Part IV covers the optional method. In this lab exercise, you will calculate the minimum size service entrance for a house using the optional method.

Materials and Equipment

House Wiring 3e textbook, 2011 *National Electrical Code®*, pencil, calculator

Procedure

1. Using the information presented in the Introduction of this lab exercise and the information found in Chapter 6 of the *House Wiring* 3e textbook, calculate the size of the residential service entrance shown in the Review section of this lab exercise.

2. Show the instructor your completed work.

3. Get the instructor to sign off upon satisfactory completion of this lab exercise.

Review

A 2000-square-foot dwelling unit; 120/240-volt, three-wire, single-phase service with the following electrical equipment:

ITEM	RATING
Electric range	12 kW @ 120/240 volts
Electric water heater	5 kW @ 240 volts
Electric clothes dryer	6 kW @ 120/240 volts
Garbage disposal	1,000 W @ 120 volts
Dishwasher	1,000 W @ 120 volts
Trash compactor	1,200 W @ 120 volts
Air conditioner	9 kW @ 240 volts
Vent fan	400 W @ 120 volts
Electric furnace	25kW @ 240 volts

Minimum Service Entrance Size: _____ AMPERE

Size Ungrounded Conductors: _____ SEU AL

_____ SEU CU

Minimum Size of the Service Neutral: _____ SEU AL

_____ SEU CU

Minimum Size of the Grounding
Electrode Conductor: _____ CU

Steps

1. Calculate the square foot area by using the outside dimensions of the dwelling. Do not include open porches, garages, and floor spaces unless they are adaptable for future living space. Include all floors of the dwelling. Refer to 220.82(B)(1).

   _____ square feet

2. Multiply the square foot area by 3 VA per square foot. Refer to 220.82(B)(1).

   _____ square feet × 3 VA/square foot = _____ VA

3. Add the load allowance for the required two small-appliance branch circuits and the one laundry circuit. These circuits are computed at 1500 VA each. Minimum total used is 4500 VA. Be sure to add in 1500 VA each for any additional small appliance or laundry circuits. Refer to 220.82(B)(2).

   _____ VA + 4500 VA + _____ VA (additional circuits) = _____ VA

4. Add the nameplate ratings of all appliances that are fastened in place, permanently connected, or located to be on a specific circuit, such as electric ranges, wall-mounted ovens, counter-mounted cooktops, electric clothes dryers, trash compactors, dishwashers, and electric water heaters.

   Example: 12 kW range = 12,000 W; a 3 kW water heater = 3000 W. Refer to 220.82(B)(3).

   _____ VA + _____ VA (Nameplate Rating Total of All Appliances) = _____ VA

Appliances	VA
Total VA	

5. Take 100% of the first 10,000 VA and 40% of the remaining load. Refer to Section 220.82(B).

    10,000 VA + [(_____ VA − 10,000VA) × 0.40] = _____ VA

6. Add the value of the largest heating and air-conditioning load from the list included in Section 220.82(C). These loads include air conditioners, heat pumps, central electric heat, and separately controlled electric space heaters, such as electric baseboard heaters. Refer to Section 220.82(C).

    _____ VA + _____ VA (largest from Section 220.82(C)) = _____ VA

7. The sum of steps 1 thru 6 is called the Total Computed Load. Divide this figure by 240 volts to find the minimum ampacity required for the ungrounded service entrance conductors. Size them according to Table 310.15(B)(7). If the total computed load in this step is less than 100 amps, Sections 230.42(B) and 230.79(C) require both the service entrance ungrounded conductors and the service disconnecting means to be rated not less than 100 amps. Refer to Table 310.15(B)(7) and Sections 230.42(B) and 230.79(C).

    _____ VA (Total Computed Load)/240 volts = _____ Amps

    Minimum Service Rating: _____ ampere based on Table 310.15(B)(7)

    Minimum Ungrounded Conductor Size: _____ Copper

    _____ Aluminum

8. The neutral size is calculated *exactly* the same as for the standard method. There is no optional method for calculating the minimum neutral size. Calculate the minimum neutral ampacity by referring to Section 220.61. Add the volt-amp load for the 120-volt general lighting, small appliance, and laundry circuits from step 4 of the standard method. Add the cooking and the drying loads and multiply the total by 70%. Now add in the loads of all appliances that operate on 120 volts. (Keep in mind that loads operating on two-wire 240 volts do not utilize a grounded neutral conductor.) Remember to apply the additional demand of 75% to the total neutral loading of the appliances if there are four or more of them. Refer to 220.61, 220.53, and 430.24.

    120-Volt Circuits Load (*step 4 in standard method*)   = _____ VA

    Cooking Load (from *step 5 in standard method*) _____ VA × 0.70 = _____ VA

    Drying Load (from *step 6 in standard method*) _____ VA × 0.70 = _____ VA

    120-Volt Appliances (from *step 8 in standard method*)

Appliances	VA

Total VA _____ × 0.75 = _____ VA (if 4 or more)

Total VA _____ × 1.00 = _____ VA (if 3 or fewer)

Largest 120-volt Motor Load _____ VA × 0.25 = _____ VA

Total _____ VA (Neutral)

9. Divide the total found in step 8 by 240 volts to find the minimum ampacity of the service neutral.

   _____ VA (step 8)/240 volts = _____ amps of neutral

10. According to Section 310.15(B)(7), the neutral may be smaller than the ungrounded service entrance conductors, but in no case can it be smaller than the grounding electrode conductor per Section 250.24(C)(1). With this in mind, size the neutral conductor based upon the ampacity found in step 9 by using Table 310.15(B)(7) or Table 310.15(B)(16) (if the neutral load is less than 100 amps). Refer to Table 310.15(B)(16) and Table 310.15(B)(7).

    Minimum Neutral Size: _____ Copper _____ Aluminum   Table 310.15(B)(7)

    Minimum Neutral Size: _____ Copper _____ Aluminum   Table 310.15(B)(16)

11. Calculate the minimum size grounding electrode conductor by using the largest size ungrounded service entrance conductor and Table 250.66.

    Minimum Grounding Electrode Conductor: _____ AWG Copper

Chapter 7

Introduction to Residential Service Entrances

Lab 7.1: Identify the Major Parts of a Residential Service Entrance

Name: _____ Date: _____ Score: _____

Introduction

The *National Electrical Code*® defines a service as the conductors and equipment for delivering electric energy from the serving electric utility to the wiring system of the premises served. There are two types of service entrances used to deliver electrical energy to a residential wiring system: an overhead service and an underground service. The advantages and disadvantages of each service type need to be considered in the preparation and planning stage of a residential electrical system. Residential electricians must recognize the differences between the two service types as well as the specific installation techniques required for each of them. In this lab exercise you will identify the major parts of both an overhead and an underground service entrance.

Materials and Equipment

House Wiring 3e textbook, pencil

Procedure

1. Identify the parts of the residential service entrance shown in Figure 7.1-1 of the Review section of this lab exercise.

2. Compare your answers with the correct terms as shown in Figure 7-3 in Chapter 7 of the *House Wiring* 3e textbook and write in the correct term for any you identified incorrectly.

3. Show the instructor your completed work.

4. Get the instructor to sign off upon satisfactory completion of this lab exercise.

Review

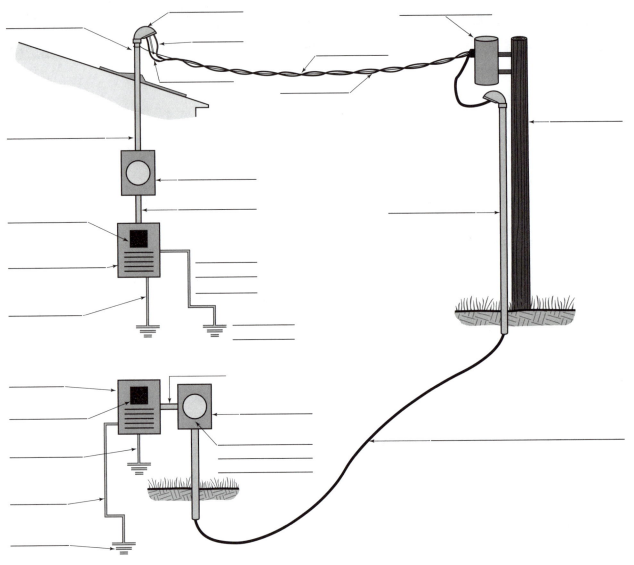

FIGURE 7.1-1

Chapter 8

Service Entrance Equipment and Installation

Lab 8.1: Install an Overhead Service Entrance Using Service Entrance Cable

Name: _____ Date: _____ Score: _____

Introduction

An electrical service is required for all buildings containing an electrical system and receiving electrical energy from a utility company. The *NEC®* describes the term *service* as the conductors required to deliver energy from the electrical supply source to the wiring system of the premises. The *NEC®* also defines the term *service-entrance conductors, overhead system* as the service conductors between the terminals of the service equipment and a point usually outside the building, clear of building walls, where they are joined by tap or splice to the service drop. An overhead service entrance installed with service entrance cable is used often in electrical work and the competent electrician must know how to install this type of service. In this lab exercise, you will install a 100-amp service entrance with Type SEU service entrance cable on your wiring mockup. The service meter socket will be mounted on the plywood side of the wiring mockup and the service loadcenter with a main circuit breaker will be mounted on the back of the wiring mockup. Before you proceed with this lab exercise, review the procedure for installing an overhead service using service entrance cable in Chapter 8 of the *House Wiring* 3e textbook.

Materials and Equipment

House Wiring 3e textbook, safety glasses, electrician's tool kit, a drill and wood bit, one weatherhead for 2 AWG AL Type SEU cable, 2 AL Type SEU cable as needed, cable straps or clips for 2 AL Type SEU, one 1¼-inch watertight connector, one 1¼-inch threaded hub, one 100-amp meter socket, one 100-amp main breaker loadcenter, one sill plate for 2 AWG AL Type SEU, 6 AWG CU grounding electrode conductor as needed, ½-inch PVC conduit as needed, two ½-inch PVC connectors, one 2-foot length of ground rod, one 2-foot length of ¾-inch EMT conduit used as a water pipe electrode, ½-inch and ¾-inch conduit hangers as needed, one ground rod clamp, one water pipe ground clamp, one 1¼-inch cable connector, one ½-inch cable connector, one intersystem bonding termination device, cable staples as needed, screws as needed, duct seal as needed, one cable hook, antioxidant as needed

Procedure

1. Put on safety glasses and observe all safety rules. Refer to Figure 8.1-1 and Figure 8.1-2 to see how this lab project should look when completed.

2. Attach a 1¼-inch threaded hub with a 1¼-inch watertight connector to the meter socket.

339

3. Mount and level the meter socket on the plywood wiring mockup so that it is 5 feet to the top of the meter socket from the floor.

4. Mount and level the 100-amp main breaker loadcenter on the other side of the wiring mockup so that it is 6 feet to the top of the enclosure from the floor. Install the loadcenter between two studs like you would in a flush-mounted situation. Have the edge of the loadcenter set out ½″ from the stud face.

5. Measure and cut the service entrance cable to a length required to go from the meter socket to the 24-inch height shown in Figure 8.1-1. Allow 2 to 3 feet extra at the weatherhead end and 1 foot extra at the meter socket end.

6. Strip 2 to 3 feet of the outside sheathing from one end of the Type SEU cable and 1 foot of outside sheathing from the other end. Twist the individual bare conductors together that make up the neutral conductor in a Type SEU cable.

7. Install the weatherhead on the end that has 2 to 3 feet of free conductor.

8. Mount the assembled service entrance cable with attached weatherhead so that it is directly above and vertically aligned with the watertight connector in the meter socket.

9. Insert the end that has 1 foot of free conductor through the watertight connector and into the meter socket. Tighten the watertight connector around the cable.

10. Make sure the cable is vertically plumb and secure the cable to the plywood with clips or straps at no more than 12 inches from the connector and weatherhead and then at intervals no more than 30 inches.

11. Install the cable hook at the location indicated in Figure 8.1-1.

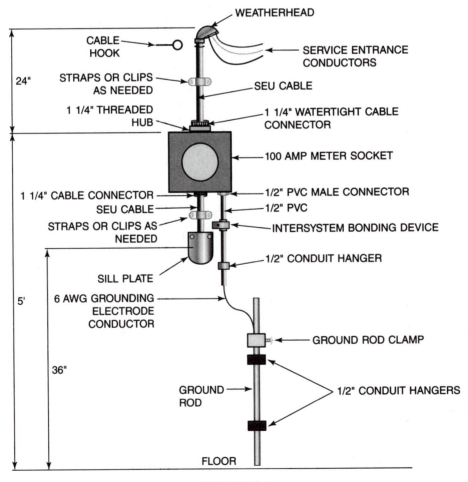

FIGURE 8.1-1

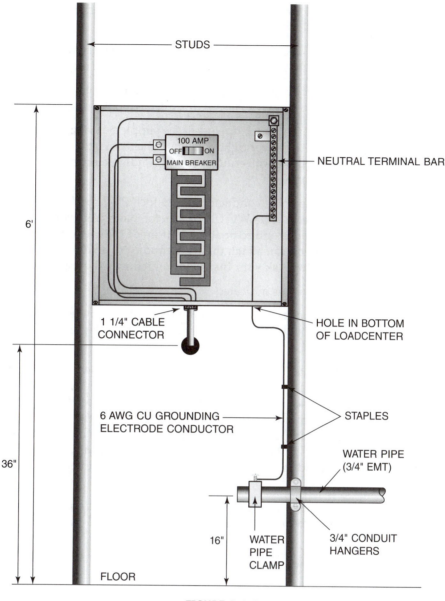

FIGURE 8.1-2

12. Now, drill a hole in the plywood wiring mockup below the center knockout in the bottom of the meter socket. The hole will need to be large enough for the Type SEU cable to go though and for this lab will be 36" to center from the floor. **Note:** This hole may already have been drilled by another student.

13. Remove the center knockout from the bottom of the meter socket and install a 1¼-inch regular cable connector in the knockout hole.

14. Measure and cut the Type SEU cable to the length required to go from the bottom of the meter socket to the loadcenter enclosure. Allow 1 foot extra at the meter socket end and enough at the loadcenter end to allow the conductors to be easily attached to the main circuit breaker and neutral lug.

15. Insert the 1-foot stripped end of service entrance cable through the cable connector on the bottom of the meter socket and tighten the connector onto the cable.

16. Insert the other end of the service entrance cable through the hole in the wall and install a sill plate over the cable at the point where the cable goes through the hole. Use duct seal around the top of the sill plate and the cable to make a watertight installation.

17. Secure the cable within 12 inches of the meter socket and the sill plate and then at intervals of no more than 30 inches.

18. Make the meter socket connections. See Figure 8-20 in the *House Wiring* 3e textbook. Check with your instructor to see if an antioxidant is to be used on the aluminum wires in this lab exercise.

19. On the other side of your wiring mockup, secure the service entrance cable to the loadcenter and make the proper connections. See Figure 8-34 in the *House Wiring* 3e textbook.

20. On the meter socket side of your wiring mock-up, install the supplemental grounding electrode ground rod in a vertical position at the lower right portion of the mockup. Support it with two ½-inch conduit hangers. This will keep it off the wall enough so a grounding electrode conductor can be attached to it with a ground rod clamp.

21. Install the 6 AWG CU grounding electrode conductor. Use ½-inch PVC conduit, supported with conduit hangers, to cover and protect the 6 AWG CU conductor. **Note: You can use the small hole designed for the grounding electrode conductor at the bottom of the meter socket and use the PVC conduit as a protective sleeve, or you can attach the PVC conduit to a ½" KO on the bottom of the meter socket using a PVC connector. Ask your instructor which way they would prefer.**

22. Install the intersystem bonding device as indicated.

23. Connect the grounding electrode conductor to the ground rod electrode by using a ground rod clamp and connect the other end to the grounding lug in the meter socket. See Figure 8-20 in the *House Wiring* 3e textbook.

24. Install the grounding electrode (water pipe) in a horizontal position below the loadcenter on the two studs to the right. Mount it at 16 inches to center from the floor. Support it with two ¾-inch conduit hangers and leave a short length under the loadcenter so that a water pipe clamp can be installed as shown in Figure 8.1-2. Install the grounding electrode conductor to the water pipe electrode. Use the small hole designed for the grounding electrode conductor at the bottom of the loadcenter. **Note:** You can also attach PVC conduit to a ½" KO on the bottom of the loadcenter using a PVC connector and install the grounding electrode conductor in the PVC conduit. Ask your instructor which way he or she would prefer.

25. Connect the grounding electrode conductor to the water pipe electrode by using a water pipe clamp and connect the other end to the grounding lug in the panel. See Figure 8-34 in the *House Wiring* 3e textbook.

26. Complete the Review for this lab exercise.

27. Show all completed work to the instructor.

28. Clean up the work area and return all tools and materials to their proper locations. **Note: Dismantle only the wiring and equipment installed on the line side of the meter socket on this lab project, including the 1¼-inch threaded hub. The load-side wiring and equipment from the load lugs in the meter socket are used in the next lab exercise.**

29. Get the instructor to sign off upon satisfactory completion of this lab exercise.

Review

1. Define a "service drop" according to the *NEC*®.

_____

_____

_____

2. Describe the protection requirements for the following grounding electrode conductors according to Article 250 of the *NEC*®.

8 AWG copper:

_____

_____

6 AWG copper:

_____

_____

4 AWG copper:

_____

_____

Lab 8.2: Install an Overhead–Mast-Type Service Entrance

Name: _____ **Date:** _____ **Score:** _____

Introduction

In some cases, an overhead service is installed that utilizes conduit rather than service entrance cable. This may be a mast-type service where conduit is used to get additional height for the point of attachment. It could also be that conduit is simply used to provide a greater degree of protection for the service entrance conductors when placed on the side of a building where physical damage could occur to the conductors. The service equipment, including the meter socket and the service loadcenter, installed in Lab 8:1 will be used for this lab exercise. Only the service equipment on the line side of the meter socket should have been removed. In this lab exercise, you will install a mast-type service entrance on the line side of the meter socket. Before you proceed with this lab, review the procedure for installing an overhead service using a mast in Chapter 8 of the *House Wiring* 3e textbook.

Materials and Equipment

House Wiring 3e textbook, safety glasses, electrician's tool kit, 24 feet or three 8-foot lengths of 2 AWG AL XHHW wire, one weatherhead for 2-inch RMC or IMC, a 5-foot length of 2-inch RMC or IMC, one roof flashing for 2-inch RMC or IMC, one rubber boot for 2-inch RMC or IMC, one porcelain standoff fitting for 2-inch RMC or IMC, one 2-inch threaded hub, three 2-inch conduit hangers for RMC or IMC, screws as needed, white electrical tape, an 8-foot step ladder, antioxidant as needed

Procedure

1. Put on safety glasses and observe all safety rules. Refer to Figure 8.2-1 to see how this lab project should look when completed.

2. Attach a 2-inch threaded hub to the meter socket.

3. Install the 2-inch conduit hangers in the locations shown in Figure 8.2-1. Make sure that they are in a line with the threaded hub on the meter socket.

4. Thread a 5-foot length of 2-inch RMC or IMC into the threaded hub and tighten it. Then secure the pipe to the wall with the conduit hangers.

5. Install three 8-foot lengths of 2 AWG AL wire into the pipe. Make sure that approximately 2 to 3 feet is coming out of the top and there is enough in the meter socket to make connections to the line lugs. Be sure to identify the service neutral conductor with white tape.

6. Install the roof flashing over the pipe and secure it to the wiring mockup with screws.

7. Install the rubber boot over the pipe and bring it down the pipe until it is against the roof flashing.

8. Attach the porcelain standoff fitting to the mast.

9. Install the weatherhead on the top of the mast, making sure to leave at least 2 to 3 feet of service conductor for the drip loop. Make sure that the service conductors exit the weatherhead through separate holes.

10. Connect the service conductors to the line lugs in the meter socket. Check with your instructor to see if an antioxidant is to be used on the aluminum wires in this lab exercise.

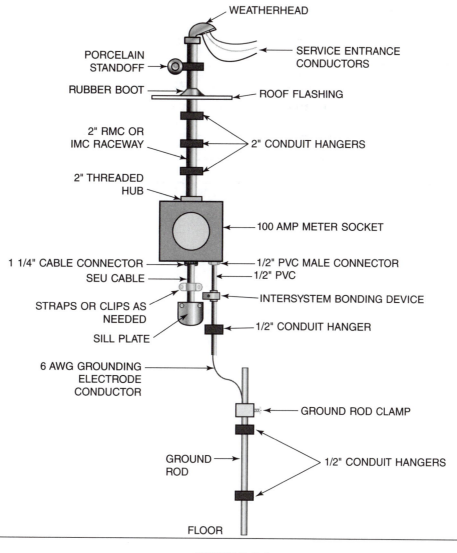

FIGURE 8.2-1

11. Complete the Review for this lab exercise.

12. Show all completed work to the instructor.

13. Clean up the work area and return all tools and materials to their proper locations.

14. Get the instructor to sign off upon satisfactory completion of this lab exercise.

Review

1. List the minimum service drop clearances required by the *NEC*® in the following residential situations.

A driveway _____ feet

A lawn _____ feet

A roof with a slope of 5/12 _____ feet

A garage roof with a slope of 3/12 _____ feet

2. What did you find to be the hardest part of this lab?

_____

_____

_____

3. Name one important thing that you learned during the completion of this lab exercise.

_____

_____

_____

Lab 9.1: Indicate the Proper Locations on a Building Plan for the Minimum Number of Receptacles and Lighting Outlets

Name: _____ Date: _____ Score: _____

Introduction

Section 210.52 in the *NEC*® tells an electrician where receptacle outlets must be installed in a dwelling unit. This information is very important for the electrician to know so that electrical boxes installed during the rough-in stage are located to meet or exceed the requirements of this section. The requirements of 210.52 apply to dwelling unit receptacles that are rated 125 volts and 15- or 20-amperes and that are not part of a lighting fixture or an appliance. Receptacles must be installed so that no point measured horizontally along the floor line in any wall space is more than 6 feet (1.8 meters) from a receptacle outlet. This means that electrical boxes for receptacles must be installed during the rough-in stage so that no point in any wall space is more than 6 feet (1.8 meters) from a receptacle. Section 210.70(A)(1) requires at least one wall-switch-controlled lighting outlet to be installed in every habitable room and bathroom. *Exception No. 1* to the general rule allows one or more receptacles controlled by a wall switch to be permitted instead of lighting outlets, but only in areas other than kitchens and bathrooms. In this lab exercise, you will be drawing the electrical symbols for the required receptacles, switches, and lighting outlets for a typical dwelling unit room shown in Figure 9.1-1. Note: This figure is not drawn to scale. Before proceeding with this lab exercise, review the electrical symbols in Chapter 5 of the *House Wiring* 3e textbook.

Materials and Equipment

House Wiring 3e textbook, pencil, ruler

Procedure

1. Draw in the electrical symbol for a duplex receptacle at the locations indicated so that the installation will meet the minimum *NEC*® requirements. Place the duplex receptacle symbol next to the dimension extension line.

2. Draw in the dimension lines between the duplex receptacle locations and indicate the maximum distance between them in feet.

3. Draw in the electrical symbol for the minimum required number of lighting outlets. Use the symbol for a ceiling-mounted incandescent lighting outlet. Also, draw in the electrical symbol for a single-pole switch that will control the lighting outlet at the proper location. Remember to draw a curved dashed line from the switch to the lighting outlet to indicate that the switch will control that outlet.

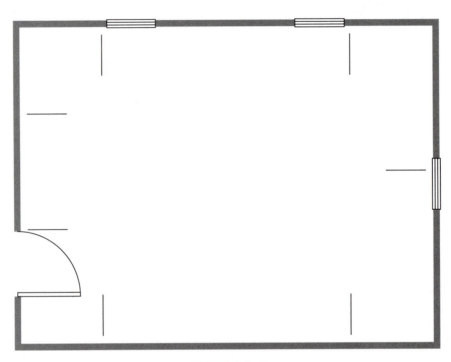

FIGURE 9.1-1

Chapter 10 — Electrical Box Installation

Lab 10.1: Installing Device Boxes

Name: _____ **Date:** _____ **Score:** _____

Introduction

Device boxes are designed to allow a device, such as a switch or receptacle, to be attached to them. Device boxes come in specific sizes—for example, a 3 × 2 × 3½-inch metal device box or a 20-cubic-inch plastic nail-on device box. Selecting the proper electrical box requires an electrician to check the electrical plans to determine what is being installed at a certain location. If the symbol for a duplex receptacle is shown on the plans, a device box will need to be installed. The determination of whether the boxes will be nonmetallic or metal will normally be left for the electrical contractor to decide. Sometimes a set of residential building plans will have specifications that call for the use of only metal boxes or only nonmetallic boxes. In today's residential construction market, most new houses are wired using nonmetallic boxes at those locations in the house where they are appropriate. A house may be wired using all nonmetallic electrical boxes or all metal electrical boxes. Usually, a house is wired using a combination of both metal and nonmetallic boxes. In this lab exercise, you will install both nonmetallic and metal device boxes on studs in the wiring mockup. Before proceeding with this lab exercise, refer to the procedure for installing device boxes in new construction in Chapter 10 of the *House Wiring* 3e textbook.

Materials and Equipment

House Wiring 3e textbook, safety glasses, electrician's tool kit, one single-gang plastic nail-on device box, one 3 × 2 × 3½-inch metal device box with a side-mounting bracket, nails and screws as needed

Procedure

1. Put on safety glasses and observe all safety rules.

2. Determine the setout for the boxes you are installing. Remember that the setout will depend on the material thickness being used for the finished wall surface. Ask your instructor for the box setout requirement for this lab exercise.

   Box setout: _____ inch

3. Determine the height from the finished floor where the boxes are to be mounted and make a mark on the stud at that location. Ask your instructor for the mounting height for the boxes used in this lab exercise. See photo A in the procedure for installing device boxes in new construction in Chapter 10 of the *House Wiring* 3e textbook.

   Mounting height: _____ inches

4. Attach the nonmetallic plastic nail-on device box to a stud with the nails provided with the box. See photo B in the procedure for installing device boxes in new construction in Chapter 10 of the *House Wiring* 3e textbook.

5. Attach the metal device box with a side-mounting bracket to a stud with either nails or screws. See photo C in the procedure for installing device boxes in new construction in Chapter 10 of the *House Wiring* 3e textbook.

6. Complete the Review of this lab exercise.

7. Show all completed work to the instructor.

8. Clean up the work area and return all tools and materials to their proper locations.

9. Get the instructor to sign off upon satisfactory completion of this lab exercise.

Review

1. List three safety items to consider when installing device boxes.

_____

_____

_____

2. What was the hardest part of this lab? Why?

_____

_____

_____

_____

Lab 10.2: Installing Outlet Boxes

Name: _____ Date: _____ Score: _____

Introduction

Outlet boxes are designed to have lighting fixtures or ceiling-suspended paddle fans attached to them. This box style is usually an octagon metal box or a round nonmetallic box. In this lab exercise, you will install both nonmetallic and metal outlet boxes on framing members in the wiring mockup. Before proceeding with this lab exercise, refer to the procedure for installing outlet boxes in Chapter 10 of the *House Wiring* 3e textbook.

Materials and Equipment

House Wiring 3e textbook, safety glasses, electrician's tool kit, one plastic nail-on round ceiling box, one 4 × 1½-inch octagon metal outlet box with a side-mounting bracket, nails and screws as needed

Procedure

1. Put on safety glasses and observe all safety rules.

2. Determine the setout for the boxes you are installing. Remember that the setout will depend on the material thickness being used for the finished wall or ceiling surface. Ask your instructor for the box setout requirement for this lab exercise.

   Box setout: ___ inch(es)

3. Determine which building framing member will provide the correct location for the outlet box as shown on an electrical plan. Your wiring mockup may not have ceiling joists on which to mount the outlet boxes. If this is the case, mount them on framing studs at 6 feet to center from the floor.

4. Make a mark on the building framing member at that location using a permanent marker or a pencil. See photo A in the procedure for installing outlet boxes with a side-mounting bracket in Chapter 10 of the *House Wiring* 3e textbook.

5. Attach the metal octagon outlet box with a side-mounting bracket to the framing member with either nails or screws. See photo A in the procedure for installing outlet boxes with a side-mounting bracket in Chapter 10 of the *House Wiring* 3e textbook.

6. Attach the plastic nail-on round ceiling box to the framing member with the nails provided with the box. See photo B in the procedure for installing outlet boxes with a side-mounting bracket in Chapter 10 of the *House Wiring* 3e textbook.

7. Complete the Review of this lab exercise.

8. Show all completed work to the instructor.

9. Clean up the work area and return all tools and materials to their proper locations.

10. Get the instructor to sign off upon satisfactory completion of this lab exercise.

Review

1. List three safety items to consider when installing outlet boxes.

   _____

   _____

   _____

2. What was the hardest part of this lab? Why?

   _____

   _____

   _____

   _____

Lab 10.3: Install Outlet Boxes with an Adjustable Bar Hanger

Name: _____ Date: _____ Score: _____

Introduction

Sometimes an outlet box that is designed to be nailed or screwed directly to the side of a stud or joist cannot be located in the proper place. If this is the case, one way to position the outlet box in the right location between two building framing members is to use an adjustable bar hanger. In this lab exercise, you will install a metal outlet box with an adjustable bar hanger between two framing members in the wiring mockup. Before proceeding with this lab exercise, refer to the procedure for installing outlet boxes with an adjustable bar hanger in Chapter 10 of the *House Wiring* 3e textbook.

Materials and Equipment

House Wiring 3e textbook, safety glasses, electrician's tool kit, one adjustable bar hanger, one 4 × 1½-inch octagon metal outlet box, nails and screws as needed

Procedure

1. Put on safety glasses and observe all safety rules.

2. Determine the setout for the boxes you are installing. Remember that the setout will depend on the material thickness being used for the finished wall or ceiling surface. Ask your instructor for the box setout requirement for this lab exercise.

   Box setout: ___ inch(es)

3. Determine which building framing members will provide the correct location for the outlet box. Your wiring mockup may not have ceiling joists on which to mount the outlet box with an adjustable bar hanger. If this is the case, mount it between two framing studs at 6 feet to center from the floor.

4. Make a mark on the building framing members at that location by using a permanent marker or a pencil.

5. Attach the metal octagon outlet box to the adjustable bar hanger with the screw(s) and fitting supplied with the bar hanger. Do not completely tighten the screw(s) that hold the box to the bar at this time. This will allow the box to be adjusted back and forth until it fits where you want it between the two framing members. See photo A in the procedure for installing outlet boxes with an adjustable bar hanger in Chapter 10 of the *House Wiring* 3e textbook.

6. Adjust the bar so it fits tightly between the two framing members and align the bar hanger with the mounting position marks.

7. While making sure that the box depth will result in a flush box position with the finished surface, nail or screw the bar hanger in place. See photo B in the procedure for installing outlet boxes with an adjustable bar hanger in Chapter 10 of the *House Wiring* 3e textbook.

8. Move the box into the desired position between the framing members and tighten the screw(s) that hold the box to the hanger. See photo C in the procedure for installing outlet boxes with an adjustable bar hanger in Chapter 10 of the *House Wiring* 3e textbook.

9. Complete the Review of this lab exercise.

10. Show all completed work to the instructor.

11. Clean up the work area and return all tools and materials to their proper locations.

12. Get the instructor to sign off upon satisfactory completion of this lab exercise.

Review

1. List three safety items to consider when installing outlet boxes with an adjustable bar hanger.

_____

_____

_____

2. What was the hardest part of this lab? Why?

_____

_____

_____

_____

Lab 11.1: Installing a Cable Run from a Loadcenter

Name: _____ Date: _____ Score: _____

Introduction

No matter which cable type is used, there are certain installation requirements that will have to be followed during the installation of the cable in the rough-in stage. These requirements are outlined in the specific article for that cable type in the *National Electrical Code*®. Installing the cable in the framework of a house is often referred to as "pulling in" the cable. The cables must be routed through or along studs, joists, and rafters to all of the receptacle, switch, or lighting outlet locations that make up a residential circuit. In this lab exercise, you will install nonmetallic sheathed cable from the loadcenter in your wiring mockup to several device and outlet boxes. See Figure 11.1-1 to see where the boxes are to be mounted and where the cable needs to be installed. Before proceeding with this lab exercise, refer to the procedure for starting a cable run from a loadcenter in Chapter 11 of the *House Wiring* 3e textbook.

Materials and Equipment

House Wiring 3e textbook, safety glasses, electrician's tool kit, ½-inch drill with a ⅞-inch auger bit, 14/2 Type NM cable as needed, two single-gang plastic nail-on device boxes, one plastic round nail-on outlet box, one 4" × 1 ½" octagon box with side mounting bracket and internal cable clamps, one 3" × 2" × 3 ½" device box with side mounting bracket and internal cable clamps, one ½-inch Type NM cable connector, Type NM cable staples as needed, nails and screws as needed, nail plates as needed

Procedure

1. Put on safety glasses and observe all safety rules.

2. Mount the boxes on the studs in the locations shown in Figure 11.1-1. Use the box types shown unless your instructor tells you otherwise.

3. Drill holes for the routing of the cable as required. Remember that any hole that is closer than 1¼ inches to the edge of a framing member requires a steel nail plate to be installed that is at least ¹⁄₁₆ inch thick.

4. Remove a KO (knockout) in the loadcenter and install a cable connector that is appropriate for the size and type of cable you are using.

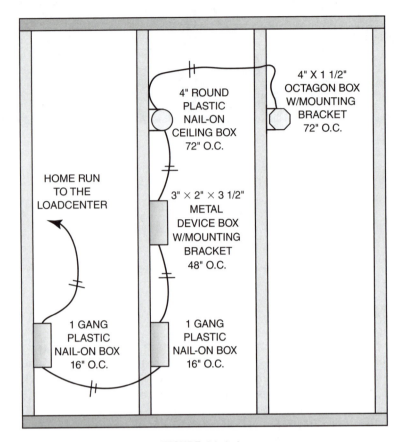

FIGURE 11.1-1

5. Place one end of the cable for the circuit you are installing through the connector and into the loadcenter. Leave enough cable in the loadcenter so the wires in the cable will have plenty of length to get to the circuit breaker, grounded neutral bar, and grounding bar connection points. It is up to each individual electrician whether the cable sheathing is stripped away at this time or during the trim-out stage of the loadcenter. The trim-out of a loadcenter is when the circuit breakers are installed and the circuit wires are properly terminated. Refer to Illustration A in the procedure for starting a cable run from a loadcenter in Chapter 11 of the *House Wiring* 3e textbook.

6. Secure the cable to the loadcenter by tightening the cable connector onto the cable.

7. Identify each circuit cable in the loadcenter with the circuit number, type, and location as it is shown on the wiring plan. This will help eliminate any confusion about which cable goes to which circuit breaker during the trim-out stage of the loadcenter. For this lab exercise, label the cable as if you are installing a branch circuit serving the master bedroom.

8. Once the cable run is started from the loadcenter, continue it along or through the framing members until you reach the first electrical box on the circuit. Secure and support the cable according to the *NEC®*.

9. At each outlet box, leave enough cable so approximately 8 inches of conductor will be available in each box for connection purposes. Remember that the Code requires at least 6 inches of free conductor in electrical boxes for connection to devices. At least 3 inches of conductor must extend from the front of each box. Refer to Illustration B in the procedure for starting a cable run from a loadcenter in Chapter 11 of the *House Wiring* 3e textbook.

Note: Some electricians install the cable into the electrical outlet box without stripping away the outside sheathing first. This technique will require the sheathing to be stripped later. Stripping the outside sheathing from a cable while it is in an electrical box can be tricky, especially with deeper boxes, and many electricians prefer to strip the outside sheathing before the cable is secured to the electrical box. There is no right or wrong way to do this. The best way to do it is the way your instructor suggests.

10. Install another cable into the electrical box and continue the cable run to the next box in the circuit. Again, be sure to leave approximately 8 inches of cable (or conductor) in the box. Refer to Illustration C in the procedure for starting a cable run from a loadcenter in Chapter 11 of the *House Wiring* 3e textbook.

11. Fold the cable (or conductors if the cable is stripped) back into the box and continue to pull the cable to the rest of the boxes in the circuit. Refer to Illustration D in the procedure for starting a cable run from a loadcenter in Chapter 11 of the *House Wiring* 3e textbook.

12. Complete the Review of this lab exercise.

13. Show all completed work to the instructor.

14. Clean up the work area and return all tools and materials to their proper locations.

15. Get the instructor to sign off upon satisfactory completion of this lab exercise.

Review

1. List three safety items to consider when installing cable.

2. What was the hardest part of this lab? Why?

Lab 12.1: Cut, Ream, and Thread Rigid Metal Conduit (RMC) by Using a Manual Threader, Reamer, and Pipe Cutter

Name: _____ Date: _____ Score: _____

Introduction

Electricians need to know how to cut to length the various conduit types used in residential wiring. Solid-length metal conduit, like rigid metal conduit, intermediate metal conduit, and electrical metallic tubing, is cut to length using a pipe cutter, a tubing cutter, a portable bandsaw, or a hacksaw. The ends of cut lengths of rigid metal conduit and intermediate metal conduit will also have to be threaded. In this lab exercise, you will cut and thread a length of ½-inch rigid metal conduit. Before proceeding with this lab exercise, review the procedure for cutting and threading in Chapter 12 of the *House Wiring* 3e textbook.

Materials and Equipment

House Wiring 3e textbook, safety glasses, gloves, electrician's tool kit, a 5-foot length of ½-inch RMC, a portable pipe vise, manual pipe threader, ½-inch thread-cutting die, cutting oil, pipe cutter, pipe reamer, metal file

Procedure

1. Observe all safety rules and be sure to wear safety glasses and gloves.

2. Secure a 5-foot length of ½-inch RMC conduit in the pipe vise and using a pipe cutter, cut off all existing threads. Refer to photo A in the procedure for cutting and threading RMC in Chapter 12 of the *House Wiring* 3e textbook.

3. Measure and clearly mark the pipe at the location you wish to make the cut. For the threading part of this lab exercise, you will need a 3-foot length of conduit. Refer to photo B in the procedure for cutting and threading RMC in Chapter 12 of the *House Wiring* 3e textbook.

4. Using a pipe cutter, cut the conduit to a 3-foot length. Refer to photo C and D in the procedure for cutting and threading RMC in Chapter 12 of the *House Wiring* 3e textbook.

5. Using a pipe reamer, ream the conduit ends. If necessary, use a file to eliminate all burrs on the inside and outside of the conduit ends. Refer to photo E in the procedure for cutting and threading RMC in Chapter 12 of the *House Wiring* 3e textbook.

6. Secure the 3-foot length of RMC in the pipe vise and thread **both** ends of the conduit using a hand threader with the proper size die. The length of the thread should be the same as the length of the cutting threads of the die. Apply a liberal amount of cutting oil during the threading process. Refer to photos F and G in the procedure for cutting and threading RMC in Chapter 12 of the *House Wiring* 3e textbook.

7. Complete the Review for this lab exercise.

8. Show all completed work to the instructor. A 3-foot length of ½-inch RMC with new threads on each end must be shown to the instructor.

9. Clean up the work area and return all tools and materials to their proper locations.

10. Get the instructor to sign off upon satisfactory completion of this lab exercise.

Review

1. Name the article in the *NEC®* that covers rigid metal conduit.

   _____

   _____

   _____

2. What does the *NEC®* have to say about reaming and threading rigid metal conduit?

   _____

   _____

   _____

3. Discuss the reasons for using cutting oil when threading rigid metal conduit.

   _____

   _____

   _____

4. List three safety items to consider when threading and cutting RMC.

   _____

   _____

   _____

Lab 12.2: Bend a 90° Stub-Up with a Hand Bender

Name: _____ Date: _____ Score: _____

Introduction

The stub-up is the most common bend that electricians have to make with a bender. Remember that the bender is marked with the "take-up" of the arc for the bender shoe. This lab exercise requires you to make two stub-ups with a rise of 14 inches. Make one bend with ½-inch EMT and the other using ¾-inch EMT. The leg length on each pipe is to be 24 inches (see Figure 12.2-1). Use a hand bender to make both bends. The height of the stub must be exact or no more than ¼ inch short of the required dimension. The leg length must be exact or no more than ¼ inch short of the required dimension. Before proceeding with this lab exercise, review the procedure for bending a 90° stub-up in Chapter 12 of the *House Wiring* 3e textbook.

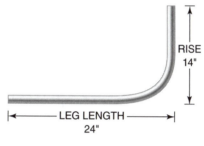

RISE
14"

|← LEG LENGTH →|
24"

FIGURE 12.2-1

Materials and Equipment

House Wiring 3e textbook, safety glasses, gloves, electrician's tool kit, hacksaw, reaming tool, pipe vise, ½-inch EMT hand bender, ¾-inch EMT hand bender, 3-foot length of ½-inch EMT, 3-foot length of ¾-inch EMT, pencil or marker

Procedure

1. Observe all safety rules and be sure to wear safety glasses and gloves.

2. Subtract the "take-up" from the finished stub height.

3. Mark this dimension clearly on the conduit. Refer to photo A in the procedure for bending a 90° stub-up in Chapter 12 of the *House Wiring* 3e textbook.

4. Place the conduit on a flat surface, such as the floor.

5. Line up the arrow on the bender with the mark on the conduit. Refer to photo B in the procedure for bending a 90° stub-up in Chapter 12 of the *House Wiring* 3e textbook.

6. Apply heavy foot pressure to the bender and bend the conduit to 90°. Refer to photo C in the procedure for bending a 90° stub-up in Chapter 12 of the *House Wiring* 3e textbook.

7. Use a torpedo level to make sure the stub-up is 90° and use a tape measure to make sure it is the correct height. Refer to photo D in the procedure for bending a 90° stub-up in Chapter 12 of the *House Wiring* 3e textbook.

8. Cut the pipe as necessary to get the dimensions required in this lab exercise. Ream all cut ends of the conduit.

9. Complete the Review for this lab exercise.

10. Show all completed work to the instructor. You should have two stub-ups to show the instructor, one with ½-inch EMT and one with ¾-inch EMT. Each will have a rise of 14 inches and a leg length of 24 inches.

11. Clean up the work area and return all tools and materials to their proper locations.

12. Get the instructor to sign off upon satisfactory completion of this lab exercise.

Review

1. Name the article of the *NEC®* that covers electrical metallic tubing.

   _____

2. Is foot pressure on the bender important? Explain.

   _____

   _____

3. How can you tell when you have bent the pipe to 90°?

   _____

4. The take-up for a ½-inch EMT bender is _____ inches and the take-up for a ¾-inch EMT bender is _____ inches.

5. What adjustment would you make if each 90° stub-up you do with a certain bender is always ¼ inch too high?

   _____

   _____

   _____

Lab 12.3: Bend a Back-to-Back Bend with a Hand Bender

Name: _____ **Date:** _____ **Score:** _____

Introduction

A back-to-back bend produces a U shape in a single length of conduit. An electrician can use the same technique for a conduit run across a floor or ceiling where the pipe turns up or down a wall. For this bend, ¾-inch EMT will be used. Leg #1 is to be 25 inches high and leg #2 is to be 30 inches high. The distance from the back of leg #1 to the back of leg #2 is 48 inches (see Figure 12.3-1). All dimensions must be within ¼ inch of the given lengths. Before proceeding with this lab exercise, review the procedure for bending a back-to-back bend in Chapter 12 of the *House Wiring* 3e textbook.

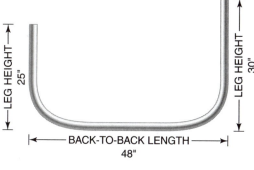

FIGURE 12.3-1

Materials and Equipment

House Wiring 3e textbook, safety glasses, gloves, electrician's tool kit, hacksaw, pipe vise, reaming tool, ¾-inch EMT bender, 10-foot length of ¾-inch EMT, pencil or marker

Procedure

1. Observe all safety rules and be sure to wear safety glasses and gloves.

2. Subtract the "take-up" from the finished stub height for leg #1.

3. Mark this dimension clearly on the conduit.

4. Place the conduit on a flat surface, such as the floor.

5. Line up the arrow on the bender with the mark on the conduit. Refer to photo A in the procedure for bending a back-to-back bend in Chapter 12 of the *House Wiring* 3e textbook.

6. Apply heavy foot pressure to the bender and bend the conduit to 90°. Refer to photo B in the procedure for bending a back-to-back bend in Chapter 12 of the *House Wiring* 3e textbook.

7. Measure to the point where the back of the second bend is to be. In this lab exercise it is 48 inches. Mark this dimension on the conduit. Refer to photo C in the procedure for bending a back-to-back bend in Chapter 12 of the *House Wiring* 3e textbook.

8. Align the mark on the conduit with the star-point and bend to 90°. Refer to photo D in the procedure for bending a back-to-back bend in Chapter 12 of the *House Wiring* 3e textbook.

9. Use a torpedo level to make sure the legs are 90° and use a tape measure to make sure they are the correct height. If necessary, cut the legs to the proper lengths with a hacksaw. Refer to photo E in the procedure for bending a back-to-back bend in Chapter 12 of the *House Wiring* 3e textbook.

10. Ream all cut ends of the conduit.

11. Complete the Review for this lab exercise.

12. Show all completed work to the instructor. You will need to show the instructor a back-to-back bend made with ¾-inch EMT having one leg 25 inches high and the other leg 30 inches high. The distance from the back of leg #1 to the back of leg #2 should be 48 inches.

13. Clean up the work area and return all tools and materials to their proper locations.

14. Get the instructor to sign off upon satisfactory completion of this lab exercise.

Review

1. Name the location on the bender head that indicates where the back of a 90° bend will be.

   _____

   _____

   _____

2. What was the hardest part for you in making a good back-to-back bend?

   _____

   _____

   _____

Lab 12.4: Bend an Offset Bend with a Hand Bender

Name: _____ **Date:** _____ **Score:** _____

Introduction

The offset bend is used when an obstruction requires a change in the conduit's plane. Before making an offset bend, you must choose the most appropriate angles for the offset. Keep in mind that shallow bends make for easier wire pulling, steeper bends conserve space. You must also consider that the conduit's overall length will shrink due to the bends you put into the pipe to form the offset. Remember to ignore shrink when working away from the obstruction, but be sure to consider it when working into it. The pipe used in this lab exercise is ½-inch EMT. The offset needs to be 4 inches high with 30° bends and have a total pipe length of 48 inches (see Figure 12.4-1). The offset must be able to clear a 4-inch obstruction without touching the object and, along with all other dimensions for this lab exercise, must be equal to or within ¼-inch greater of the given parameters. Before proceeding with this lab exercise, review the procedure for bending an offset bend in Chapter 12 of the *House Wiring* 3e textbook.

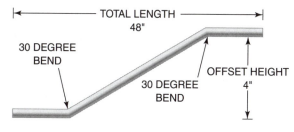

| TOTAL LENGTH 48" |
| 30 DEGREE BEND |
| 30 DEGREE BEND |
| OFFSET HEIGHT 4" |

FIGURE 12.4-1

Materials and Equipment

House Wiring 3e textbook, safety glasses, gloves, electrician's tool kit, hacksaw, pipe vise, reaming tool, ½-inch EMT bender, 5-foot length of ½-inch EMT, pencil or marker

Procedure

1. Observe all safety rules and be sure to wear safety glasses and gloves.

2. Secure a length of ½-inch EMT in a pipe vise and cut it to a 5-foot length. Refer to photo A in the procedure for bending an offset bend in Chapter 12 of the *House Wiring* 3e textbook.

3. Ream the ends of the conduit after cutting it to length. Refer to photo B in the procedure for bending an offset bend in Chapter 12 of the *House Wiring* 3e textbook.

4. Determine the center of the 5-foot length of conduit. Refer to photo C in the procedure for bending an offset bend in Chapter 12 of the *House Wiring* 3e textbook.

5. Determine the distance to be marked on the conduit. Refer to the offset bending chart shown in part D of the procedure for bending an offset bend in Chapter 12 of the *House Wiring* 3e textbook. Remember, you are using 30° angles.

6. Mark the dimensions determined in the previous step on the conduit. Do this by making a mark at an equal distance on each side of center. The distance from the center of each mark is found by dividing the dimension by two. Refer to photo E in the procedure for bending an offset bend in Chapter 12 of the *House Wiring* 3e textbook.

7. Place the conduit on a flat surface, such as the floor.

8. Line up the arrow on the bender with the first mark on the conduit.

9. Apply heavy foot pressure to the bender and bend the conduit to 30°. Note that when the bender handle is straight up and down you have a 30° bend. (This is true for most, but not necessarily all hand benders.) Refer to photo F in the procedure for bending an offset bend in Chapter 12 of the *House Wiring* 3e textbook.

10. Invert the bender, rotate the conduit 180°, and align the arrow on the next mark. Then while standing up, bend the conduit to 30°. Make sure to keep the bends in the same plane. Refer to photo G in the procedure for bending an offset bend in Chapter 12 of the *House Wiring* 3e textbook.

11. Check the offset bend for "wows" or "dog legs" and fine-tune as necessary.

12. Check to be sure that the offset amount is correct. If it is not enough or is too much, add or subtract some angle from the bends. Refer to photo H in the procedure for bending an offset bend in Chapter 12 of the *House Wiring* 3e textbook.

13. Cut the conduit to the proper length required in this lab exercise and ream all cut ends.

14. Complete the Review for this lab exercise.

15. Show all completed work to the instructor. You will need to show the instructor a 48-inch length of ½-inch EMT with a 4-inch offset.

16. Clean up the work area and return all tools and materials to their proper locations.

17. Get the instructor to sign off upon satisfactory completion of this lab exercise.

Review

1. Name the location on the bender that aligns with the marks on the pipe when making an offset bend.

_____

_____

_____

2. Give an example of where it would be necessary for an electrician to put an offset bend in a pipe.

_____

_____

_____

3. What was the hardest part for you in making a good offset bend?

_____

_____

_____

Lab 12.5: Bend a Three-Point Saddle Bend with a Hand Bender

Name: _____   Date: _____   Score: _____

Introduction

The saddle bend is similar to the offset bend, but in this case the same plane is resumed. It is used most often when another pipe is encountered that runs perpendicular to your conduit path and results in you having to go over it. Most common is a 45° center bend and two 22.5° outer bends, but you can use a 60° center bend and two 30° outer bends. In this lab exercise, you will use a 45° center bend and 22.5° outer bends. Use the same calculation for either set of angles. For this lab exercise, an electrician encounters a 3¾-inch O.D. pipe that needs to be saddled over. The conduit used in this lab exercise is ½-inch EMT. The final length for the conduit is to be 58 inches (see Figure 12.5-1). The dimension for the saddle must be equal to or within ¼ inch greater than 4 inches. All other dimensions must be equal to or within ¼ inch less than the given parameters. Before proceeding with this lab exercise, review the procedure for bending a three-point saddle in Chapter 12 of the *House Wiring* 3e textbook.

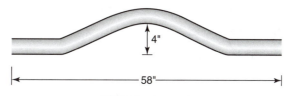

4"

58"

FIGURE 12.5-1

Materials and Equipment

House Wiring 3e textbook, safety glasses, gloves, electrician's tool kit, hacksaw, pipe vise, reaming tool, ½-inch EMT bender, 5-foot length of ½-inch EMT, pencil or marker

Procedure

1. Observe all safety rules and be sure to wear safety glasses and gloves.

2. Secure a length of ½-inch EMT in a pipe vise and cut it to a 5-foot length. Ream the conduit ends after cutting.

3. Determine the mark spacing for the 4-inch three-point saddle using the Saddle Bending Marks table in part B of the procedure for bending a three-point saddle in Chapter 12 of the *House Wiring* 3e textbook. (**Note:** The multiplier is 2.5 per inch of saddle height for all conduit sizes when bending the 45°/22.5° three-point saddle).

4. Determine the center of the three-point saddle and mark it on the conduit. In this lab exercise the center will be the center point of the 5-foot length of EMT you are working with. Refer to photo A in the procedure for bending a three-point saddle in Chapter 12 of the *House Wiring* 3e textbook.

5. Determine the two outer marks from the center mark. Mark these two dimensions on the conduit at an equal distance on each side of the center mark. Refer to photo C in the procedure for bending a three-point saddle in Chapter 12 of the *House Wiring* 3e textbook.

6. Place the conduit on a flat surface, such as the floor.

7. Line up the rim notch or saddle mark on the bender with the saddle center mark on the conduit. Refer to photo D in the procedure for bending a three-point saddle in Chapter 12 of the *House Wiring* 3e textbook.

8. While keeping the rim notch and the center mark aligned, place the conduit on the floor. Apply heavy foot pressure to the bender and bend the conduit to 45° degrees.

9. Invert the bender and rotate the conduit 180°. Refer to photo E in the procedure for bending a three-point saddle in Chapter 12 of the *House Wiring* 3e textbook.

10. Slide the conduit ahead and align the arrow with the first outer mark. Refer to photo F in the procedure for bending a three-point saddle in Chapter 12 of the *House Wiring* 3e textbook.

11. While standing up, bend the conduit to 22.5°. Refer to photo G in the procedure for bending a three-point saddle in Chapter 12 of the *House Wiring* 3e textbook.

12. Remove and reverse the conduit end for end.

13. Align the remaining outside mark at the arrow, and make another 22.5° bend. Refer to photo H in the procedure for bending a three-point saddle in Chapter 12 of the *House Wiring* 3e textbook.

14. Check the saddle bend for "wows" or "dog legs" and fine-tune as necessary. Be sure to have all bends in the same plane. Refer to photo I in the procedure for bending a three-point saddle in Chapter 12 of the *House Wiring* 3e textbook.

15. Cut off any excess conduit from each end. Be sure to ream any end that is cut.

16. Complete the Review for this lab exercise.

17. Show all completed work to the instructor. You will need to show the instructor a 58-inch length of ½-inch EMT with a 4-inch three-point saddle in the middle.

18. Clean up the work area and return all tools and materials to their proper locations.

19. Get the instructor to sign off upon satisfactory completion of this lab exercise.

Review

1. Name the location on the bender that is used to make the first bend in a three-point saddle bend.

   _____

   _____

   _____

2. Name the location on the bender that is used to make the second and third bends in a three-point saddle bend.

   _____

   _____

   _____

3. What was the hardest part for you in making a good three-point saddle?

   _____

   _____

   _____

Lab 12.6: Bend Box Offsets with a Hand Bender

Name: _____ Date: _____ Score: _____

Introduction

Box offsets are used where a piece of conduit is to be attached to an electrical box or enclosure. This bend can be accomplished through the use of a hand bender or the use of a box offset tool called the Little Kicker®. An electrician must become proficient in the use of a hand bender to make box offsets. For this lab exercise, you will have to install box offsets on each end of a ½-inch EMT pipe that fits between two utility boxes. The utility boxes will be 36 inches apart (see Figure 12.6-1). All dimensions must be equal to or within ¼ inch less than the given parameters. It is not practicable to lay out box offsets in smaller pipe sizes the same way as normal offset bends. Therefore, this lab exercise is a *suggested* way to do box offsets until you are proficient at bending them. In the case of box offsets, practice is really the only way to get good at bending them. Before proceeding with this lab exercise, review the procedure for bending box offsets in Chapter 12 of the *House Wiring* 3e textbook.

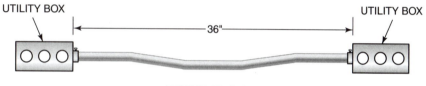

UTILITY BOX   36"   UTILITY BOX

FIGURE 12.6-1

Materials and Equipment

House Wiring 3e textbook, safety glasses, gloves, electrician's tool kit, hacksaw, pipe vise, reaming tool, ½-inch EMT bender, 5-foot length of ½-inch EMT, two utility boxes with ½-inch KOs, two ½-inch EMT set-screw connectors, pencil or marker

Procedure

1. Observe all safety rules and be sure to wear safety glasses and gloves.

2. Cut the 5-foot length of ½-inch EMT to a length of 36 inches with a hacksaw. Make sure to ream the ends.

3. Mark the conduit at 2 inches and 8 inches from each end of the pipe. Refer to photo A in the procedure for box offsets in Chapter 12 of the *House Wiring* 3e textbook.

4. Invert the bender with the handle on the floor and, placing the 2-inch mark at the arrow, make a small bend (approximately 5°). Refer to photo B in the procedure for box offsets in Chapter 12 of the *House Wiring* 3e textbook.

5. Rotate the pipe 180°, and keeping the bender in the same position, slide the pipe ahead so that the 8-inch mark now lines up with the arrow. Refer to photo C in the procedure for box offsets in Chapter 12 of the *House Wiring* 3e textbook.

6. Make a small bend that is equal to the first bend made (approximately 5°).

7. Check this offset by placing a utility box with a ½-inch set-screw connector installed on a flat surface and see if the pipe will now slide into the connector opening without hitting the sides. If it does not fit correctly, fine-tune the conduit by adding or taking away offset height until the pipe easily fits into the connector opening. Refer to photo D in the procedure for box offsets in Chapter 12 of the *House Wiring* 3e textbook.

8. Make a box offset on the other end of the pipe by following steps 4, 5, and 6. Make sure that the offsets are in the same plane and that there are no "wows" or "dog legs."

9. Check this offset by placing a utility box with a ½-inch set-screw connector installed on a flat surface and see if the pipe will now slide into the connector opening without hitting the sides. If it does not fit correctly, fine-tune the conduit by adding or taking away offset height until the pipe easily fits into the connector opening. Refer to photo D in the procedure for box offsets in Chapter 12 of the *House Wiring* 3e textbook.

10. Complete the Review for this lab exercise.

11. Show all completed work to the instructor. You will have a 36-inch length of ½-inch EMT with box offsets on each end.

12. Clean up the work area and return all tools and materials to their proper locations.

13. Get the instructor to sign off upon satisfactory completion of this lab exercise.

Review

1. What was the hardest part for you in making box offsets?

_____

_____

_____

2. Name a tool other than a hand bender that electricians can use to make box offsets in ½-inch and ¾-inch EMT.

_____

_____

_____

Lab 12.7: Bend and Install EMT as Part of a Conduit System

Name: _____ **Date:** _____ **Score:** _____

Introduction

When electricians install a conduit system, there are usually several bends that have to be made. Couplings are used to connect two lengths of conduit and conduit connectors are used to connect the conduit to electrical boxes and enclosures. Install the conduit system shown in Figure 12.7-1. Use a hand bender to make the bends. Boxes A, B, and C are 4 × 1½-inch square with ½-inch knockouts. The conduit used is ½-inch EMT. Box offsets must be made at each box. Use the knockouts as indicated. All dimensions must be equal to or within ¼ inch less than the given parameters. Before proceeding with this lab exercise, review the conduit bending information and bending procedures in Chapter 12 of the *House Wiring* 3e textbook.

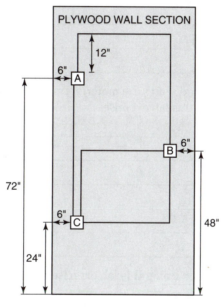

FIGURE 12.7-1

Materials and Equipment

House Wiring 3e textbook, safety glasses, gloves, electrician's tool kit, hacksaw, pipe vise, reaming tool, ½-inch EMT bender, ½-inch EMT as needed, ½-inch EMT set screw connectors as needed, three 4 × 1½-inch square boxes with ½-inch knock-outs, ½-inch one-hole straps for EMT as needed, #10 × ¾-inch sheet metal screws as needed

Procedure

1. Observe all safety rules and be sure to wear safety glasses and gloves.

2. Mount the 4-inch square boxes at the specified location and heights as shown in Figure 12.7-1. Use two screws to secure each box.

3. Cut and bend the conduit into the shapes shown in Figure 12.7-1. Remember to ream each cut end.

4. Connect the EMT conduit to the boxes using set-screw connectors.

5. Secure and support the EMT conduit as required by the *NEC®*.

6. Complete the Review for this lab exercise.

7. Show all completed work to the instructor.

8. Clean up the work area and return all tools and materials to their proper locations.

9. Get the instructor to sign off upon satisfactory completion of this lab exercise.

Review

1. Make out a complete material list for this lab. Include a description of the item used and the quantity. Be sure to include *all* items used.

_____

_____

_____

_____

_____

_____

2. Based on an hourly rate of $50.00 and using the material price list located in Appendix B, calculate the cost of doing the electrical work in this lab exercise.

_____

_____

_____

_____

_____

_____

3. What was the hardest part for you in doing this lab exercise?

_____

_____

_____

Chapter 13   Switching Circuit Installation

Lab 13.1: Install a Single-Pole Switch for a Lighting Load with the Power Source Feeding the Switch

Name: _____   Date: _____   Score: _____

Introduction

The most often used switching arrangement is one where a lighting fixture is to be controlled from one location. Usually the electrical feed is brought to the electrical box housing the switch and the proper connections are made for control of the light fixture. This switching circuit is one that all electricians must know. Before proceeding with this lab exercise, review the information on single-pole switching circuit 1 in Chapter 13 of the *House Wiring* 3e textbook.

Materials and Equipment

House Wiring 3e textbook, safety glasses, electrician's tool kit, power cord made in Lab 3.8, 14/2 Type NM as needed, one plastic single-gang nail-on device box, one plastic nail-on 4-inch round ceiling box, one 15-amp 120-volt single-pole switch, cable staples, grounding screw, keyless lamp holder, 60-watt lamp, wirenuts as needed, 6-32 and 8-32 machine screws as needed

Procedure

1. Put on safety glasses and observe all safety rules.

2. On the wiring mock-up, mount the electrical boxes as shown in Figure 13.1-1. When securing the boxes, hammer the nails in only far enough to securely fasten the box to the stud. It will be easier to take off the box if the nails are not all of the way in. Assume that ½-inch wallboard will be installed later. Refer to the procedures for installing electrical boxes in new construction in Chapter 10 of the *House Wiring* 3e textbook.

3. Install 14/2 Type NM cable between the two boxes and secure with staples as required by the *NEC*®. Follow the installation practices as outlined in Chapters 9 and 11 of the *House Wiring* 3e textbook.

4. Install the power cord into the switch box and secure with staples as required by the *NEC*®. Install the power cord as if it were Type NM cable.

5. Connect the conductors to the devices according to Figure 13-9 in the *House Wiring* 3e textbook. Remember to ground all metal electrical boxes. When making the conductor connections, review the following procedures in the textbook:

   – Connecting wires together with a wirenut in Chapter 2.

   – Using terminal loops to connect circuit conductors to terminal screws on receptacles or switches in Chapter 18.

   – Installing receptacles (or switches) in a nonmetallic electrical box in Chapter 18.

   – Installing receptacles (or switches) in a metal electrical box in Chapter 18.

   – Installation steps for installing a light fixture directly to an outlet box in Chapter 17.

6. Show all completed work to the instructor. Do not secure the switch or the lighting fixture to the electrical boxes until the instructor has reviewed your work to this point.

7. Once the instructor has approved your work, secure the devices to the boxes and install a light bulb in the lamp holder. Do not install switch covers unless your instructor tells you to.

8. Test the circuit by plugging in the power cord and activating the switch. The lamp should come on. If it does not, disconnect the power, lock it out, and troubleshoot the circuit.

9. Complete the Review for this lab exercise.

10. Show all completed work to the instructor.

11. Clean up the work area and return all tools and materials to their proper locations.

12. Get the instructor to sign off upon satisfactory completion of this lab exercise.

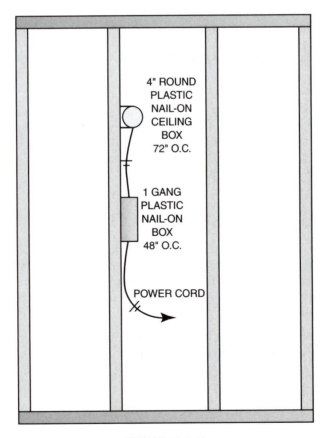

FIGURE 13.1-1

Review

1. Complete the wiring diagram shown below.

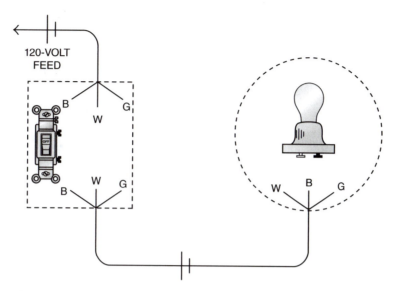

2. Using symbols from Chapter 5 in the *House Wiring* 3e textbook, draw a cabling diagram for the circuit wired in this lab exercise.

Lab 13.2: Install a Single-Pole Switch for a Lighting Load with the Power Source Feeding the Lighting Outlet (Switch Loop)

Name: _____ Date: _____ Score: _____

Introduction

There are times when it is easier to feed the lighting outlet first and use a switch loop to control the lighting outlet. A white insulated wire in a cable wiring method used in a single-pole switch loop is permitted in Section 200.7(C)(2) of the *NEC®* but must be reidentified. Before proceeding with this lab exercise, review the information on single-pole switching circuit 2 in Chapter 13 of the *House Wiring* 3e textbook.

Materials and Equipment

House Wiring 3e textbook, safety glasses, electrician's tool kit, power cord made in Lab 3.8, 14/2 Type NM as needed, one plastic single-gang nail-on device box, one plastic nail-on 4-inch round ceiling box, one 15-amp 120-volt single-pole switch, cable staples, grounding screw, keyless lamp holder, 60-watt lamp, wirenuts as needed, 6-32 and 8-32 machine screws as needed, black electrical tape

Procedure

1. Put on safety glasses and observe all safety rules.

   Note: You are allowed to use the same box setup used in Lab 13.1 and skip to step 4. If the setup for Lab 13.1 has been taken down, follow the steps as outlined below.

2. On the wiring mock-up, mount the electrical boxes as shown in Figure 13.2-1. When securing the boxes, hammer the nails in only far enough to securely fasten the box to the stud. It will be easier to take off the box if the nails are not all of the way in. Assume that ½-inch wallboard will be installed later. Refer to the procedures for installing electrical boxes in new construction in Chapter 10 of the *House Wiring* 3e textbook.

3. Install 14/2 Type NM cable between the two boxes and secure with staples as required by the *NEC®*. Follow the installation practices as outlined in Chapters 9 and 11 of the *House Wiring* 3e textbook.

4. Install the power cord into the lighting outlet box and secure with staples as required by the *NEC®*. Install the power cord as if it were Type NM cable.

5. Connect the conductors to the devices according to Figure 13-10 in the *House Wiring* 3e textbook. Remember to ground all metal electrical boxes. When making the conductor connections, review the following procedures in the textbook:

   – Connecting wires together with a wirenut in Chapter 2.

   – Using terminal loops to connect circuit conductors to terminal screws on receptacles or switches in Chapter 18.

   – Installing receptacles (or switches) in a nonmetallic electrical box in Chapter 18.

– Installing receptacles (or switches) in a metal electrical box in Chapter 18.

– Installation steps for installing a light fixture directly to an outlet box in Chapter 17.

6. Show all completed work to the instructor. Do not secure the switch or the lighting fixture to the electrical boxes until the instructor has reviewed your work to this point.

7. Once the instructor has approved your work, secure the devices to the boxes and install a light bulb in the lamp holder. Do not install switch covers unless your instructor tells you to.

8. Test the circuit by plugging in the power cord and activating the switch. The lamp should come on. If it does not, disconnect the power, lock it out, and troubleshoot the circuit.

9. Complete the Review for this lab exercise.

10. Show all completed work to the instructor.

11. Clean up the work area and return all tools and materials to their proper locations.

12. Get the instructor to sign off upon satisfactory completion of this lab exercise.

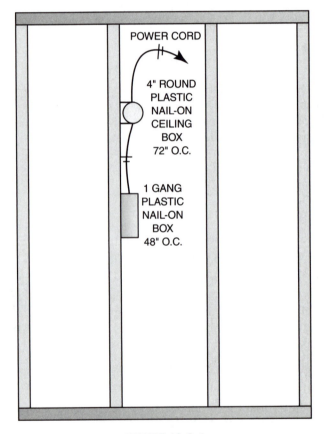

FIGURE 13.2-1

Review

1. Complete the wiring diagram shown below.

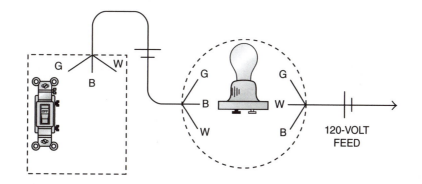

2. Using symbols from Chapter 5 in the *House Wiring* 3e textbook, draw a cabling diagram for the circuit wired in this lab exercise.

Lab 13.3: Install a Three-Way Switching Circuit with the Power Source Feeding the First Three-Way Switch Location

Name: _____ Date: _____ Score: _____

Introduction

The need for three-way switches occurs in rooms that have entry and exit from more than one location. An example would be a living room with a doorway to a kitchen as well as a doorway to the outside of the house. Another example would be controlling a light or lights from the top and bottom of a stairway. Before proceeding with this lab exercise, review the information on three-way switching circuit 1 in Chapter 13 of the *House Wiring* 3e textbook.

Materials and Equipment

House Wiring 3e textbook, safety glasses, electrician's tool kit, ½-inch drill with a 7/8-inch auger bit, power cord made in Lab 3.8, 14/2 Type NM as needed, 14/3 Type NM as needed, one 4-inch round plastic nail-on ceiling box, two single-gang plastic nail-on device boxes, two 15-amp 120-volt three-way switches, staples, grounding screws, one keyless lamp holder, one 60-watt lamp, wirenuts as needed, 6-32 and 8-32 machine screws as needed, nail plates as needed

Procedure

1. Put on safety glasses and observe all safety rules.

2. On the wiring mock-up, mount the electrical boxes as shown in Figure 13.3-1. When securing the boxes, hammer the nails in only far enough to securely fasten the box to the stud. It will be easier to take off the box if the nails are not all of the way in. Assume that ½-inch wallboard will be installed later. Refer to the procedure for installing electrical boxes in new construction in Chapter 10 of the *House Wiring* 3e textbook.

3. Install 14/2 Type NM cable between the 4-inch plastic ceiling box and one of the single-gang plastic device boxes as shown in Figure 13.3-1 and secure with staples as required by the *NEC®*. Follow the installation practices for cables as outlined in Chapters 9 and 11 of the *House Wiring* 3e textbook.

4. Drill the mock-up framing members where needed, install 14/3 Type NM cable between the two single-gang plastic device boxes as shown in Figure 13.3-1 and secure with staples as required by the *NEC®*. Follow the installation practices for cables as outlined in Chapters 9 and 11 of the *House Wiring* 3e textbook.

5. Install the power cord into the single-gang plastic device box that has only the 14/3 Type NM cable and secure it with staples as required by the *NEC®*. Install the power cord as if it were Type NM cable.

6. Connect the conductors to the devices according to Figure 13-13 in the *House Wiring* 3e textbook. Remember to ground all metal electrical boxes. When making the conductor connections, review the following procedures in the textbook:

   – Connecting wires together with a wirenut in Chapter 2.

– Using terminal loops to connect circuit conductors to terminal screws on receptacles or switches in Chapter 18.

– Installing receptacles (or switches) in a nonmetallic electrical box in Chapter 18.

– Installing receptacles (or switches) in a metal electrical box in Chapter 18.

– Installation steps for installing a light fixture directly to an outlet box in Chapter 17.

7. Show all completed work to the instructor. Do not secure the switch or the lighting fixture to the electrical boxes until the instructor has reviewed your work to this point.

8. Once the instructor has approved your work, secure the devices to the boxes and install a light bulb in the lamp holder. Do not install switch covers unless your instructor tells you to.

9. Test the circuit by plugging in the power cord and following the procedure on testing a standard three-way switching arrangement in Chapter 20 of the *House Wiring* 3e textbook. The lamp should be able to be controlled from both switching locations. If not, disconnect the power, lock it out, and troubleshoot the circuit.

10. Complete the Review for this lab exercise.

11. Show all completed work to the instructor.

12. Clean up the work area and return all tools and materials to their proper locations.

13. Get the instructor to sign off upon satisfactory completion of this lab exercise.

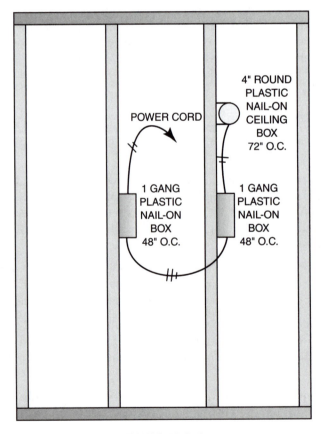

FIGURE 13.3-1

Review

1. Complete the wiring diagram shown below.

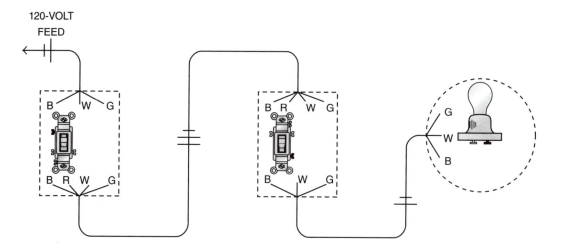

2. Using symbols from Chapter 5 in the *House Wiring* 3e textbook, draw a cabling diagram for the circuit wired in this lab exercise.

Lab 13.4: Install a Three-Way Switching Circuit with the Power Source Feeding the Lighting Outlet Location

Name: _____ Date: _____ Score: _____

Introduction

The need for three-way switches occurs in rooms that have entry and exit from more than one location. An example would be a living room with a doorway to a kitchen as well as a doorway to the outside of the house. Another example would be controlling a light or lights from the top and bottom of a stairway. Sometimes it is necessary to feed a lighting outlet instead of a switch. The wiring arrangement is similar to a single-pole switch loop. Before proceeding with this lab exercise, review the information on three-way switching circuit 2 in Chapter 13 of the *House Wiring* 3e textbook.

Materials and Equipment

House Wiring 3e textbook, safety glasses, electrician's tool kit, ½-inch drill with a 7⁄8-inch auger bit, power cord made in Lab 3.8, 14/2 Type NM as needed, 14/3 Type NM as needed, one plastic 4-inch round nail-on ceiling box, two single-gang plastic nail-on device boxes, two 15-amp 120-volt three-way switches, staples, grounding screws, one keyless lamp holder, one 60-watt lamp, wirenuts as needed, 6-32 and 8-32 machine screws as needed, nail plates as needed

Procedure

1. Put on safety glasses and observe all safety rules.

   Note: You are allowed to use the same box setup used in Lab 13.3 and skip to step 5. If the setup for Lab 13.3 has been taken down, follow all of the steps below.

2. On the wiring mock-up, mount the electrical boxes as shown in Figure 13.4-1. When securing the boxes, hammer the nails in only far enough to securely fasten the box to the stud. It will be easier to take off the box if the nails are not all of the way in. Assume that ½-inch wallboard will be installed later. Refer to the procedure for installing device boxes in new construction in Chapter 10 of the *House Wiring* 3e textbook.

3. Install 14/2 Type NM cable between the 4-inch plastic ceiling box and one of the single-gang plastic device boxes as shown in Figure 13.4-1 and secure with staples as required by the *NEC®*. Follow the installation practices for cables as outlined in Chapters 9 and 11 of the *House Wiring* 3e textbook.

4. Drilling the mock-up framing members where needed, install 14/3 Type NM cable between the two single-gang plastic device boxes as shown in Figure 13.4-1 and secure with staples as required by the *NEC®*. Follow the installation practices for cables as outlined in Chapters 9 and 11 of the *House Wiring* 3e textbook.

5. Install the power cord into the ceiling box and secure with staples as required. Install the power cord as if it were Type NM cable.

6. Connect the conductors to the devices according to Figure 13-14 in the *House Wiring* 3e textbook. Remember to ground all metal electrical boxes. When making the conductor connections, review the following procedures in the textbook:

   – Connecting wires together with a wirenut in Chapter 2.

   – Using terminal loops to connect circuit conductors to terminal screws on receptacles or switches in Chapter 18.

   – Installing receptacles (or switches) in a nonmetallic electrical box in Chapter 18.

   – Installing receptacles (or switches) in a metal electrical box in Chapter 18.

   – Installation steps for installing a light fixture directly to an outlet box in Chapter 17.

7. Show all completed work to the instructor. Do not secure the switch or the lighting fixture to the electrical boxes until the instructor has reviewed your work to this point.

8. Once the instructor has approved your work, secure the devices to the boxes and install a light bulb in the lamp holder. Do not install switch covers unless your instructor tells you to.

9. Test the circuit by plugging in the power cord and following the procedure on testing a standard three-way switching arrangement in Chapter 20 of the *House Wiring* 3e textbook. The lamp should be able to be controlled from both switching locations. If not, disconnect the power and troubleshoot the circuit.

10. Complete the Review for this lab exercise.

11. Show all completed work to the instructor.

12. Clean up the work area and return all tools and materials to their proper locations.

13. Get the instructor to sign off upon satisfactory completion of this lab exercise.

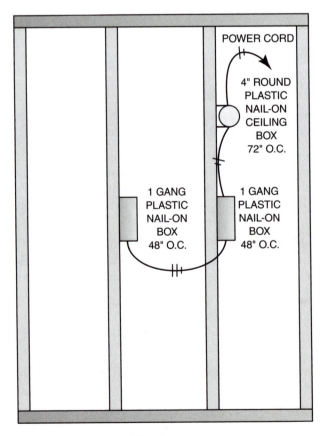

FIGURE 13.4-1

Review

1. Complete the wiring diagram shown below.

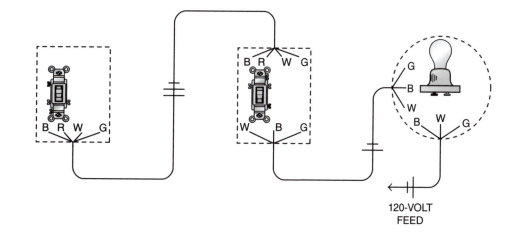

120-VOLT
FEED

2. Using symbols from Chapter 5 in the *House Wiring* 3e textbook, draw a cabling diagram for the circuit wired in this lab exercise.

Lab 13.5: Install a Three-Way Switching Circuit with the Power Source Feeding the Lighting Outlet Location and with Three-Wire Cable Run from the Lighting Outlet to Each Three-Way Switch Location

Name: _____ Date: _____ Score: _____

Introduction

The need for three-way switches occurs in rooms that have entry and exit from more than one location. An example would be a living room with a doorway to a kitchen as well as a doorway to the outside of the house. Another example would be controlling a light or lights from the top and bottom of a stairway. Sometimes it is necessary to feed a lighting outlet instead of a switch. However, in this situation the three-wire cable is run from the lighting outlet box to each switch box. Care must be taken to make the proper connections in the lighting outlet box. Before proceeding with this lab exercise, review the information on three-way switching circuit 3 in Chapter 13 of the *House Wiring* 3e textbook.

Materials and Equipment

House Wiring 3e textbook, safety glasses, electrician's tool kit, ½-inch drill with a 7⁄8-inch auger bit, power cord made in Lab 3.8, 14/2 Type NM as needed, 14/3 Type NM as needed, one plastic 4-inch ceiling box, two single-gang plastic nail-on device boxes, two 15-amp 120-volt three-way switches, staples, one keyless lamp holder, one 60-watt lamp, wirenuts as needed, 6-32 and 8-32 machine screws as needed, nail plates as needed

Procedure

1. Put on safety glasses and observe all safety rules.

2. On the wiring mock-up, mount the electrical boxes as shown in Figure 13.5-1. When securing the boxes, hammer the nails in only far enough to securely fasten the box to the stud. It will be easier to take the boxes off if the nails are not all of the way in. Assume that ½-inch wallboard will be installed later on this wall. Refer to the procedures for installing electrical boxes in new construction in Chapter 10 of the *House Wiring* 3e textbook.

3. Drilling the mock-up framing members where needed, install 14/3 Type NM cable from the 4-inch round plastic ceiling box to each of the single-gang plastic nail-on device boxes and secure with staples as required by the *NEC*®. Follow the installation practices for cables as outlined in Chapters 9 and 11 of the *House Wiring* 3e textbook.

4. Install the power cord into the 4-inch round plastic ceiling box and secure with staples as required by the *NEC*®. Install the power cord as if it were Type NM cable.

5. Connect the conductors to the devices according to Figure 13-15 in the *House Wiring* 3e textbook. Remember to ground all metal electrical boxes. When making the conductor connections, review the following procedures in the textbook:

   – Connecting wires together with a wirenut in Chapter 2.

   – Using terminal loops to connect circuit conductors to terminal screws on receptacles or switches in Chapter 18.

   – Installing receptacles (or switches) in a nonmetallic electrical box in Chapter 18.

   – Installing receptacles (or switches) in a metal electrical box in Chapter 18.

   – Installation steps for installing a light fixture directly to an outlet box in Chapter 17.

6. Show all completed work to the instructor. Do not secure the switch or the lighting fixture to the electrical boxes until the instructor has reviewed your work to this point.

7. Once the instructor has approved your work, secure the devices to the boxes and install a light bulb in the lamp holder. Do not install switch covers unless your instructor tells you to.

8. Test the circuit by plugging in the power cord and following the procedure on testing a standard three-way switching arrangement in Chapter 20 of the *House Wiring* 3e textbook. The lamp should be able to be controlled from both switching locations. If not, disconnect the power, lock it out, and troubleshoot the circuit.

9. Complete the Review for this lab exercise.

10. Show all completed work to the instructor.

11. Clean up the work area and return all tools and materials to their proper locations.

12. Get the instructor to sign off upon satisfactory completion of this lab exercise.

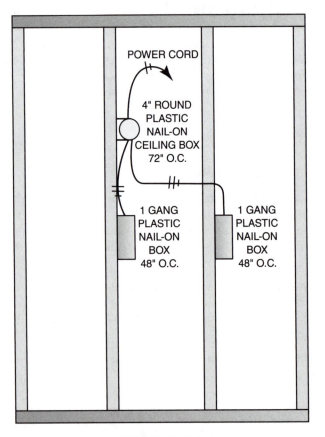

FIGURE 13.5-1

Review

1. Complete the wiring diagram shown below.

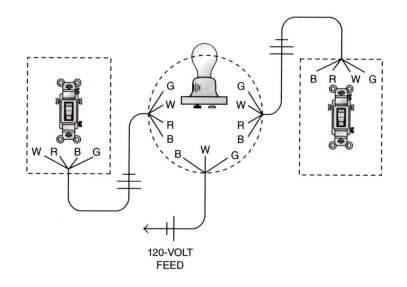

120-VOLT
FEED

2. Using symbols from Chapter 5 in the *House Wiring* 3e textbook, draw a cabling diagram for the circuit wired in this lab exercise.

Lab 13.6: Install a Four-Way Switching Circuit with the Power Source Feeding the First Three-Way Switch Location

Name: _____ Date: _____ Score: _____

Introduction

Four-way switches are used when a lighting load must be controlled from more than two switching points. To accomplish this, three-way switches are connected to the source and to the load. The switches at *all* other control points must be four-way switches. An example for an application of four-way switches might be a living room with an outside entrance, a doorway to a hall, and another doorway to a kitchen area. Switch control of a lighting outlet is usually installed at all three of these door locations. Before proceeding with this lab exercise, review the information on four-way switching circuit 1 in Chapter 13 of the *House Wiring* 3e textbook.

Materials and Equipment

House Wiring 3e textbook, safety glasses, electrician's tool kit, ½-inch drill with a 7/8-inch auger bit, power cord made in Lab 3.8, 14/2 Type NM as needed, 14/3 Type NM as needed, one 4-inch round plastic ceiling box, two single-gang plastic nail-on device boxes, one 4 × 1½-inch square box with mounting bracket and internal cable clamps, one 4 × ½-inch square single-gang raised plaster ring, two 15-amp 120-volt three-way switches, one 15-amp 120-volt four-way switch, staples as needed, one keyless lamp holder, one 60-watt lamp, wirenuts as needed, 6-32 and 8-32 machine screws as needed, nail plates as needed

Procedure

1. Put on safety glasses and observe all safety rules.

2. On the wiring mock-up, mount the 4-inch square box with side-mounting bracket as shown in Figure 13.6-1. Use staples or screws to secure the box. They will be easier to remove than nails, which would be used in the field. Assume that ½-inch wallboard will be installed later on this wall. Refer to the procedure for installing outlet boxes with a side-mounting bracket in new construction in Chapter 10 of the *House Wiring* 3e textbook.

3. Mount the two plastic nail-on single-gang switch boxes and the 4-inch round plastic ceiling box as shown in Figure 13.6-1. Hammer the nails in only far enough to securely fasten the boxes to the studs. It will be easier to take the boxes off if the nails are not all of the way in. Assume that ½-inch wallboard will be installed later on this wall. Refer to the procedure for installing device boxes in new construction in Chapter 10 of the *House Wiring* 3e textbook.

4. Drilling the mock-up framing members where needed, install 14/3 Type NM cable between each of the plastic device boxes and the 4-inch square box. Secure the cables with staples as required by the *NEC*®. Follow the installation practices for cables as outlined in Chapters 9 and 11 of the *House Wiring* 3e textbook.

5. Drilling the mock-up framing members where needed, install 14/2 Type NM cable between the *right*-hand switch box and the lighting outlet box. Secure with staples as required by the *NEC*®. Follow the installation practices for cables as outlined in Chapters 9 and 11 of the *House Wiring* 3e textbook.

6. Install the power cord into the *left*-hand device box and secure with staples as required by the *NEC*®. Install the power cord as if it were Type NM cable.

7. Connect the conductors to the devices according to Figure 13-17 in the *House Wiring* 3e textbook. Remember to ground all metal electrical boxes. When making the conductor connections, review the following procedures in the textbook:

   – Connecting wires together with a wirenut in Chapter 2.

   – Using terminal loops to connect circuit conductors to terminal screws on receptacles or switches in Chapter 18.

   – Installing receptacles (or switches) in a nonmetallic electrical box in Chapter 18.

   – Installing receptacles (or switches) in a metal electrical box in Chapter 18.

   – Installation steps for installing a light fixture directly to an outlet box in Chapter 17.

8. Show all completed work to the instructor. Do not secure the switch or the lighting fixture to the electrical boxes until the instructor has reviewed your work to this point.

9. Once the instructor has approved your work, secure the devices to the boxes and install a light bulb in the lamp holder. Do not install switch covers unless your instructor tells you to.

10. Test the circuit by plugging in the power cord and toggling any switch. The lamp should come on. (**Note:** The lamp may already be on when you energize the circuit based upon the toggle position of the installed switches.) Toggle another switch and the lamp should go off. Now toggle the third switch. The lamp should come on again. If the lamp does not function properly, disconnect the power, lock it out, and troubleshoot the circuit.

11. Complete the Review for this lab exercise.

12. Show all completed work to the instructor.

13. Clean up the work area and return all tools and materials to their proper locations.

14. Get the instructor to sign off upon satisfactory completion of this lab exercise.

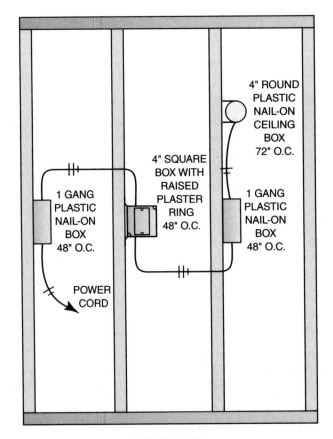

FIGURE 13.6-1

Review

1. Complete the wiring diagram shown below.

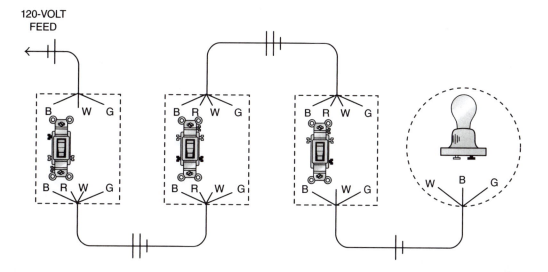

2. Using symbols from Chapter 5 in the *House Wiring* 3e textbook, draw a cabling diagram for the circuit wired in this lab exercise.

Lab 13.7: Install a Four-Way Switching Circuit with the Power Source Feeding the Lighting Outlet Location

Name: _____ Date: _____ Score: _____

Introduction

Four-way switches are used when a lighting load must be controlled from more than two switching points. To accomplish this, three-way switches are connected to the source and to the load. The switches at *all* other control points must be four-way switches. An example for an application of four-way switches might be a living room with an outside entrance, a doorway to a hall, and another doorway to a kitchen area. Switch control of a lighting outlet is usually installed at all three of these door locations. In Lab 13.6 the power source was brought to the first three-way switch in the circuit. In this lab exercise, the power source will be brought to the lighting outlet location. This is considered a "switch loop" wiring installation. Before proceeding with this lab exercise, review the information on four-way switching circuit 2 in Chapter 13 of the *House Wiring* 3e textbook.

Materials and Equipment

House Wiring 3e textbook, safety glasses, electrician's tool kit, ½-inch drill with a 7/8-inch auger bit, power cord made in Lab 3.8, 14/2 Type NM as needed, 14/3 Type NM as needed, one 4 -inch round plastic ceiling box, two single-gang plastic nail-on device boxes, one 4 × 1½-inch square box with mounting bracket and internal cable clamps, one 4 × ½-inch square single-gang raised plaster ring, two 15-amp 120-volt three-way switches, one 15-amp 120-volt four-way switch, staples as needed, one keyless lamp holder, one 60-watt lamp, wirenuts as needed, 6-32 and 8-32 machine screws as needed, nail plates as needed

Procedure

1. Put on safety glasses and observe all safety rules.

   Note: You are allowed to use the same box setup used in Lab 13.6 and skip to step 6. If the setup for Lab 13.6 has been taken down, follow all of the steps below.

2. On the wiring mock-up, mount the 4-inch square box with side-mounting bracket as shown in Figure 13.7-1. Use staples or screws to secure the box. They will be easier to remove than nails, which would be used in the field. Assume that ½-inch wallboard will be installed later on this wall. Refer to the procedure for installing device boxes with a side-mounting bracket in new construction in Chapter 10 of the *House Wiring* 3e textbook.

3. Mount the two plastic nail-on single-gang switch boxes and the 4-inch round plastic ceiling box as shown in Figure 13.7-1. Hammer the nails in only far enough to securely fasten the box to the stud. It will be easier to take the boxes off if the nails are not all of the way in. Assume that ½-inch wallboard will be installed later on this wall. Refer to the procedure for installing device boxes in new construction in Chapter 10 of the *House Wiring* 3e textbook.

4. Drilling the mock-up framing members where needed, install 14/3 Type NM cable between <u>each</u> of the plastic device boxes <u>and</u> the 4-inch square box. Secure the cables with staples as required by the *NEC®*. Follow the installation practices for cables as outlined in Chapters 9 and 11 of the *House Wiring* 3e textbook.

5. Drilling the mock-up framing members where needed, install 14/2 Type NM cable between the right-hand switch box and the lighting outlet box. Secure with staples as required by the *NEC®*. Follow the installation practices for cables as outlined in Chapters 9 and 11 of the *House Wiring* 3e textbook.

6. Install the power cord into the lighting outlet box and secure with staples as required. Install the power cord as if it were Type NM cable.

7. Connect the conductors to the devices according to Figure 13-18 in the *House Wiring* 3e textbook. Remember to ground all metal electrical boxes. When making the conductor connections, review the following procedures in the textbook:

   – Connecting wires together with a wirenut in Chapter 2.

   – Using terminal loops to connect circuit conductors to terminal screws on receptacles or switches in Chapter 18.

   – Installing receptacles (or switches) in a nonmetallic electrical box in Chapter 18.

   – Installing receptacles (or switches) in a metal electrical box in Chapter 18.

   – Installation steps for installing a light fixture directly to an outlet box in Chapter 17.

8. Show all completed work to the instructor. Do not secure the switch or the lighting fixture to the electrical boxes until the instructor has reviewed your work to this point.

9. Once the instructor has approved your work, secure the devices to the boxes and install a light bulb in the lamp holder. Do not install switch covers unless your instructor tells you to.

10. Test the circuit by plugging in the power cord and toggling any switch. The lamp should come on. (**Note:** The lamp may already be on when you energize the circuit based upon the toggle position of the installed switches.) Toggle another switch and the lamp should go off. Now toggle the third switch. The lamp should now come on again. If the lamp does not function properly, disconnect the power, lock it out, and troubleshoot the circuit.

11. Complete the Review for this lab exercise.

12. Show all completed work to the instructor.

13. Clean up the work area and return all tools and materials to their proper locations.

14. Get the instructor to sign off upon satisfactory completion of this lab exercise.

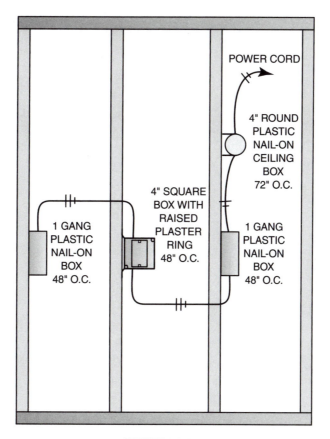

FIGURE 13.7-1

Review

1. Complete the wiring diagram shown below.

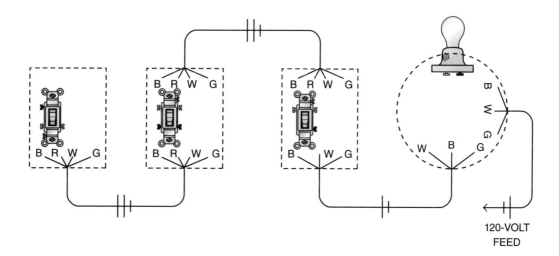

2. Using symbols from Chapter 5 in the *House Wiring* 3e textbook, draw a cabling diagram for the circuit wired in this lab exercise.

Lab 13.8: Install Split-Duplex Receptacles so That the Bottom Half Is Controlled by a Single-Pole Switch and the Top Half Is "Hot" at All Times

Name: _____ Date: _____ Score: _____

Introduction

In many new homes the designer has elected not to have actual lighting fixtures installed in rooms such as a bedroom or a living room. Instead, split-duplex receptacles will be used so that the bottom half of the duplex receptacle is switched and the top half is "hot" at all times. This will allow for table or floor lamps to be plugged into the bottom half of the receptacle and controlled from a switch located at the entrance or exit to the room. The top half of the receptacle will allow for items such as a TV or stereo to be plugged in and have power available at all times, regardless of the switch position. Electricians must be familiar with the proper wiring practices associated with switching split-duplex receptacles. In this lab exercise, you will wire a circuit so that a single-pole switch will control the bottom half of three split-duplex receptacles while the top half of the receptacles will be "hot" at all times. Before proceeding with this lab exercise, review the information on switched duplex receptacle circuit 3 in Chapter 13 of the *House Wiring* 3e textbook.

Materials and Equipment

House Wiring 3e textbook, safety glasses, electrician's tool kit, ½-inch drill with a 7⁄8-inch auger bit, power cord made in Lab 3.8, 14/3 Type NM as needed, staples as needed, wirenuts as needed, three 15-amp 125-volt split-duplex receptacles, one 15-amp 120-volt single-pole switch, four plastic nail-on single-gang device boxes, 6-32 and 8-32 machine screws as needed, nail plates as needed

Procedure

1. Put on safety glasses and observe all safety rules.

2. On the wiring mock-up, mount four single-gang plastic nail-on boxes as shown in Figure 13.8-1. Assume the use of ½-inch wallboard on this wall. Refer to the procedure for installing device boxes in new construction in Chapter 10 of the *House Wiring* 3e textbook.

3. Drilling the mock-up framing members where needed, install 14/3 Type NM cable between all of the boxes. Secure the cables with staples as required by the *NEC®*. Follow the installation practices as outlined in Chapters 9 and 11 of the *House Wiring* 3e textbook.

4. Drilling the mock-up framing members where needed, install the power cord into the single-gang plastic nail-on device box that is located 48 inches to center from the floor and secure with staples as required by the *NEC®*. Install the power cord as if it were Type NM cable.

5. Connect the conductors to the devices according to Figure 13-21 in the *House Wiring* 3e textbook. Remember to ground all metal electrical boxes. When making the conductor connections, review the following procedures in the textbook:

   – Connecting wires together with a wirenut in Chapter 2.

   – Using terminal loops to connect circuit conductors to terminal screws on receptacles or switches in Chapter 18.

   – Installing receptacles (or switches) in a nonmetallic electrical box in Chapter 18.

   – Installing receptacles (or switches) in a metal electrical box in Chapter 18.

   – Installation steps for installing a light fixture directly to an outlet box in Chapter 17.

6. Show all completed work to the instructor. Do not secure the switch or the lighting fixture to the electrical boxes until the instructor has reviewed your work to this point.

7. Once the instructor has approved your work, secure the devices to the boxes. Do not install switch and receptacle covers unless your instructor tells you to.

8. Test the circuit by plugging in the power cord. Using a voltage tester, check the top half of each receptacle for proper polarity and grounding. Activate the switch to power the bottom half of the split receptacles. If a problem is found, disconnect the power, lock it out, and troubleshoot the circuit.

9. Complete the Review for this lab exercise.

10. Show all completed work to the instructor.

11. Clean up the work area and return all tools and materials to their proper locations.

12. Get the instructor to sign off upon satisfactory completion of this lab exercise.

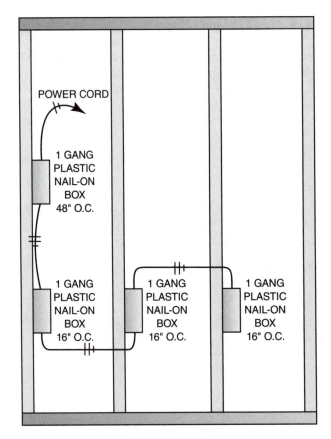

FIGURE 13.8-1

Review

1. Complete the wiring diagram shown below.

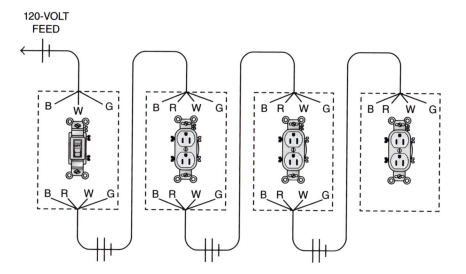

2. Using symbols from Chapter 5 in the *House Wiring* 3e textbook, draw a cabling diagram for the circuit wired in this lab exercise.

Lab 13.9: Install a Single-Pole Switch Controlling a Lighting Load with a Continuously "Hot" Receptacle Located Downstream of the Lighting Outlet—The Power Source Will Feed the Switch

Name: _____   Date: _____   Score: _____

Introduction

There are several applications for single-pole switches. An interesting wiring arrangement is one where the feed comes to a single-pole switch, which controls a lighting outlet, and there is a need for a continuously energized duplex receptacle downstream from the lighting outlet. Before proceeding with this lab exercise, review the information on switched duplex receptacle circuit 4 in Chapter 13 of the *House Wiring* 3e textbook.

Materials and Equipment

House Wiring 3e textbook, safety glasses, electrician's tool kit, ½-inch drill with a 7⁄8-inch auger bit, power cord made in Lab 3.8, 14/2 Type NM as needed, 14/3 Type NM as needed, one plastic nail-on ceiling box, one 3 × 2 × 3½-inch metal device box with clamps and mounting bracket, one single-gang plastic nail-on device box, one 15-amp 120-volt single-pole switch, one 15-amp 125-volt duplex receptacle, staples, grounding screw, keyless lamp holder, 60-watt lamp, wirenuts as needed, 6-32 and 8-32 machine screws as needed, nail plates as needed

Procedure

1. Put on safety glasses and observe all safety rules.

2. On the wiring mock-up, mount the 3 × 2 × 3½-inch device box as shown in Figure 13.9-1. Assume that ½-inch wallboard will be installed later. Use staples or screws to secure the box. They will be easier to remove than nails, which would be used in the field. Refer to the procedure for installing device boxes in new construction in Chapter 10 of the *House Wiring* 3e textbook.

3. Mount the 4-inch plastic nail-on ceiling box at 72 inches to center from the floor as shown in Figure 13.9-1. Refer to the procedure for installing outlet boxes in new construction in Chapter 10 of the *House Wiring* 3e textbook.

4. Mount the single-gang plastic nail-on device box as shown in Figure 13.9-1. Refer to the procedure for installing device boxes in new construction in Chapter 10 of the *House Wiring* 3e textbook.

5. Drilling the mock-up framing members where needed, install 14/2 Type NM cable between the 4-inch round plastic nail-on ceiling box and the plastic single-gang nail-on box and secure with staples as required by the *NEC®*. Follow the installation practices as outlined in Chapters 9 and 11 of the *House Wiring* 3e textbook.

6. Drilling the mock-up framing members where needed, install 14/3 Type NM cable between the 4-inch round plastic nail-on ceiling box and the 3 × 2 × 3½-inch metal device box and secure with staples as required by the *NEC®*. Follow the installation practices as outlined in Chapters 9 and 11 of the *House Wiring* 3e textbook.

7. Install the power cord into the 3 × 2 × 3½-inch metal device box and secure with staples as required by the *NEC*®. Install the power cord as if it were Type NM cable.

8. Connect the conductors to the devices according to Figure 13-22 in the *House Wiring* 3e textbook. Remember to ground all metal electrical boxes. When making the conductor connections, review the following procedures in the textbook:

   – Connecting wires together with a wirenut in Chapter 2.

   – Using terminal loops to connect circuit conductors to terminal screws on receptacles or switches in Chapter 18.

   – Installing receptacles (or switches) in a nonmetallic electrical box in Chapter 18.

   – Installing receptacles (or switches) in a metal electrical box in Chapter 18.

   – Installation steps for installing a light fixture directly to an outlet box in Chapter 17.

9. Show all completed work to the instructor. Do not secure the switch or the lighting fixture to the electrical boxes until the instructor has reviewed your work to this point.

10. Once the instructor has approved your work, secure the devices to the boxes and install a light bulb in the lamp holder. Do not install switch and receptacle covers unless your instructor tells you to.

11. Test the circuit by plugging in the power cord and activating the switch. The lamp should come on. If it does not, disconnect power, lock it out, and troubleshoot the circuit. The receptacle should be "hot" at all times regardless of the position of the switch.

12. Complete the Review for this lab exercise.

13. Show all completed work to the instructor.

14. Clean up the work area and return all tools and materials to their proper locations.

15. Get the instructor to sign off upon satisfactory completion of this lab exercise.

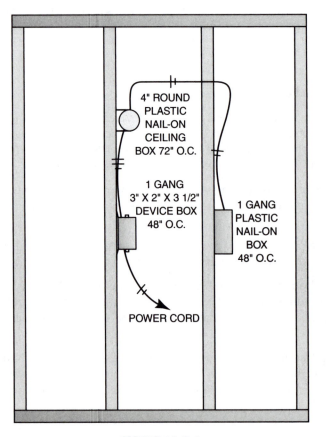

FIGURE 13.9-1

Review

1. Complete the wiring diagram shown below.

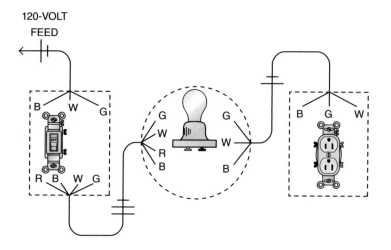

2. Using symbols from Chapter 5 in the *House Wiring* 3e textbook, draw a cabling diagram for the circuit wired in this lab exercise.

Chapter 14 Branch-Circuit Installation

Lab 14.1: Install a General Lighting Branch Circuit

Name: _____ Date: _____ Score: _____

Introduction

A general lighting circuit is a branch circuit that has both lighting and receptacle loads connected to it. A good example of this circuit type would be a bedroom branch circuit that has several receptacles and a ceiling-mounted lighting fixture controlled by a single-pole switch connected to it. This type of circuit makes up the majority of the branch circuits found in residential wiring. Electricians must properly install single and duplex receptacles when installing a residential general lighting circuit. It is very important for the proper polarity to be observed when connecting conductors to the receptacles—that the silver terminals (long slot) have the grounded conductor(s) (white) attached to them and the brass terminals (short slot) have the ungrounded (hot) conductor(s) (typically black or red) attached to them. It is equally important to observe proper grounding procedures when installing devices such as receptacles and switches. Failure to ground the device properly can result in a safety hazard that could prove fatal.

According to Section 210.12 of the *NEC*®, branch circuits supplying 120-volt, 15- and 20-amp outlets in family rooms, dining rooms, living rooms, parlors, libraries, dens, bedrooms, sunrooms, recreation rooms, closets, hallways, and similar areas of a house must be arc fault circuit interrupter (AFCI) protected with a combination AFCI device. AFCI devices are designed to trip when they sense rapid fluctuations in the current flow that is typical of arcing conditions. They are set up to recognize the "signature" of dangerous arcs and trip the circuit off when one occurs. AFCIs can distinguish between dangerous arcs and the operational arcs that occur when a plug is inserted or removed from a receptacle or a switch is turned on or off. An AFCI circuit breaker is installed in the service panel to protect the entire circuit. The *NEC*® is clear that the objective is to provide protection of the entire branch circuit. This means that for an electrician to comply with Section 210.12, an AFCI circuit breaker will need to be used so that the entire branch circuit, from the overcurrent protective device to the last outlet installed on the circuit, is AFCI protected. AFCI circuit breakers look very similar to GFCI circuit breakers. The AFCI breaker has a Push-to-Test button but it is typically a different color than the Push-to-Test button on a GFCI breaker.

Section 406.12 of the 2011 *NEC*® requires receptacles in all areas specified in Section 210.52 that are of the non-locking-type and rated at 125-volts, 15 and 20 amps to be tamper-resistant. That would include receptacles located on most general lighting branch circuits.

In this lab exercise, a homerun will be run from the circuit breaker panel to the first box in the circuit. Two duplex receptacles are in the circuit and are "hot" at all times. A single-pole switch controls the one lighting outlet on the circuit. Before proceeding with this lab exercise, review the information on installing a general lighting branch circuit in Chapter 14 of the *House Wiring* 3e textbook.

Materials and Equipment

House Wiring 3e textbook, safety glasses, electrician's tool kit, voltage tester, ½-inch drill with a 7/8-inch auger bit, step ladder, one 15-amp 120-volt AFCI single-pole circuit breaker, 14/2 Type NM cable as needed, staples as needed, wirenuts as needed, grounding screws as needed, 6-32 and 8-32 machine screws as needed, cable connectors as needed, two 15-amp 125-volt tamper-resistant duplex receptacles, one 15-amp 120-volt single-pole switch, one single-gang plastic nail-on device box, one 4-inch round plastic ceiling box, one 3 × 2 × 3½-inch device box with internal clamps and mounting bracket, one 3 × 2 × 2¾-inch device box with internal clamps and mounting bracket, one 60-watt lamp, one keyless lamp holder, nail plates as needed

Procedure

1. Put on safety glasses and observe all safety rules.

2. On the studded wiring mock-up, mount the electrical boxes as shown in Figure 14.1-1. Assume the use of ½-inch wallboard on this wall.

3. Drill the framing members as needed and install 14/2 Type NM cable from the circuit breaker panel to the single-gang plastic nail-on box. Secure and support the cable with staples as required by the *NEC*®. Use nail plates if necessary.

4. Drill the framing members as needed and install 14/2 Type NM cable between all of the other boxes in the circuit. Secure and support the cables with staples as required by the *NEC*®. Use nail plates if necessary.

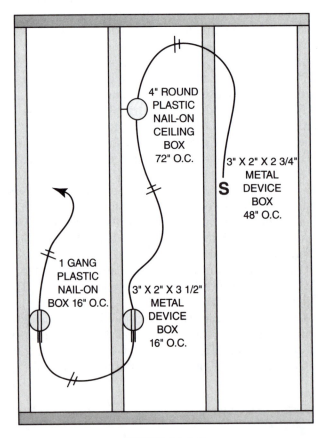

FIGURE 14.1-1

5. Connect the conductors to the devices in the boxes so that the duplex receptacles are "hot" at all times and the single-pole switch controls the lighting outlet. Use Chapter 13 and Chapter 18 in the *House Wiring* 3e textbook for help if necessary. When making the conductor connections, review the following procedures:

   – Connecting wires together with a wirenut in Chapter 2 of the textbook.

   – Using terminal loops to connect circuit conductors to terminal screws on receptacles or switches in Chapter 18 of the textbook.

   – Installing receptacles (or switches) in a nonmetallic electrical box in Chapter 18 of the textbook.

   – Installing receptacles (or switches) in a metal electrical box in Chapter 18 of the textbook.

   – Installation steps for installing a light fixture directly to an outlet box in Chapter 17 of the textbook.

6. Secure the devices to the boxes and install a 60-watt lamp in the lamp holder. Do not install switch or receptacle covers unless your instructor tells you to.

7. Install a 15-amp AFCI circuit breaker in the mock-up loadcenter and make the proper connections for the homerun wiring. Refer to the procedure on installing a single-pole AFCI circuit breaker in Chapter 19 of the *House Wiring* 3e textbook.

8. Test the circuit by:

   • Unlocking the electrical power supply and turning on the circuit breaker.

   • Follow the procedure on testing 120-volt receptacles with a voltage tester in Chapter 20 of the *House Wiring* 3e textbook and check each receptacle for proper voltage, polarity, and grounding.

   • Toggle the switch on and off and observe if the light fixture is working correctly.

   • If a problem is found, disconnect the power using a lock-out/tag-out procedure like the one in Chapter 1 of the *House Wiring* 3e textbook, and troubleshoot the circuit.

9. Show all completed work to the instructor.

10. Lock out the power supply, clean up the work area, and return all tools and materials to their proper locations.

11. Complete the Review for this lab exercise.

12. Get the instructor to sign off upon satisfactory completion of this lab exercise.

Review

1. Make out a complete material list for this lab. Include a description of the item used and the quantity. Be sure to include *all* items used.

_____

_____

_____

_____

_____

_____

2. Based on an hourly rate of $50.00 and using the material price list located in Appendix B, calculate the cost of doing the electrical work in this lab exercise.

_____

_____

_____

_____

_____

_____

Lab 14.2: Install a Small-Appliance Branch Circuit

Name: _____ **Date:** _____ **Score:** _____

Introduction

The *NEC*® requires a minimum of two small-appliance branch circuits in each house. They are 20-amp-rated circuits and are normally wired with 12 AWG conductors. They feed the kitchen and dining room areas. Receptacles that serve the kitchen countertop must be GFCI protected. A GFCI is a device intended for the protection of personnel that functions to de-energize a whole circuit or portion of a circuit within an established period of time when a current to ground exceeds some predetermined value that is less than that required to operate the overcurrent device of the supply circuit. A feed-through GFCI receptacle will provide GFCI protection for its own outlet and several other receptacle outlets connected downstream. A GFCI circuit breaker is used to provide ground fault protection for a complete circuit. It is installed in the circuit breaker panel where the homerun of the circuit to be protected is terminated.

Section 406.12 of the 2011 *NEC*® requires receptacles in all areas specified in Section 210.52 that are of the non-locking-type and rated at 125-volts, 15 and 20 amps to be tamper-resistant. This includes GFCI receptacles. However, tamper-resistant GFCI receptacles will not be required in this lab exercise because tamper-resistant receptacles are not required in kitchens.

In this lab exercise, you will use both a GFCI circuit breaker and a feed-through GFCI receptacle to provide GFCI protection to several duplex receptacles. Before proceeding with this lab exercise, review the information on installing a small-appliance branch circuit in Chapter 14 of the *House Wiring* 3e textbook.

Materials and Equipment

House Wiring 3e textbook, safety glasses, electrician's tool kit, voltage tester, ½-inch drill with a 7/8-inch auger bit, four plastic nail-on single-gang boxes, 12/2 Type NM cable as needed, four 15-amp 125-volt duplex receptacles, staples as needed, wirenuts as needed, ground screws as needed, one 15-amp 125-volt feed-through GFCI duplex receptacle, 6-32 and 8-32 machine screws as needed, one 20-amp single-pole GFCI circuit breaker, one 20-amp single-pole regular circuit breaker, nail plates as needed

Procedure

1. Put on safety glasses and observe all safety rules.

2. On the wiring mock-up, mount the electrical boxes as shown in Figure 14.2-1. Assume the use of ½-inch wallboard on this wall.

3. Drill the framing members as needed and install a 12/2 Type NM cable homerun from the circuit breaker panel to the single-gang plastic nail-on box indicated in Figure 14.2-1. Secure and support the cable with staples as required by the *NEC*®. Use nail plates if necessary.

4. Drill the framing members as needed and install 12/2 Type NM cable between all four device boxes. Secure and support the cables according to the *NEC*®. Use nail plates if necessary.

5. Install a GFCI feed-through *receptacle* in the first box on the left and connect it so that the other three receptacle outlets are GFCI protected. Refer to the procedure for installing feed-through GFCI duplex receptacles in Chapter 18 of the *House Wiring* 3e textbook. Do not install a cover.

6. Install *regular* duplex receptacles in the other three boxes. Refer to the procedures for installing duplex receptacles in Chapter 18 of the *House Wiring* 3e textbook. Do not install covers.

7. Install a 20-amp single-pole circuit breaker in the circuit breaker panel and make the proper connections for the homerun wiring. Refer to the procedure for installing a single-pole circuit breaker in Chapter 19 of the *House Wiring* 3e textbook.

8. Test the circuit by:

   • Unlocking the electrical power supply and turning on the circuit breaker.

   • Push the Reset button on the GFCI receptacle.

   • Follow the procedure for testing 120-volt receptacles with a voltage tester in Chapter 20 of the *House Wiring* 3e textbook. Check each receptacle for proper voltage, polarity, and grounding. **Note:** When using a voltage tester to check for voltage from the short (ungrounded) slot to the U-shape slot (ground), the GFCI will trip. This will verify that the GFCI is working correctly. Reset the GFCI receptacle and continue to check each receptacle. The GFCI will not trip if you use a digital multimeter for this test.

   • If a problem is found, disconnect the power using a lock-out/tag-out procedure like the one in Chapter 1 of the *House Wiring* 3e textbook, and troubleshoot the circuit.

9. Show all completed work to the instructor.

10. De-energize the circuit and lock it out using a lock-out/tag-out procedure like the one in Chapter 1 of the *House Wiring* 3e textbook.

11. In the first box on the left, replace the GFCI feed-through receptacle with a *regular* duplex receptacle. Do not install a cover.

12. In the circuit breaker panel, replace the 20-amp single-pole circuit breaker with a *20-amp GFCI circuit breaker*. Follow the procedure for installing a GFCI circuit breaker in Chapter 19 of the *House Wiring* 3e textbook.

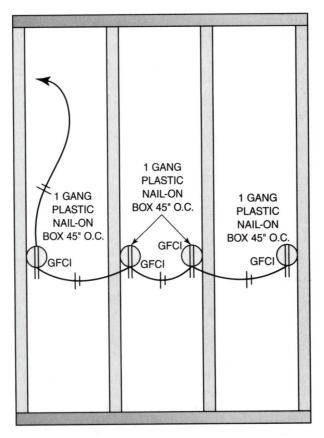

FIGURE 14.2-1

13. Test the circuit by:

   - Unlocking the electrical power supply and turning on the GFCI circuit breaker.

   - Push the Reset button on the GFCI circuit breaker.

   - Follow the procedure for testing 120-volt receptacles with a voltage tester in Chapter 20 of the *House Wiring* 3e textbook. Check each receptacle for proper voltage, polarity, and grounding. **Note:** When using a voltage tester to check for voltage from the short (ungrounded) slot to the U-shape slot (ground), the GFCI will trip. This will verify that the GFCI is working correctly. Reset the GFCI circuit breaker and continue to check each receptacle. The GFCI will not trip if you use a digital multimeter for this test.

   - If a problem is found, disconnect the power using a lock-out/tag-out procedure like the one in Chapter 1 of the *House Wiring* 3e textbook, and troubleshoot the circuit.

14. Show all completed work to the instructor.

15. Lock out the power supply, clean up the work area, and return all tools and materials to their proper locations.

16. Complete the Review for this lab exercise.

17. Get the instructor to sign off upon satisfactory completion of this lab exercise.

Review

1. Make out a complete material list for this lab. Include a description of the item used and the quantity. Be sure to include *all* items used.

_____

_____

_____

_____

_____

_____

2. Based on an hourly rate of $50.00 and using the material price list located in Appendix B, calculate the cost of doing the electrical work in this lab exercise.

_____

_____

_____

_____

_____

_____

Lab 14.3: Install an Electric Range Branch Circuit

Name: _____ Date: _____ Score: _____

Introduction

Some circuits are intended to feed only one piece of equipment. If the piece of equipment is to be cord-and-plug connected, a special receptacle is required. A stationary appliance, such as an electric range, requires large amounts of current and is connected to receptacles that are designed specifically for the amperage and voltage that this appliance needs to operate. Special receptacles are available in a flush-mount style and a surface-mount style. Flush mount is used in new construction and requires first mounting an electrical box and then installing the flush-mount receptacle in the box. It gets its name because it is "flush," or even, with the finished wall when the installation is complete. Surface-mount receptacles are not attached to an electrical box. It is a self-contained piece of equipment and is installed by attaching a back plate to the floor or wall surface, connecting the wiring to the proper terminals, and securing a plastic cover over the installation. It can be used in new construction but is most often used in remodel work when an electrical box is not easily installed in a wall. An electric range requires a heavy-duty, 50-amp, 250-volt rated receptacle and attachment plug for its installation. The receptacle and plug have special letter designations that you should know. The letter G indicates the location of the equipment grounding conductor. The letter W indicates the location of the white grounded conductor. The letters X and Y indicate the location of the ungrounded conductors. In this lab exercise, you will install both a flush-mounted range receptacle and a surface-mounted range receptacle. Before proceeding with this lab exercise, review the information on installing an electric range branch circuit in Chapter 14 of the *House Wiring* 3e textbook.

Materials and Equipment

House Wiring 3e textbook, safety glasses, electrician's tool kit, voltage tester, ½-inch drill with a 7/8-inch auger bit, 8/3 Type NM cable as needed, ¾-inch cable connectors as needed, one 4 × 2⅛-inch square box with mounting bracket and ½-inch and ¾-inch knockouts, one 4 × ½-inch two-gang raised plaster ring, one 50-amp 125/250-volt flush-mount receptacle, one 50-amp 125/250-volt surface-mount receptacle, one 40-amp two-pole circuit breaker, staples as needed, #10 × ¾-inch sheet metal screws as needed, 6-32 and 8-32 machine screws as needed, nail plates as needed

Procedure

1. Put on safety glasses and observe all safety rules.

2. Install the square box at 48 inches to center on a wiring mock-up stud as indicated in Figure 14.3-1. Assume that ½-inch wallboard will be used on the walls.

3. Drill the framing members as needed and install an 8/3 Type NM cable homerun from the circuit breaker panel to the square box. Secure and support the cable according to the *NEC*®. Make sure to leave enough free conductor for connections in the circuit breaker panel and the square box. Use nail plates if necessary.

4. Install a two-gang raised plaster ring on the square box and then make the connections to the *flush-mount* range receptacle. Secure the receptacle to the box. Do not install a receptacle cover. Refer to Figure 14-3 in the *House Wiring* 3e textbook.

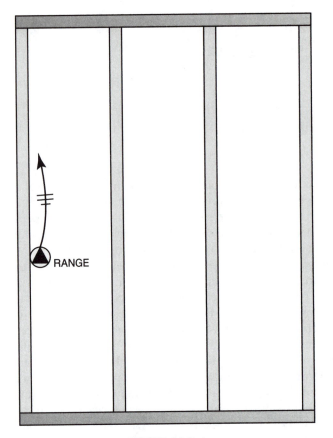

FIGURE 14.3-1

5. Install a 40-amp double-pole circuit breaker in the panel and connect the "homerun" conductors to it. Refer to the procedure for installing a two-pole circuit breaker for a 120/240-volt branch circuit in Chapter 19 of the *House Wiring* 3e textbook.

6. Test the circuit by:

    • Unlocking the electrical power supply and turning on the circuit breaker.

    • Follow the procedure for testing 120/240-volt range and dryer receptacles with a voltage tester in Chapter 20 of the *House Wiring* 3e textbook.

    • If a problem is found, disconnect the power using a lock-out/tag-out procedure like the one in Chapter 1 of the *House Wiring* 3e textbook, and troubleshoot the circuit.

7. Show all completed work to the instructor.

8. De-energize the branch circuit and lock out the power supply using a lock-out/tag-out procedure like the one in Chapter 1 of the *House Wiring* 3e textbook.

9. Remove the *flush-mount* receptacle and Type NM cable from the square box. Take the square box off the stud but leave the homerun wiring.

10. Install a *surface-mount* range receptacle on the plywood backing of the wiring mock-up or on the face of a wiring mock-up stud in approximately the same location where the square box was.

11. Attach the cable to the *surface-mount* receptacle and connect the wires. Refer to Figure 14-3 in the *House Wiring* 3e textbook.

12. Test the circuit by:

- Unlocking the electrical power supply and turning on the circuit breaker.

- Follow the procedure for testing 120/240-volt range and dryer receptacles with a voltage tester in Chapter 20 of the *House Wiring* 3e textbook.

- If a problem is found, disconnect the power using a lock-out/tag-out procedure like the one in Chapter 1 of the *House Wiring* 3e textbook, and troubleshoot the circuit.

13. Show all completed work to the instructor.

14. Lock out the power supply, clean up the work area, and return all tools and materials to their proper locations.

15. Complete the Review of this lab exercise.

16. Get the instructor to sign off upon satisfactory completion of this lab exercise.

Review

1. Make out a complete material list for this lab exercise. Include a description of the item used and the quantity. Be sure to include *all* items used.

_____

_____

_____

_____

_____

_____

2. Based on an hourly rate of $50.00 and using the material price list located in Appendix B, calculate the cost of doing the electrical work in this lab exercise.

_____

_____

_____

_____

_____

_____

Lab 14.4: Install an Electric Clothes Dryer Branch Circuit

Name: _____ **Date:** _____ **Score:** _____

Introduction

A stationary appliance, such as an electric clothes dryer, requires large amounts of current and is connected to receptacles that are designed specifically for the amperage and voltage that this appliance needs to operate. Special receptacles are available in a flush-mount style and a surface-mount style. Flush mount is used in new construction and requires first mounting an electrical box and then installing the flush-mount receptacle in the box. It gets its name because it is flush or even with the finished wall when the installation is complete. A surface-mount receptacle is not attached to an electrical box. It is a self-contained piece of equipment and is installed by attaching a back plate to the floor or wall surface, connecting the wiring to the proper terminals, and securing a plastic cover over the installation. It can be used in new construction but is most often used in remodel work when an electrical box is not easily installed in a wall. An electric dryer requires a heavy-duty 30-amp, 250-volt rated receptacle and attachment plug. The dryer receptacle and plug have special letter designations that you should know. The letter G indicates the location of the equipment grounding conductor. The letter W indicates the location of the white grounded conductor. The letters X and Y indicate the location of the ungrounded conductors. In this lab exercise, you will install both a flush-mounted dryer receptacle and a surface-mounted dryer receptacle. Before proceeding with this lab exercise, review the information on installing an electric clothes dryer branch circuit in Chapter 14 of the *House Wiring* 3e textbook.

Materials and Equipment

House Wiring 3e textbook, safety glasses, electrician's tool kit, voltage tester, ½-inch drill with a 7/8-inch auger bit, 10/3 Type NM cable as needed, ¾-inch cable connectors as needed, ½-inch cable connectors as needed, one two-gang plastic nail-on device box, one 30-amp 125/250-volt flush-mount receptacle, one 30-amp 125/250-volt surface-mount receptacle, one 30-amp two-pole circuit breaker, staples as needed, #10 × ¾-inch sheet metal screws as needed, 6-32 and 8-32 machine screws as needed, nail plates as needed

Procedure

1. Put on safety glasses and observe all safety rules.

2. Install the two-gang plastic nail-on device box at 48 inches to center on a wiring mock-up stud as indicated in Figure 14.4-1. Assume that ½-inch wallboard will be used on the walls.

3. Drill the framing members as needed and install a 10/3 Type NM cable homerun from the circuit breaker panel to the device box. Secure and support the cable according to the *NEC®*. Make sure to leave enough free conductor for connections in the circuit breaker panel and the device box. Use nail plates if necessary.

4. Install the *flush-mount* dryer receptacle and secure it to the device box. Do not install a receptacle cover. Refer to Figure 14-17 in the *House Wiring* 3e textbook.

5. Install a 30-amp double-pole circuit breaker in the circuit breaker panel and connect the homerun conductors to it. Refer to the procedure for installing a two-pole circuit breaker for a 120/240-volt branch circuit in Chapter 19 of the *House Wiring* 3e textbook.

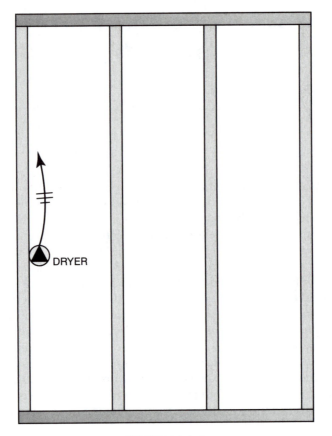

FIGURE 14.4-1

6. Test the circuit by:

   - Unlocking the electrical power supply and turning on the circuit breaker.

   - Follow the procedure for testing 120/240-volt range and dryer receptacles with a voltage tester in Chapter 20 of the *House Wiring* 3e textbook.

   - If a problem is found, disconnect the power using a lock-out/tag-out procedure like the one in Chapter 1 of the *House Wiring* 3e textbook, and troubleshoot the circuit.

7. Show all completed work to the instructor.

8. De-energize the branch circuit and lock out the power supply using a lock-out/tag-out procedure like the one in Chapter 1 of the *House Wiring* 3e textbook.

9. Remove the *flush-mount* receptacle and Type NM cable from the device box. Take the device box off the stud but leave the homerun wiring.

10. Install a *surface-mount* dryer receptacle on the plywood backing of the wiring mock-up or on the face of a wiring mock-up stud in approximately the same location where the two-gang device box was.

11. Attach the cable to the *surface-mount* receptacle and connect the wires. Refer to Figure 14-17 in the *House Wiring* 3e textbook.

12. Test the circuit by:

    - Unlocking the electrical power supply and turning on the circuit breaker.

    - Follow the procedure for testing 120/240-volt range and dryer receptacles with a voltage tester in Chapter 20 of the *House Wiring* 3e textbook.

    - If a problem is found, disconnect the power using a lock-out/tag-out procedure like the one in Chapter 1 of the *House Wiring* 3e textbook, and troubleshoot the circuit.

13. Show all completed work to the instructor.

14. Lock out the power supply, clean up the work area, and return all tools and materials to their proper locations.

15. Complete the Review of this lab exercise.

16. Get the instructor to sign off upon satisfactory completion of this lab exercise.

Review

1. Make out a complete material list for this lab. Include a description of the item used and the quantity. Be sure to include *all* items used.

_____

_____

_____

_____

_____

_____

2. Based on an hourly rate of $50.00 and using the material price list located in Appendix B, calculate the cost of doing the electrical work in this lab exercise.

_____

_____

_____

_____

_____

_____

Lab 14.5: Install an Electric Baseboard Heating Branch Circuit

Name: _____ Date: _____ Score: _____

Introduction

There are many different styles of electric heating available to a home owner and electricians should be familiar with the most common types and installation techniques. Electric furnaces used to heat an entire house, individually controlled baseboard electric heaters used to heat specific rooms, and unit heaters that are used to heat a specific area (such as a basement or garage) are the most common types installed. Electric heat has many advantages. It is very convenient to control because thermostats are easily installed in each room. It is considered safer than using fossil fuels such as oil or gas because there is no storage of fuel on the premises and the explosion and fire hazard is certainly not as great. Electric heat is very quiet because there are few, if any, moving parts. It is relatively inexpensive and easy to install. Once it is installed, there is little maintenance required to keep it working. And one last advantage to consider is that no chimney is required to exhaust the combustion gasses into the outside air. This results in a more economical heating installation because the home owner does not have to pay for a chimney installation, as well as an environmentally cleaner heating system. You would think that with all of the advantages of electric heat that the result would be electric heating installations in all homes. However, one big disadvantage of heating with electricity, the relatively high cost of electricity, severely limits the electric heating installations done by electricians. In this lab exercise, you will install electric baseboard heating controlled by a line voltage thermostat. Before proceeding with this lab exercise, review the information on installing an electric heating branch circuit in Chapter 14 of the *House Wiring* 3e textbook.

Materials and Equipment

House Wiring 3e textbook, safety glasses, electrician's tool kit, ½-inch drill with a 7/8-inch auger bit, 12/2 Type NM cable as needed, one 20-amp two-pole circuit breaker, one two-pole line voltage thermostat, one single-gang plastic nail-on wall box, one 4-foot section of electric baseboard heat, cable connectors as needed, wirenuts as needed, fasteners as needed, staples as needed, 6-32 and 8-32 machine screws as needed, nail plates as needed

Procedure

1. Put on safety glasses and observe all safety rules.

2. Install the single-gang plastic nail-on box at 60 inches to center from the floor at the location shown in Figure 14.5-1.

3. Drill the framing members as needed and install the 12/2 Type NM homerun from the circuit breaker panel to the thermostat box location. Secure and support as needed, making sure to leave enough free conductor for connections in the panel and at the device box. Use nail plates if necessary.

4. Drill the framing members as needed and install 12/2 Type NM cable from the thermostat box to the location of the built-in junction box on the left end of the heating unit. Secure and support the cable as required by the *NEC®*. Leave enough cable for making connections in the baseboard heater's junction box. Use nail plates if necessary.

5. Install a 4-foot section of electric baseboard heat on the studded wall section of the wiring mock-up, as shown in Figure 14.5-1. Keep the bottom of the heating unit 1 inch above the floor. This will allow the installation of carpeting or other finish floor material.

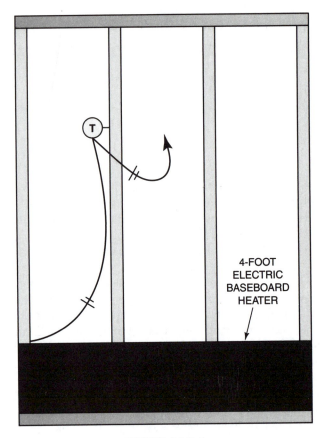

FIGURE 14.5-1

6. Attach the 12/2 Type NM cable to the baseboard heater's junction box with a cable connector or the factory-supplied internal cable clamps. Make sure to leave enough free conductor in the junction box for the required connections.

7. Referring to Figure 14-30 in the *House Wiring* 3e textbook, connect the thermostat and secure it to the device box in the proper manner. Be sure to reidentify the white circuit conductors as "hot" by using black electrical tape.

8. Referring to Figure 14-30 in the *House Wiring* 3e textbook, connect the wires in the heater junction box to the feed wires from the thermostat making sure to mark the white conductor as a "hot" conductor by using black tape.

9. Referring to the procedure for installing a two-pole circuit breaker for a 240-volt branch circuit in Chapter 19 of the *House Wiring* 3e textbook, install a 20-amp two-pole circuit breaker in the panel and make the required connections. Remember that the electric baseboard heater unit requires 240 volts, not 120/240 volts.

10. Test the circuit by:

    - Unlocking the electrical power supply and turning on the circuit breaker.

    - Turn the thermostat on and adjust it until you hear a click. After a short time, the electric baseboard heater will begin to heat up.

    - If a problem is found, disconnect the power using a lock-out/tag-out procedure like the one in Chapter 1 of the *House Wiring* 3e textbook, and troubleshoot the circuit.

11. Show all completed work to the instructor.

12. Lock out the power supply, clean up the work area, and return all tools and materials to their proper locations.

13. Complete the Review of this lab exercise.

14. Get the instructor to sign off upon satisfactory completion of this lab exercise.

Review

1. Make out a complete material list for this lab. Include a description of the item used and the quantity. Be sure to include *all* items used.

   _____

   _____

   _____

   _____

   _____

2. Based on an hourly rate of $50.00 and using the material price list located in Appendix B, calculate the cost of doing the electrical work in this lab exercise.

   _____

   _____

   _____

   _____

   _____

   _____

Lab 14.6: Install a Smoke Detector Branch Circuit

Name: _____ Date: _____ Score: _____

Introduction

Because fire in homes is the third leading cause of accidental death, some type of fire alarm system is required in all new home construction. The National Fire Protection Association (NFPA) publishes the National Fire Alarm Code, called NFPA 72. NFPA 72 outlines the minimum requirements for the selection, installation, and maintenance of fire alarm equipment. Chapter 2 of NFPA 72 covers residential fire alarm systems. It defines a household fire alarm system as a system of devices that produces an alarm signal in the house for the purpose of notifying the occupants of the house of a fire so that they will evacuate the house. NFPA 72 is usually adopted by the building codes that must be followed when building a house, and as such, must be followed by electricians when installing a residential electrical system. The most common fire warning device used in a house is a smoke detector. When wiring the smoke detector circuit for a new house, a detector unit should be placed in each bedroom and in the area just outside of the bedroom areas. They also should be installed on each level of a house. The smoke detectors must be installed in new residential construction so that when one detector is operated; all other detectors in the house will also operate. This is called "interconnecting." When wiring smoke detectors, electricians will often run a two-wire cable (or two conductors in a raceway) to the first smoke detector location. A three-wire cable (or three wires in a raceway) is then run to each of the other smoke detector locations. The black and white conductors in the circuit are used to provide 120 volts to each smoke detector. The red conductor (or third wire in a raceway) is used as the interconnection between all of the smoke detectors. Each smoke detector has a yellow wire that is used to connect to the red wire in a three-wire cable. The maximum number of interconnected smoke detectors varies by manufacturer but a good rule of thumb is to not connect more than 10 smoke detectors on circuit. In this lab exercise, you will wire the circuit so that if smoke is detected at any one of the smoke detectors, all three of them will sound an audible alarm signal. You will be using an AFCI circuit breaker in this lab exercise because in an actual house wiring situation, the smoke detector branch circuit would be supplying electrical power to those areas of a house that need to be protected by an AFCI circuit breaker as required by Section 210.12 in the *NEC*®. Before proceeding with this lab exercise, review the information on installing a smoke detector branch circuit in Chapter 14 of the *House Wiring* 3e textbook.

Materials and Equipment

House Wiring 3e textbook, safety glasses, electrician's tool kit, ½-inch drill with a 7/8-inch auger bit, step ladder, 14/2 and 14/3 Type NM cable as needed, one 15-amp single-pole AFCI circuit breaker, three interconnecting type 120-volt smoke detectors, three plastic nail-on outlet boxes, connectors as needed, wirenuts as needed, staples as needed, 6-32 and 8-32 machine screws as needed, nail plates as needed

Procedure

1. Put on safety glasses and observe all safety rules.

2. Install the plastic nail-on outlet boxes on the studs in the wiring mock-up as indicated in Figure 14.6-1 at 72 inches to the center from the floor.

3. Drill the framing members as needed and install the 14/2 Type NM cable homerun from the panel to the first smoke detector box on the left. Secure and support as required by the *NEC*®. Make sure to leave enough free conductor for connections in the panel and at the smoke detector box. Use nail plates if necessary.

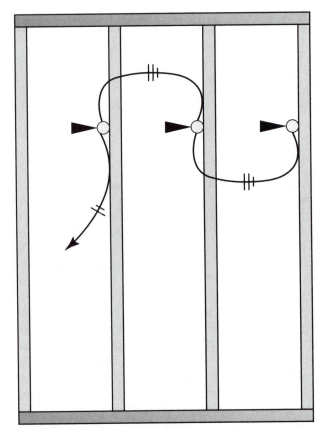

FIGURE 14.6-1

4. Drill the framing members as needed and install the 14/3 Type NM cable between each of the smoke detector boxes. Secure and support the cable as required by the *NEC®*. Use nail plates if necessary.

5. Connect the smoke detectors and secure them to the boxes in the proper manner. Use the red conductor in the three-wire cable as the interconnect wire between smoke detectors. Refer to Figure 14-47 in the *House Wiring* 3e textbook for help with making the proper connections.

6. Install a 15-amp AFCI circuit breaker in the mock-up loadcenter and make the proper connections for the homerun wiring. Refer to the procedure on installing a single-pole AFCI circuit breaker in Chapter 19 of the *House Wiring* 3e textbook.

7. Test the circuit by:

   • Unlocking the electrical power supply and turning on the circuit breaker.

   • Push the test button on any of the smoke detectors. All three detectors should sound an alarm.

   • Have the instructor provide you with a smoke source (such as canned smoke) and test the detectors with smoke. Do this by directing some smoke around any one of the three smoke detectors. All three detectors should sound an alarm.

   • If a problem is found, disconnect the power using a lock-out/tag-out procedure like the one in Chapter 1 of the *House Wiring* 3e textbook, and troubleshoot the circuit.

8. Show all completed work to the instructor.

9. Lock out the power supply, clean up the work area, and return all tools and materials to their proper locations.

10. Complete the Review of this lab exercise.

11. Get the instructor to sign off upon satisfactory completion of this lab exercise.

Review

1. Make out a complete material list for this lab. Include a description of the item used and the quantity. Be sure to include *all* items used.

   _____

   _____

   _____

   _____

   _____

   _____

2. Based on an hourly rate of $50.00 and using the material price list located in Appendix B, calculate the cost of doing the electrical work in this lab exercise.

   _____

   _____

   _____

   _____

   _____

   _____

Lab 14.7: Install a Low-Voltage Chime Circuit

Name: _____ **Date:** _____ **Score:** _____

Introduction

A chime system is used in homes to signal when somebody is at the front or rear door. A chime system will consist of the chime, the momentary contact switch buttons, a transformer, and the wire used to connect the system together. This system is often referred to as the doorbell system even though bells (and buzzers) have not been used in homes for many years. However, the term "doorbell" has stayed with us and many electricians still refer to this part of the electrical system installation as "installing the doorbell." A chime system sounds a musical note or a series of notes when a button located next to an outside entrance or exit door is pushed. The signal from the door buttons is delivered to the chime itself through low-voltage Class 2 wiring. The chime should be located in an area of the house where once it sounds a tone, the people in the house can hear it. Sometimes chimes are located on each floor of a larger home to make sure that everybody in the house will hear the chime when a tone is sounded. A chime button will need to be located next to the front door and rear door of a house. Chime buttons at any other exterior doorways may be included in the chime system if desired. Chime buttons are of the momentary contact type. This means that when someone's finger pushes the button, contacts are closed and current can pass through the switch to the chime, causing a tone to be sounded. When finger pressure on the button is taken away, springs cause the contacts to come apart and the switch will not pass current. This de-energizes the chime and the tone stops. In other words, current can pass through the switch only as long as someone is holding the button in. The chime transformer is used to transform the normal residential electrical system voltage of 120 volts down to the value that a chime system will work on, usually around 16 volts. These transformers have built-in thermal overload protection and no additional overcurrent protection is required to be installed for them. The transformer is installed in a separate metal electrical box or right at the service entrance panel or sub-panel. A ½-inch KO is removed and the high-voltage side of the transformer (120 volts) fits through the KO. A set screw or locknut is used to hold the transformer in place. The high-voltage side has two black pigtails and a green grounding pigtail. The two black pigtail wires are connected across a 120-volt source and the green pigtail is connected to an equipment grounding conductor. The body of the transformer with the low-voltage side is located on the outside of the electrical box or panel. There are two screws on the low-voltage side (16 volts) of the transformer that are used to connect the chime wiring to the transformer. The wire used to connect the button's chime and transformer is often called "bell wire" or simply "thermostat wire" because it is the same type as that used to wire heating and cooling system thermostats. It is usually 16 AWG or 18 AWG solid copper with an insulation type that limits it to use on 30-volt or less circuits. It comes in a cable assembly with two, three, or more single conductors covered with a protective outer sheathing. Each conductor is color coded in the cable assembly. The cable is run through bored holes in the building framing members or on the surface. It is supported with small insulated staples or cleats. It can also be installed in a raceway for added protection from physical damage. There are two common wiring schemes for a chime circuit. The first is to run a two-wire thermostat cable from each doorbell button location and from the transformer location to the chime. The second scheme that electricians often use is to install a two-wire thermostat cable from each doorbell button location to the transformer location. Next, install a three-wire thermostat cable from the chime location to the transformer location. In this lab exercise, you will wire the circuit so that either of two push buttons will activate the chime. The first wiring scheme described above will be used. Before proceeding with this lab exercise, review the information on installing a low-voltage chime circuit in Chapter 14 of the *House Wiring* 3e textbook.

Materials and Equipment

House Wiring 3e textbook, safety glasses, electrician's tool kit, pistol-grip drill with a ¼-inch wood drill bit, ½-inch drill with a 7/8-inch auger bit, step ladder, 18/2 chime wire as needed, 14/2 Type NM cable as needed, one 15-amp single-pole AFCI circuit breaker, one 4 × 1½-inch octagon box with side-mounting bracket and internal clamps, one 4-inch octagon flat blank cover, one two-tone chime, two chime push buttons (surface-mount), one-chime transformer, connectors as needed, wirenuts as needed, fasteners as needed, 6-32 and 8-32 machine screws as needed, small insulated staples as needed, nail plates as needed

Procedure

1. Put on safety glasses and observe all safety rules.

2. Install the 4-inch octagon box and the chime on the wiring mock-up studs indicated in Figure 14.7-1. They should be placed at 72 inches to the center from the floor.

3. Drill the framing members as needed and install a length of 14/2 Type NM from the circuit breaker panel to the octagon box. Use nail plates if necessary.

4. Install the chime transformer into a ½-inch KO on the side or bottom of the 4-inch octagon box so that the primary leads are inside of the box and connect the transformer primary conductors to the 14 AWG conductors in the box. Place a blank cover on the octagon box. **Note:** Usually, there are two black wires on the transformer primary. Connect one of them to the black 14 AWG and the other to the white 14 AWG.

5. Drill a small hole (¼ inch) at 45 inches from the floor on the face of the studs that require doorbell buttons to be mounted on them. The hole must be drilled at an angle from the stud face to the inside of the studs.

6. Drill the framing members as needed and install the 18/2 chime wire cable between the switches, transformer, and the chime as indicated in Figure 14.7-1. Support with small insulated staples where required. Extend the cable through the small drilled holes at the switch locations by approximately 4 inches and leave approximately 12 inches of cable at all other locations. Use nail plates if necessary.

7. Connect the conductors in the chime cable to the push buttons and mount the doorbell push buttons at 45 inches to center from the floor over the drilled holes on the studs indicated, pushing any extra chime cable back through the holes.

8. Connect the wires to the chime and secondary side of the transformer as shown in Figure 14-54 in the *House Wiring* 3e textbook.

9. Install a 15-amp AFCI circuit breaker in the mock-up loadcenter and make the proper connections for the homerun wiring. Refer to the procedure on installing a single-pole AFCI circuit breaker in Chapter 19 of the *House Wiring* 3e textbook.

10. Test the circuit by:

    - Unlocking the electrical power supply and turning on the circuit breaker.

    - Push each of the doorbell buttons. The chime should sound an audible tone.

    - If a problem is found, disconnect the power using a lock-out/tag-out procedure like the one in Chapter 1 of the *House Wiring* 3e textbook, and troubleshoot the circuit.

11. Show all completed work to the instructor.

12. Lock out the power supply, clean up the work area, and return all tools and materials to their proper locations.

13. Complete the Review of this lab exercise.

14. Get the instructor to sign off upon satisfactory completion of this lab exercise.

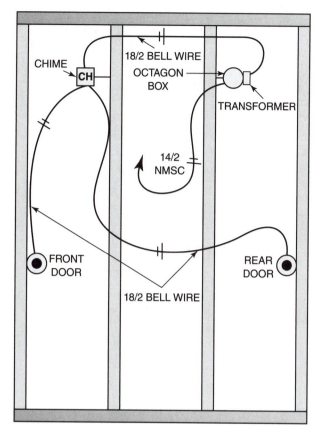

FIGURE 14.7-1

Review

1. Make out a complete material list for this lab. Include a description of the item used and the quantity. Be sure to include *all* items used.

_____

_____

_____

_____

_____

_____

2. Based on an hourly rate of $50.00 and using the material price list located in Appendix B, calculate the cost of doing the electrical work in this lab exercise.

_____

_____

_____

_____

_____

_____

Chapter 15

Special Residential Wiring Situations

Lab 15.1: Install a Garage Branch Circuit

Name: _____ **Date:** _____ **Score:** _____

Introduction

Garages are either attached to the main house during the construction process or built detached at some distance away from the main house. If the garage is attached, the *NEC®* requires at least one 15- or 20-amp, 120-volt receptacle and at least one wall switch-controlled lighting outlet. While these minimum requirements may be satisfactory for a small attached garage, most garages are large enough to require several lighting outlets and several receptacle outlets. There is no *NEC®* requirement concerning the maximum distance between receptacle outlets in a garage. Therefore, it is up to the electrician as to how many and how far apart the receptacles are located. The branch circuits supplying the receptacle(s) and lighting outlet(s) may be on the same branch circuit and can be rated either 15 or 20 amps. Section 210.8 of the *NEC®* requires all of the 15- or 20-amp, 120-volt receptacles installed in a garage to be GFCI protected. In this lab exercise, you will wire the circuits that are shown in Figure 15.1-1 on the stud section of the wiring mock-up. Circuit #1 is a 15-amp branch circuit wired with 14 AWG wire. Wire circuit #1 so that lighting outlets #1, #2, and #3 are controlled by switches #5, #6, and #7. Switch #8 controls lighting outlet #4. Circuit #2 is a 20-amp branch circuit wired with 12 AWG wire. Receptacle #11 is a feed-through GFCI receptacle that also protects receptacles #9 and #10. Special purpose outlet #12 is a garage door opener receptacle that is "hot" at all times and is GFCI protected. You will need to choose a box type and size that will satisfy *NEC®* requirements for each box location. Before proceeding with this lab exercise, review the material on installing garage feeders and branch circuits in Chapter 15 of the *House Wiring* 3e textbook.

Materials and Equipment

House Wiring 3e textbook, safety glasses, electrician's tool kit, voltage tester, step ladder, ½-inch drill with a ⅞-inch auger bit, 14 AWG and 12 AWG Type NM cable as needed, electrical boxes as needed, three duplex receptacles, one GFCI feed-through duplex receptacle, one 15-amp single-pole circuit breaker, one 20-amp single-pole circuit breaker, four keyless lamp holders, four 60-watt medium-base lamps, connectors, wirenuts, staples, nail plates as needed

Procedure

1. Put on safety glasses and observe all safety rules.

2. Install the device boxes for receptacles #9, #10, and #11 at 24 inches to center on the sides of the studs indicated in Figure 15.1-1. The box for receptacle #12 is to be located at 84 inches to center on the stud indicated. Assume that ½-inch wallboard will be used on the walls and ceiling.

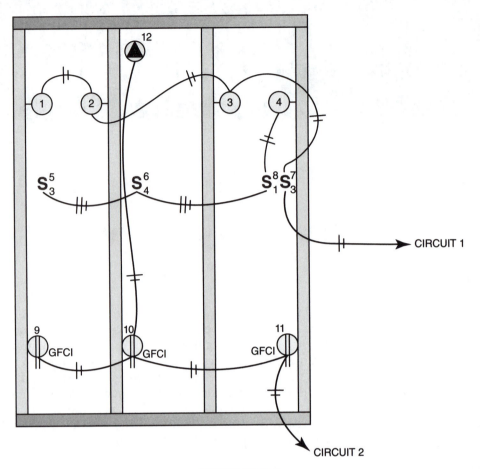

FIGURE 15.1-1

3. Install single-gang device boxes for switches #5 and #6 and a two-gang device box for switches #7 and #8 at 48 inches to center on the studs indicated in Figure 15.1-1.

4. Install the outlet boxes for lighting outlets #1, #2, #3, and #4 at 72 inches to center on the studs as indicated in Figure 15.1-1.

5. Drill the framing members as needed and install 14/2, 14/3, or 12/2 Type NM cable between all boxes as indicated in Figure 15.1-1. Secure the cable according to the requirements of the *NEC*®. Use nail plates when required.

6. Drill the framing members as needed and install the 15-amp-rated homerun from the two-gang switch box to the circuit breaker panel and secure as needed, making sure to leave enough free conductor length for connections in the panel and in the device box. Use nail plates when required.

7. Drill the framing members as needed and install the 20-amp-rated homerun from receptacle box #11 to the circuit breaker panel and secure as needed, making sure to leave enough free conductor length for connections in the panel and in the device box. Use nail plates when required.

8. Connect the conductors to the devices in the boxes. Use Chapters 13 and 18 in the *House Wiring* 3e textbook for help if necessary. When making the conductor connections, review the following procedures:

   – Connecting wires together with a wirenut in Chapter 2 of the *House Wiring* 3e textbook.

   – Using terminal loops to connect circuit conductors to terminal screws on receptacles or switches in Chapter 18 of the *House Wiring* 3e textbook.

   – Installing receptacles (or switches) in a nonmetallic electrical box in Chapter 18 of the *House Wiring* 3e textbook.

– Installing receptacles (or switches) in a metal electrical box in Chapter 18 of the *House Wiring* 3e textbook.

– Installation steps for installing a light fixture directly to an outlet box in Chapter 17 of the *House Wiring* 3e textbook.

9. Secure the devices to the boxes and install 60-watt lamps in the lamp holders. Do not install switch or receptacle covers unless your instructor tells you to.

10. Install a 15- and a 20-amp single-pole circuit breaker in the circuit breaker panel. Connect the grounding wire of the homerun to the grounding bar, the grounded conductor to the neutral bar, and the ungrounded wire to the breaker. Refer to the procedure on installing a single-pole circuit breaker in Chapter 19 of the *House Wiring* 3e textbook.

11. Test the circuit by:

   • Unlocking the electrical power supply and turning on the circuit breakers.

   • Follow the procedure on testing 120-volt receptacles with a voltage tester in Chapter 20 of the *House Wiring* 3e textbook and check each receptacle for proper voltage, polarity, and grounding.

   • Test the GFCI receptacle to make sure it trips when it is supposed to.

   • Toggle the switches on and off and observe if the lighting fixtures are working correctly.

   • If a problem is found, disconnect the power using a lock-out/tag-out procedure like the one in Chapter 1 of the *House Wiring* 3e textbook, and troubleshoot the circuit.

12. Complete the Review of this lab exercise.

13. Show all completed work to the instructor.

14. Lock out the power supply, clean up the work area, and return all tools and materials to their proper locations.

15. Get the instructor to sign off upon satisfactory completion of this lab exercise.

Review

1. Make out a complete material list for this lab. Include a description of the item used and the quantity. Be sure to include *all* items used.

_____

_____

_____

_____

_____

_____

2. Based on an hourly rate of $50.00 and using the material price list located in Appendix B, calculate the cost of doing the electrical work in this lab exercise.

_____

_____

_____

_____

_____

_____

Chapter 16 Video, Voice, and Data Wiring Installation

Lab 16.1: Install Crimp Style F-Type Connectors on RG-6 Coaxial Cable

Name: _____ Date: _____ Score: _____

Introduction

Electricians install coaxial cables for video signals to video jacks throughout a building. EIA/TIA 570 recommends using RG-6, 75-ohm coaxial cables. It is also recommended that two runs of coaxial cable be installed from a central video distribution center to each TV outlet location. The two runs of cable will allow for video distribution from any video source as well as distribution to a video monitor at each outlet location. Also, coaxial cable is becoming more popular as a data transmission line for high-speed Internet connections and the extra cable could be used to serve a computer located close to the video outlet location. At each end of the coaxial cable, male F-Type connectors are installed. Threaded F-Type connectors are recommended to reduce signal interference. Push-on fittings are not recommended and should be avoided. At the wall-mounted TV outlet end, a female-to-female F-Type coupler is used. At the service center end the F-Type connector is threaded onto the proper fitting. At the TV outlet end, connection to a video device is done using a 75-ohm RG-6 coaxial patch cord. Once the system has been completely installed and cable television or satellite television service is brought to the distribution center, connections of individual televisions and other video devices to the wall jacks is accomplished by simply connecting them to any video outlet in the building. In this lab exercise, you will construct a 6-foot-long RG-6 patch cord with an F-Type connector on each end. Before proceeding with this lab exercise, review the procedure for installing an F-Type connector on an RG-6 coaxial cable in Chapter 16 of the *House Wiring* 3e textbook.

Materials and Equipment

House Wiring 3e textbook, safety glasses, electrician's tool kit, 6-foot length of RG-6 coaxial cable, coaxial cable stripper for RG-6 coaxial cable, two F-Type crimp-style connectors with ferrule for RG-6 coaxial cable, crimping tool with the appropriate die for RG-6 cable, a measuring device, a TV and a video player for testing

Procedure

1. Put on safety glasses and observe all applicable safety rules.

2. Measure and mark ¾ inch from the end of the cable.

3. Using the coaxial cable stripper, remove ¾ inch of the outside sheathing of the cable. **Note:** Some coaxial stripper types have two blades, an inside blade that makes a shallow cut and an outside blade that makes a deeper cut. Refer to photo A of the procedure for installing an F-Type connector on a RG-6 coaxial cable in Chapter 16 of the *House Wiring* 3e textbook.

4. Using the coaxial stripper, remove ½ inch of the white dielectric insulation from the center conductor. The end of the cable should now look similar to Figure 16.1-1.

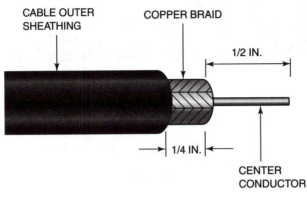

CABLE OUTER SHEATHING

COPPER BRAID

1/2 IN.

1/4 IN.

CENTER CONDUCTOR

FIGURE 16.1-1

5. Fold the copper braid back over the outer sheathing and push the RG-6 F-Type connector onto the end of the cable until the white dielectric insulation is even with the hole on the inside of the connector. Refer to Illustration C of the procedure for installing an F-Type connector on a RG-6 coaxial cable in Chapter 16 of the *House Wiring* 3e textbook.

6. Using a proper sized crimping tool like the one shown in Figure 16.1-2, crimp the connector ferrule onto the cable. Make sure the crimp is made toward the cable side of the connector. Too close to the ferrule head will pinch the crimp.

DIE FOR RG-59 AND RG-6 F-TYPE COAXIAL CABLE CONNECTORS

FIGURE 16.1-2

7. Install an F-Type connector on the other end of the coaxial cable by following steps 2–6.

8. Test the RG-6 coaxial patch cord constructed in this lab exercise by connecting it between a video player and a TV. Turn on the video player and the TV and observe whether the coaxial cable successfully carries the video signal. If it does not, troubleshoot the connections to determine the cause of the malfunction. If necessary, cut off the existing F-Type connectors, install new connectors, and retest.

9. Complete the Review for this lab exercise.

10. Show all completed work to the instructor.

11. Clean up the work area and return all tools and materials to their proper locations.

12. Get the instructor to sign off upon satisfactory completion of this lab exercise.

Review

1. What did you find to be the hardest part of this job?

_____

_____

_____

2. List at least two safety procedures that must be followed when doing this job.

_____

_____

_____

Lab 16.2: Install Compression Style F-Type Connectors on an RG-6, 75-Ohm Coaxial Cable

Name: _____ Date: _____ Score: _____

Introduction

In some installations, it is necessary to use a weatherproof F-Type coaxial cable connector. The Thomas and Betts Corporation has developed a system called Snap-N-Seal®. It is considered by many to be the best system of its type available to electricians. Other companies, such as IDEAL Industries, now offer a compression F-Type coaxial cable connector system. These systems work so well that many electricians use them for all of their coaxial cable terminations that require an F-Type connector, whether the installation is outside or inside a building. Compression coaxial cable fittings are now the preferred style of F-Type connector used in many parts of the country. In this lab exercise, you will install a compression F-Type coaxial cable connector on each end of a 6-foot length of RG-6 cable. Then you will test the cable to determine if it carries a video signal as it should. Before proceeding with this lab exercise, review the material on installing coaxial cable connectors in Chapter 16 of the *House Wiring* 3e textbook.

Materials and Equipment

House Wiring 3e textbook, safety glasses, electrician's tool kit, 6-foot length of RG-6 coaxial cable, coaxial cable stripper for RG-6 coaxial cable, two F-Type compression style connectors with ferrule for RG-6 coaxial cable, a measuring device, a video player and TV for testing, a compression connector tool

Procedure

1. Put on safety glasses and observe all applicable safety rules.

2. Follow the instructions shown in Figure 16.2-1 and install a compression style F-Type coaxial cable connector on each end of the 6-foot length of RG-6 coaxial cable.

3. Show all completed work to the instructor before you proceed.

4. Test the RG-6 coaxial patch cord constructed in this lab exercise by connecting it between a video player and a TV. Turn the video player and the TV on and observe whether the coaxial cable successfully carries the video signal. If it does not, troubleshoot the connections to determine the cause of the malfunction. If necessary, cut off the existing F-Type connectors and install new connectors; then retest.

5. Complete the Review for this lab exercise.

6. Show all completed work to the instructor.

7. Clean up the work area and return all tools and materials to their proper locations.

8. Get the instructor to sign off upon satisfactory completion of this lab exercise.

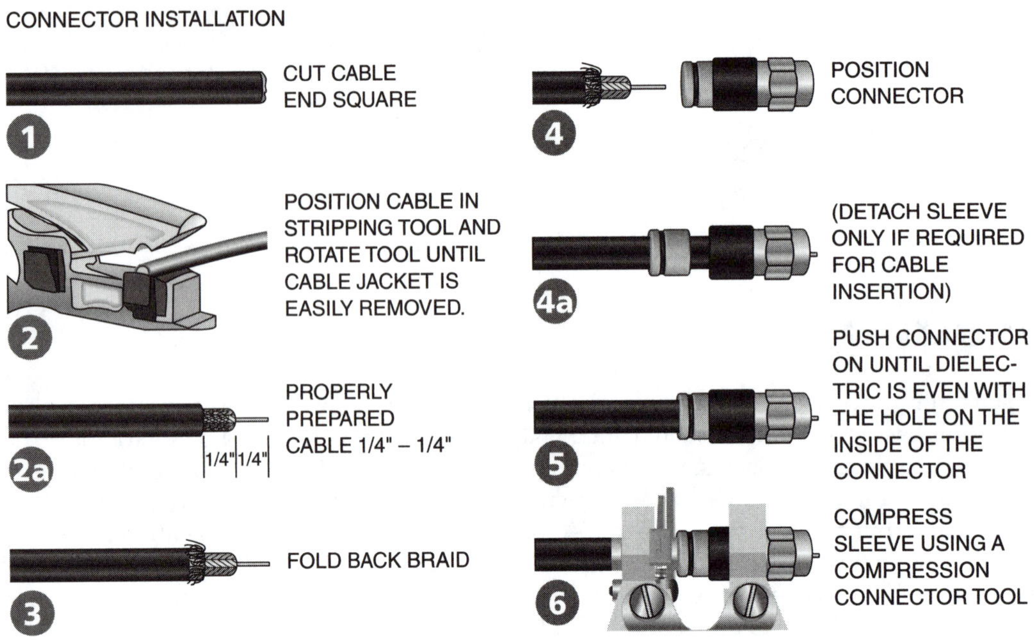

CONNECTOR INSTALLATION

1. CUT CABLE END SQUARE

2. POSITION CABLE IN STRIPPING TOOL AND ROTATE TOOL UNTIL CABLE JACKET IS EASILY REMOVED.

2a. PROPERLY PREPARED CABLE 1/4" – 1/4"

3. FOLD BACK BRAID

4. POSITION CONNECTOR

4a. (DETACH SLEEVE ONLY IF REQUIRED FOR CABLE INSERTION)

5. PUSH CONNECTOR ON UNTIL DIELEC-TRIC IS EVEN WITH THE HOLE ON THE INSIDE OF THE CONNECTOR

6. COMPRESS SLEEVE USING A COMPRESSION CONNECTOR TOOL

FIGURE 16.2-1

Review

1. What did you find to be the hardest part of this job?

_____

_____

_____

2. List at least two safety procedures that must be followed when doing this job.

_____

_____

_____

Lab 16.3: Install RJ-45 Jacks on the Ends of a Four-Pair Category 5e UTP Cable

Name: _____ Date: _____ Score: _____

Introduction

In a hard-wired structured cabling system, four-pair UTP cables such as Cat 5e are used for both voice and data applications. After the homeruns are installed from the video/voice/data distribution center to the various outlet locations in a house, the electrician must then terminate the cables on each end. Typically, an RJ-45 8-pin jack is used at the outlet location and the voice/video/data distribution center end is terminated on a punch-down bar in the distribution center. On both ends the terminating is accomplished by using a punchdown tool rather than placing wires under a binding post. The termination sequence will be either EIA/TIA T568A or EIA/TIA T568B. In this lab exercise, you will use the EIA/TIA 568B scheme. Additional information on installing a structured cabling system can be found in Chapter 16 of the *House Wiring* 3e textbook. In this lab exercise, you will install an RJ-45 8-pin jack on each end of the Cat 5e UTP cable. Before proceeding with this lab exercise, review the procedure for installing an RJ-45 jack on a four-pair category 5e cable in Chapter 16 of the *House Wiring* 3e textbook.

Materials and Equipment

House Wiring 3e textbook, safety glasses, electrician's tool kit, 6-foot length of category 5e UTP cable, Cat 5e cable stripper, two RJ-45 jacks, a 110 punchdown tool, a measuring device, UTP cable tester

Procedure

1. Put on safety glasses and observe all applicable safety rules.

2. Using a UTP cable jacket stripping tool, remove about 2 inches of jacket from the cable end. Refer to photo A of the procedure for installing an RJ-45 jack on a four-pair category 5e cable in Chapter 16 of the *House Wiring* 3e textbook.

3. Determine which wiring scheme you will be using (T568A or T568B) and note the color coding and pin numbers on the jack. **Note:** Most manufacturers of RJ-45 jacks will have color coding and pin numbering on their products. Refer to Illustration B of the procedure for installing an RJ-45 jack on a four-pair category 5e cable in Chapter 16 of the *House Wiring* 3e textbook.

4. Route the conductors for termination. Terminate one pair at a time starting from the rear of the jack. Terminating each pair after placement will prevent crushing the inside pairs with the punchdown tool. **Note:** The cable should be placed so that the cable jacket touches the rear of the jack housing, as shown. Refer to Illustration C of the procedure for installing an RJ-45 jack on a four-pair category 5e cable in Chapter 16 of the *House Wiring* 3e textbook.

5. Using a 110-style punchdown tool, seat each conductor into the proper IDC slot. Be sure to keep the twists within ½ inch of the IDC slot. The punchdown tool, if used properly, will trim any excess wire off flush with the device body. Refer to Illustration D of the procedure for installing an RJ-45 jack on a four-pair category 5e cable in Chapter 16 of the *House Wiring* 3e textbook.

6. Place the cap that comes with the RJ-45 device over the terminated wires and press it into place. This cap will provide a more secure connection of the wires to the IDC slots as well as provide some additional strain relief. The jack can now be inserted into a wall plate assembly and secured to an electrical box or mud ring placed at the desired location. Refer to Illustration E of the procedure for installing an RJ-45 jack on a four-pair category 5e cable in Chapter 16 of the *House Wiring* 3e textbook.

7. Follow steps 2–6 and install an RJ-45 jack on the other end of the Cat 5e cable length.

8. Test the cable with a UTP cable tester. The tester will indicate whether there is continuity from one end to the other through the RJ-45 jacks. It will also indicate whether the correct wiring scheme (T568A or T568B) has been followed on each end. If a problem is indicated, troubleshoot the connections to determine the cause of the malfunction and fix the problem.

9. Complete the Review for this lab exercise.

10. Show all completed work to the instructor.

11. Clean up the work area and return all tools and materials to their proper locations.

12. Get the instructor to sign off upon satisfactory completion of this lab exercise.

Review

1. The most often used category rated UTP cable for wiring from various voice/data outlets is _____.

2. Describe the difference between a 568A and a 568B connection.

   _____

   _____

   _____

   _____

3. Name the document that contains the installation requirements for a structured cabling system that must be followed.

   _____

4. The popular name given to an 8-pin connector or jack used to terminate UTP cable is _____.

5. The standard color coding for a four-pair UTP cable is:
   - Pair 1: Tip is _____; ring is _____
   - Pair 2: Tip is _____; ring is _____
   - Pair 3: Tip is _____; ring is _____
   - Pair 4: Tip is _____; ring is _____

Lab 16.4: Install RJ-45 Modular Plugs on the Ends of Category 5e UTP Cable

Name: _____ **Date:** _____ **Score:** _____

Introduction

A patch cord connects the voice/data outlet to the piece of equipment (like a computer) that sends and receives the voice or data signals. Sometimes patch cords are simply purchased already made. Other times, the installer has to make the patch cords. In this lab exercise, you will make a patch cord by using Cat 5e UTP cable with a modular RJ-45 plug installed on each end. You will use the EIA/TIA 568B wiring scheme. Additional information on installing a structured cabling system can be found in Chapter 16 of the *House Wiring* 3e textbook. Before proceeding with this lab exercise, review the procedure for assembling a patch cord with RJ-45 plugs and category 5e cable in Chapter 16 of the *House Wiring* 3e textbook.

Materials and Equipment

House Wiring 3e textbook, safety glasses, electrician's tool kit, 6-foot length of category 5e UTP cable, Cat 5e cable stripper, one ratchet-type crimping tool with changeable dies, one die set for RJ-45 modular plugs, two RJ-45 modular plugs, a measuring device, UTP cable tester

Procedure

1. Put on safety glasses and observe all applicable safety rules.

2. Determine the length of the patch cord and using a pair of cable cutters, cut the desired length from a spool or box of the category 5e UTP cable. For this lab exercise, you will be using a 6-foot length.

3. Using a UTP cable stripping tool, strip the cable jacket back about 3⁄4 inch from each end of the cable.

4. Determine whether the connection will be done to the EIA/TIA 568A color scheme or to the EIA/TIA 568B color scheme. Remember that for this lab exercise, you will be using the 568B scheme.

5. On one end of the patch cord, sort the pairs out so they fit into the plug in the order shown in Figure 16-14 of the *House Wiring* 3e textbook. Again, for this lab exercise you will be using the 568B scheme. Refer to Illustration A (Patch Cord B) of the procedure for assembling a patch cord with RJ-45 plugs and category 5e cable in Chapter 16 of the *House Wiring* 3e textbook.

6. Insert the wire pairs into the plug.

7. Using a crimping tool with the proper crimping die, crimp the pins. Refer to Illustration B of the procedure for assembling a patch cord with RJ-45 plugs and category 5e cable in Chapter 16 of the *House Wiring* 3e textbook.

8. Following steps 3–7, install a plug on the other end of the UTP cable.

9. Test the cable with a UTP cable tester. The tester will indicate whether there is continuity from one end to the other through the RJ-45 plugs. It will also indicate whether the correct wiring scheme (T568A or T568B) has been followed on each end. If a problem is indicated, troubleshoot the connections to determine the cause of the malfunction and fix the problem.

10. Complete the Review for this lab exercise.

11. Show all completed work to the instructor.

12. Clean up the work area and return all tools and materials to their proper locations.

13. Get the instructor to sign off upon satisfactory completion of this lab exercise.

Review

1. What do the letters IDC stand for?

   _____

2. A _____ is the short length of cable with an RJ-45 plug on either end used to connect a computer to the data outlet.

Chapter 17

Lighting Fixture Installation

Lab 17.1: Install a Light Fixture to an Electrical Box with a Strap

Name: _____ **Date:** _____ **Score:** _____

Introduction

In previous lab exercises, you used the direct connection method to attach a lamp holder directly to a lighting outlet box. The installation process for attaching a lighting fixture to an outlet box with a strap is very similar to the direct connection method. The main difference is that a metal strap is connected to the lighting outlet box and the fixture is attached to the metal strap with headless bolts and decorative nuts or with long screws. The metal strap must be connected to the system grounding conductor. For this type of installation, it is important to read, understand, and follow the manufacturer's instructions. Before proceeding with this lab exercise, review the procedure for installing a lighting fixture to an outlet box with a strap in Chapter 17 of the *House Wiring* 3e textbook.

Materials and Equipment

House Wiring 3e textbook, safety glasses, electrician's tool kit, step ladder, ½-inch drill with a 7/8-inch auger bit, one lighting fixture that uses a strap to connect to the outlet box, one plastic nail-on ceiling outlet box, one single-gang plastic nail-on device box, 14/2 Type NM cable as needed, one 15-amp 120-volt single-pole switch, one 15-amp single-pole circuit breaker, 60-watt lamp(s) for the lighting fixture, Type NM cable connectors as needed, wirenuts as needed, staples as needed, nail plates as needed

Procedure

1. Put on safety glasses and observe all safety rules.

2. On the wiring mockup, install a single-gang plastic nail-on device box on a stud at 48 inches to center from the floor. Assume the use of ½-inch wallboard on this wall.

3. On the wiring mockup, install a plastic nail-on ceiling outlet box on a joist that lines up with the stud that the device box is attached to. Assume the use of ½-inch wallboard on this ceiling.

4. Drill the framing members as needed and install 14/2 Type NM cable from the circuit breaker panel into the single-gang plastic nail-on device box. Secure and support the cable with staples as required by the *NEC®*. Use nail plates where required.

5. Drill the framing members as needed and install 14/2 Type NM cable between the single-gang plastic nail-on device box and the plastic nail-on ceiling box. Secure and support the cables with staples as required by the *NEC®*. Use nail plates where required.

6. Connect the conductors to the switch so that it controls the lighting outlet. Use Chapters 13 and 18 in the *House Wiring* 3e textbook for help if necessary. When making the conductor connections, review the following procedures:

   – Connecting wires together with a wirenut in Chapter 2 of the *House Wiring* 3e textbook.

   – Using terminal loops to connect circuit conductors to terminal screws on receptacles or switches in Chapter 18 of the *House Wiring* 3e textbook.

   – Installing receptacles (or switches) in a nonmetallic electrical box in Chapter 18 of the *House Wiring* 3e textbook.

7. Install the lighting fixture to the outlet box. Refer to Illustration A in the procedure for installing a lighting fixture to an outlet box with a strap in Chapter 17 of the *House Wiring* 3e textbook.

8. Secure the switch to the box and install the lamps in the lighting fixture. Install any lamp cover(s) at this time. Do not install a switch cover unless your instructor tells you to.

9. Insert a 15-amp single-pole circuit breaker in the mock-up loadcenter and make the proper connections for the "homerun" wiring. Refer to the procedure on installing a single-pole circuit breaker in Chapter 19 of the *House Wiring* 3e textbook.

10. Test the circuit by:

    – Unlocking the electrical power supply and turning on the circuit breaker.

    – Toggling the switch on and off and observe if the light fixture is working correctly.

    – If a problem is found, disconnect the power using a lock-out/tag-out procedure like the one in Chapter 1 of the *House Wiring* 3e textbook and troubleshoot the circuit.

11. Complete the Review for this lab exercise.

12. Show all completed work to the instructor.

13. Lock out the power supply, clean up the work area, and return all tools and materials to their proper locations.

14. Get the instructor to sign off upon satisfactory completion of this lab exercise.

Review

1. What was the hardest part of this lab exercise?

   _____

   _____

   _____

2. List two safety rules to follow when installing a lighting fixture to an outlet box with a strap.

   _____

   _____

   _____

Lab 17.2: Install a Light Fixture to an Electrical Box with a Stud and Strap

Name: _____ Date: _____ Score: _____

Introduction

Larger and heavier types of light fixtures use the stud and strap method of installation. Hanging fixtures, like a chandelier or pendant fixture, often require extra mounting support when compared to smaller light fixtures. It is extremely important to read and follow the manufacturer's instructions because there are some slight variations with this type of installation. Before proceeding with this lab exercise, review the procedure for installing a lighting fixture to a lighting outlet box using a stud and strap in Chapter 17 of the *House Wiring* 3e textbook.

Materials and Equipment

House Wiring 3e textbook, safety glasses, electrician's tool kit, step ladder, ½-inch drill with a 7/8-inch auger bit, one lighting fixture that uses a stud and strap to connect to the outlet box, one plastic nail-on ceiling outlet box, one single-gang plastic nail-on device box, 14/2 Type NM cable as needed, one 15-amp 120-volt single-pole switch, one 15-amp single-pole circuit breaker, 60-watt lamp(s) for the lighting fixture, Type NM cable connectors as needed, wirenuts as needed, staples as needed, nail plates as needed

Procedure

1. Put on safety glasses and observe all safety rules.

   Note: You are allowed to use the same box setup used in Lab 17.1 and skip to step 7. If the setup for Lab 17.1 has been taken down, follow the steps as outlined below.

2. On the wiring mock-up, install a single-gang plastic nail-on device box on a stud at 48 inches to center from the floor. Assume the use of ½-inch wallboard on this wall.

3. On the wiring mock-up, install a plastic nail-on ceiling outlet box on a joist that lines up with the stud that the device box is attached to. Assume the use of ½-inch wallboard on this ceiling.

4. Drill the framing members as needed and install 14/2 Type NM cable from the circuit breaker panel into the single-gang plastic nail-on device box. Secure and support the cable with staples as required by the *NEC®*. Use nail plates where required.

5. Drill the framing members as needed and install 14/2 Type NM cable between the single-gang plastic nail-on device box and the plastic nail-on ceiling box. Secure and support the cables with staples as required by the *NEC®*. Use nail plates where required.

6. Connect the conductors to the switch so that it controls the lighting outlet. Use Chapters 13 and 18 in the *House Wiring* 3e textbook for help if necessary. When making the conductor connections, review the following procedures:

   – Connecting wires together with a wirenut in Chapter 2 of the *House Wiring* 3e textbook.

   – Using terminal loops to connect circuit conductors to terminal screws on receptacles or switches in Chapter 18 of the *House Wiring* 3e textbook.

   – Installing receptacles (or switches) in a nonmetallic electrical box in Chapter 18 of the *House Wiring* 3e textbook.

7. Install the lighting fixture to the outlet box. Refer to Illustration A in the procedure for installing a lighting fixture to a lighting outlet box using a stud and strap in Chapter 17 of the *House Wiring* 3e textbook.

8. Secure the switch to the box and install the lamps in the lighting fixture. Install any lamp cover(s) at this time. Do not install a switch cover unless your instructor tells you to.

9. Insert a 15-amp single-pole circuit breaker in the mock-up loadcenter and make the proper connections for the "homerun" wiring. Refer to the procedure on installing a single-pole circuit breaker in Chapter 19 of the *House Wiring* 3e textbook.

10. Test the circuit by:

    – Unlocking the electrical power supply and turning on the circuit breaker.

    – Toggling the switch on and off and observe if the light fixture is working correctly.

    – If a problem is found, disconnect the power using a lock-out/tag-out procedure like the one in Chapter 1 of the *House Wiring* 3e textbook and troubleshoot the circuit.

11. Complete the Review for this lab exercise.

12. Show all completed work to the instructor.

13. Lock out the power supply, clean up the work area, and return all tools and materials to their proper locations.

14. Get the instructor to sign off upon satisfactory completion of this lab exercise.

Review

1. What was the hardest part of this lab?

   _____

   _____

   _____

2. List two safety rules to follow when installing a lighting fixture to an outlet box with a stud and strap.

   _____

   _____

   _____

Chapter 18   Device Installation

Lab 18.1: Identify Common Receptacle and Plug Configurations

Name: _____   **Date:** _____   **Score:** _____

Introduction

A receptacle is an electrical device that allows the home owner to access the electrical system with cord-and-plug-connected equipment such as lamps, computers, kitchen appliances, and tools. The *NEC®* defines a receptacle as a contact device installed at an outlet for the connection of an attachment plug. An attachment plug is defined as a device that, by insertion in a receptacle, establishes a connection between the conductors of the attached flexible cord and the conductors connected permanently to the receptacle. In residential applications, general use receptacles are usually 125-volt, 15- or 20-amp devices and have specific slot configurations. For appliances that require just 240 volts, such as a room air conditioner, receptacles are made with special blade configurations to meet the requirements of the appliance. For appliances that require 120/240 volts, such as electric clothes dryers and electric ranges, special blade configurations to meet the requirements of the appliance are also available. The National Electrical Manufacturers Association (NEMA) has developed a chart of the receptacle and plug configurations that are available. In this lab exercise, you will identify receptacle and plug configurations commonly found in residential wiring.

Materials and Equipment

House Wiring 3e textbook, pencil

Procedure

1. Using Table 2-4 in the *House Wiring* 3e textbook, identify the NEMA non-locking designation for the receptacles and plugs needed for the applications listed in the Review section of this lab exercise.

2. Show the instructor your completed work.

3. Get the instructor to sign off upon satisfactory completion of this lab exercise.

Review

Residential Application	NEMA Receptacle Designation	NEMA Plug Designation
Duplex receptacles used on 15-amp, 120-volt general lighting branch circuits		
A single receptacle used on a 20-amp, 120-volt laundry branch circuit		
Duplex receptacles used on 20-amp, 120-volt small-appliance branch circuits		
Duplex receptacles used on a 20-amp, 120-volt bathroom branch circuit		
A newly installed 30-amp, 120/240-volt electric clothes dryer individual branch circuit		
A newly installed 40-amp, 120/240-volt electric range individual branch circuit		
A 30-amp, 120/240-volt electric clothes dryer individual branch circuit installed prior to 1996		
A 40-amp, 120/240-volt electric range individual branch circuit installed prior to 1996		
A 15-amp, 240-volt room air conditioner branch circuit		
A 30-amp, 240-volt room air conditioner branch circuit		

Chapter 19 — Service Panel Trim-Out

Lab 19.1: Install and Trim Out a Single-Phase Circuit Breaker Loadcenter

Name: _____ Date: _____ Score: _____

Introduction

An important piece of electrical equipment that is part of an electrical service is the circuit breaker loadcenter. The loadcenter usually will include the service main circuit breaker and the branch-circuit overcurrent protection devices. When a loadcenter is used as a sub-panel, it will contain only branch-circuit overcurrent protection devices. Installation of a loadcenter in a neat and workmanlike manner is an important job duty for an electrician. A loadcenter must always be trimmed out in as neat a manner as possible. Before you proceed with this lab exercise, review the material on trimming out a service panel in Chapter 19 of the *House Wiring* 3e textbook. **Note:** Check with your instructor to see if you need to use AFCI (Arc-Fault Circuit Interrupter) circuit breakers where required in this lab exercise.

Materials and Equipment

House Wiring 3e textbook, safety glasses, electrician's tool kit, one 1¼-inch cable connector, four #10 × ¾-inch sheet metal screws, one 100-amp main breaker loadcenter, 3 feet of ¾-inch EMT, two ¾-inch conduit hangers, one water pipe grounding clamp, 2 feet of 2 AWG AL Type SEU cable, ½-inch Type NM cable connectors as needed, ¾-inch Type NM cable connectors as needed, 20-amp single-pole circuit breakers as needed, 15-amp single-pole circuit breakers as needed, 30-amp double-pole circuit breakers as needed, 40-amp double-pole circuit breakers as needed, 20-amp double-pole circuit breakers as needed, AFCI circuit breakers as needed if required by the lab instructor, 12/2 Type NM cable as needed, 14/2 Type NM cable as needed, 10/3 Type NM cable as needed, 8/3 Type NM cable as needed, 6 AWG bare CU grounding wire as needed, staples as needed, antioxidant as needed, ½-inch PVC as needed, one ½-inch PVC connector, two ½-inch conduit hangers

Procedure

1. Put on safety glasses and observe all safety rules. Refer to Figure 19.1-1 to see how this lab project should look when completed.

2. Wire the 100-amp main circuit breaker loadcenter on the 4 × 8-foot plywood part of the wiring mock-up. Do *not* use antioxidant on the aluminum conductor terminations unless your instructor tells you to do so. The circuits in this loadcenter are as follows:

   five 15-amp, 120-volt general lighting branch circuits

   two 20-amp, 120-volt small-appliance branch circuits

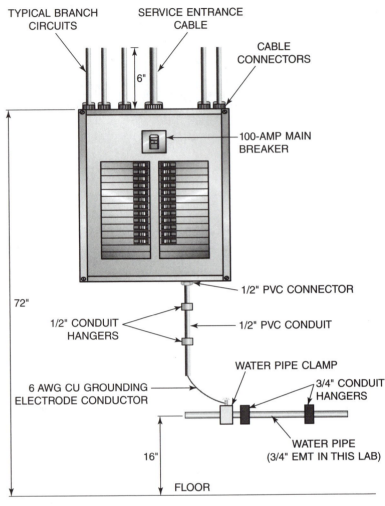

TYPICAL BRANCH
CIRCUITS

SERVICE ENTRANCE
CABLE

CABLE
CONNECTORS

6"

100-AMP MAIN
BREAKER

72"

1/2" PVC CONNECTOR

1/2" CONDUIT
HANGERS

1/2" PVC CONDUIT

WATER PIPE CLAMP

6 AWG CU GROUNDING
ELECTRODE CONDUCTOR

3/4" CONDUIT
HANGERS

16"

WATER PIPE
(3/4" EMT IN THIS LAB)

FLOOR

FIGURE 19.1-1

one 20-amp, 120-volt laundry branch circuit

one 20-amp, 120-volt bathroom branch circuit

one 20-amp, 120-volt garbage disposal circuit

one 20-amp, 240-volt water pump circuit

one 40-amp, 120/240-volt electric range branch circuit

one 30-amp, 120/240-volt electric clothes dryer branch circuit

one 30-amp, 240-volt electric water heater branch circuit

3. Using four sheet metal screws, mount and level the loadcenter on the plywood at 72 inches from the floor to the top of the loadcenter. Mount the loadcenter so the right side of the loadcenter is 12 inches in from the right edge of the plywood wall.

4. Strip back 12 inches of outer sheathing from the 2 AWG AL Type SEU cable and, using the 1¼-inch cable connector, connect it to the middle knockout on the top of the loadcenter. Leave approximately 6 inches of Type SEU cable outside the loadcenter.

5. In the loadcenter, twist the grounded neutral conductor together and connect it to the largest neutral bar lug. Then, strip back and connect the ungrounded Type SEU cable conductors to the line lugs in the loadcenter. These lugs are typically on the main circuit breaker. Refer to Figure 8-34 in the *House Wiring* 3e textbook.

6. Install the grounding electrode (water pipe) in a horizontal position at the lower right portion of the mock-up at 16 inches to center from the floor. Support it with two ¾-inch conduit hangers. For this lab exercise, a 3-foot length of ¾-inch EMT will serve as the water pipe grounding electrode.

7. Install the 6 AWG bare CU grounding electrode conductor. Use ½-inch PVC conduit, supported with conduit hangers, to cover and protect the 6 AWG CU conductor.

8. Connect the grounding electrode conductor to the water pipe electrode by using a water pipe clamp. Connect the other end to the neutral terminal bar in the loadcenter. Refer to Figure 8-34 in the *House Wiring* 3e textbook.

9. Attach the required cables for the circuits described in step 2 of this lab exercise to the loadcenter with the proper size cable connector. Make sure to leave approximately 6 inches of sheathed cable on the outside and enough free conductors on the inside of the loadcenter to make connections at each circuit breaker.

10. Install the required circuit breakers and connect the bare grounding wires, the white grounded wires, and the ungrounded wires to the proper circuit breakers. Follow the procedures for installing circuit breakers in Chapter 19 of the *House Wiring* 3e textbook. **Note:** If any white conductors are to be used as ungrounded conductors, be sure to reidentify them with black tape.

11. Complete the Review for this lab exercise.

12. Show all completed work to the instructor.

13. Clean up the work area.

   Note: Leave everything you installed in this lab exercise mounted on the wall. It will be needed to complete the next lab exercise.

14. Get the instructor to sign off upon satisfactory completion of this lab exercise.

Review

1. Complete the circuit directory shown below for the loadcenter you trimmed out in this lab exercise. A permanent and legible circuit directory is required by Section 408.4 of the *NEC®* at each loadcenter. Indicate when double-pole circuit breakers are used by drawing a dashed line between two of the circuit breaker handles.

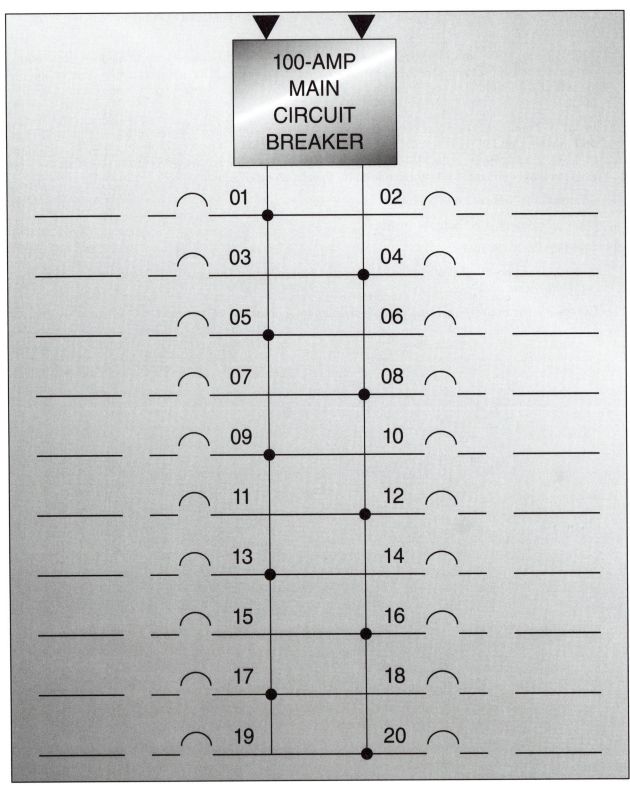

Lab 19.2: Install and Trim Out a Single-Phase Main Lug Only Sub-Panel

Name: _____ Date: _____ Score: _____

Introduction

When the main service panel is located some distance from those areas having many circuits and a heavy load concentration, it is recommended that sub-panels be installed closer to where the majority of the electrical load is located. The individual branch-circuit conductors are run to the sub-panel, not back to the main panel. Thus, the branch-circuit runs are shorter, and line losses are less than if the circuits had been run all the way back to the main panel. In this lab exercise, you will install a sub-panel with three 15-amp general lighting circuits. Before you proceed with this lab exercise, review the material on trimming out a service panel in Chapter 19 of the *House Wiring* 3e textbook. **Note:** Check with your instructor to see if you need to use AFCI circuit breakers where required in this lab exercise.

Materials and Equipment

House Wiring 3e textbook, safety glasses, electrician's tool kit, one 125-amp MLO six- or eight-circuit sub-panel, #10 × ¾-inch sheet metal screws as needed, 6 AWG AL Type SER cable or 8/3 CU Type NM cable as needed, 14/2 Type NM cable as needed, two 1-inch cable connectors, three ½-inch Type NM cable connectors, one 40-amp double-pole circuit breaker, three 15-amp single-pole circuit breakers, AFCI circuit breakers as needed if required by the lab instructor, one grounding bar kit for the sub-panel, antioxidant as needed

Procedure

1. Put on safety glasses and observe all safety rules. Refer to Figure 19.2-1 to see how this lab project should look when completed.

2. Using the sheet metal screws, mount and level the sub-panel 12 inches from the left edge of the 100-amp main breaker loadcenter installed in Lab 19.1. The top of the sub-panel should be even with the top of the main breaker loadcenter. Install a grounding bar kit in the sub-panel unless there already is one installed.

3. Install the Type SER cable between the sub-panel and the main breaker loadcenter, making sure to strip the outer sheathing to provide enough free conductor to make the connections in the loadcenters. Find out from your instructor whether you will be using 6 AWG AL Type SER or 8/3 CU Type NM cable. Do *not* use antioxidant on the aluminum terminations for this lab unless your instructor tells you to do so.

4. In the sub-panel, connect the grounding conductor to the grounding bar. Refer to Figure 8-37 in the *House Wiring* 3e textbook.

5. In the sub-panel, connect the grounded neutral wire to the grounded neutral bar. Refer to Figure 8-37 in the *House Wiring* 3e textbook.

6. In the sub-panel, connect the ungrounded conductors to the main lugs. Refer to Figure 8-37 in the *House Wiring* 3e textbook.

7. In the main breaker loadcenter, connect the grounding conductor and the grounded neutral conductor to the grounded neutral bar.

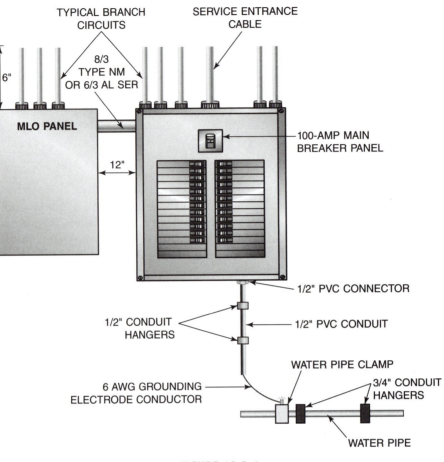

FIGURE 19.2-1

8. In the main breaker loadcenter, install a 40-amp two-pole circuit breaker and connect the ungrounded conductors of the Type SER cable to the breaker.

9. In the sub-panel, attach the three required 14/2 Type NM cables for the circuits to be installed using the proper size cable connector. Make sure to leave approximately 6 inches of sheathed cable on the outside and enough free conductors on the inside of the sub-panel to make connections at each circuit breaker.

10. In the sub-panel, install three 15-amp general-purpose circuits in the manner used in Lab 19.1. Remember to separate the bare grounding conductors and the white grounded conductors in the sub-panel. Install the required circuit breakers and connect the following:

   • Bare grounding wires to the equipment grounding bar

   • White grounded wires to the grounded neutral bar

   • Ungrounded wires to the proper circuit breakers

Follow the procedures for installing circuit breakers in Chapter 19 of the *House Wiring* 3e textbook.

Note: If any white conductors are to be used as ungrounded conductors, be sure to reidentify them with black tape.

11. Complete the Review for this lab exercise.

12. Show all completed work to the instructor.

13. Clean up the work area and return all tools and materials to their proper locations.

14. Get the instructor to sign off upon satisfactory completion of this lab exercise.

Review

1. Section 250.142(B) prohibits connecting the grounded neutral to any equipment on the load-side of the main service disconnect. Describe what this means for installations involving sub-panels.

_____

_____

_____

_____

_____

2. Describe the difference between a Type SEU cable and a Type SER cable.

_____

_____

_____

_____

_____

Lab 20.1: Complete a Successful Service Call

Name: _____ **Date:** _____ **Score:** _____

Introduction

Once a residential electrical system has been installed, it is checked out and any problems found are fixed. However, this does not mean that an electrician will never visit that house again to work on the electrical system. After the home owners move into the house, they may discover that something is not working correctly. It could be as simple as a light bulb not working because it is burned out or something more complicated, such as a circuit breaker that keeps tripping whenever they use their microwave oven and the garbage disposal at the same time. Typically, the home owner will call the company that installed the electrical system and request a service call. A service call is required if a home owner finds something wrong with the house electrical system. The company will get some information about the problem over the phone and then send an electrician on a service call. In this lab exercise, you will analyze an electrical problem that a home owner is having and then discuss the steps you would take to make a successful service call to fix the problem. Before you proceed with this lab exercise, review the material on service calls in Chapter 20 of the *House Wiring* 3e textbook.

Materials and Equipment

House Wiring 3e textbook, pencil

Procedure

1. Using the information presented throughout the *House Wiring* 3e textbook and the information presented on service calls in Chapter 20, analyze the problem described in the Review section and answer the questions in a way that will result in a successful completion of a service call.

2. Get the instructor to sign off upon satisfactory completion of this lab exercise.

Review

A home owner calls your company office and explains that a circuit breaker in the home's electrical panel has tripped and cannot be reset. The branch circuit that the circuit breaker protects serves several receptacles in the kitchen. As a result, these receptacles are not energized and anything plugged into them will not work.

Your boss decides to send you over to the house to troubleshoot the problem and fix it. For this lab exercise, we will assume that the electrical licensing rules in your area will allow you to do a service call on your own.

1. When you get to the house on the service call, what is the first thing you need to consider?

   _____

   _____

   _____

2. Once you knock on the door and the home owner opens it, what should you do?

   _____

   _____

   _____

3. You look at the problem and determine that one of the small-appliance branch circuit's 20-amp single-pole circuit breakers has tripped and cannot be reset. List some possible causes for this situation.

   _____

   _____

   _____

   _____

4. In this example, many electricians would begin to suspect that a ground fault in one of the receptacle outlet boxes on the circuit is causing the circuit breaker to trip. Often, a bare grounding wire from a Type NM cable will be touching an ungrounded conductor receptacle terminal screw in the box. This can happen when the receptacle and wires are not carefully pushed back into the box during the trim-out stage. To determine if a ground fault in one of the boxes is the cause of the problem, you will need to troubleshoot the branch circuit. Begin by determining which receptacles in the kitchen/dining area are on the circuit and make sure that there are no small appliances plugged into any of the receptacles on the circuit. Then, using the procedure for troubleshooting a receptacle circuit with a ground fault using a continuity tester in Chapter 20 of the *House Wiring* 3e textbook, you would determine which receptacle box has the fault. Review the procedure in Chapter 20 at this time. List the first two steps of this procedure.

   _____

   _____

   _____

   _____

5. Once you have fixed the problem, there are a few things to do to complete the service call. Name them.

   _____

   _____

   _____

   _____

6. Before completing the service call and finally leaving the home, there is one final thing to do. Describe what it is.

   _____

   _____

   _____

   _____

Chapter 21   Green Wiring Practices

Lab 21.1: Install a Charging Station

Name: _____ **Date:** _____ **Score:** _____

Introduction

Some appliances and electric devices consume electricity even when they are not performing their intended function. Cable TV boxes, Internet routers, microwave ovens, and televisions are just a few examples of devices that use electricity even when turned "off." Televisions, video recorders, and stereo equipment consume electricity in "standby mode" waiting for a remote control single to turn them on. Battery chargers for cell phones and laptop computers often use electricity even if the gadgets they power are fully charged or not even connected. Power converters for an Internet router and cordless phones are other examples of devices that consume electricity even when they are not being used. The total electrical load of all these devices can add up to 5% or more of a monthly electric bill. This load is called the "phantom load" because it is not obvious the devices are using electricity.

When wiring a green home, an electrician can help the building team locate and install special circuits to reduce electrical loads from sucking electricity needlessly. Special circuits with switch or timer control allow the home owner to completely turn off electricity to the loads either at will or automatically when the loads are not needed.

A charging station for battery-operated loads can be designed and installed in a house. Several receptacles are ganged together in a closet or cabinet and controlled by an automatic timer. All the battery chargers can be left plugged into the station, but the timer will only supply electricity for a short period of time during the night to do the charging.

In this lab exercise, a charging station controlled by a timer will be installed. A homerun will be run from the circuit breaker panel to the device box housing the timer. The timer will control three duplex receptacles located in a three-gang plastic nail-on device box. Before proceeding with this lab exercise, review the information on green wiring practices in Chapter 21 of the *House Wiring* 3e textbook.

Materials and Equipment

House Wiring 3e textbook, safety glasses, electrician's tool kit, voltage tester, ½-inch drill with a 7/8-inch auger bit, step ladder, one 15-amp 120-volt AFCI single-pole circuit breaker, 14/2 Type NM cable as needed, staples as needed, wirenuts as needed, grounding screws as needed, 6-32 and 8-32 machine screws as needed, cable connectors as needed, three 15-amp 125-volt tamper-resistant duplex receptacles, one 15- or 20-amp 120-volt single-pole timer, one single-gang plastic nail-on device box, one three-gang plastic nail-on device box, nail plates as needed

Procedure

1. Put on safety glasses and observe all safety rules.

2. On the studded wiring mock-up, mount the electrical boxes as shown in Figure 21.1-1. Assume the use of ½-inch wallboard on this wall.

3. Drill the framing members as needed and install the homerun with 14/2 Type NM cable from the circuit breaker panel to the single-gang plastic nail-on box. Secure and support the cable with staples as required by the *NEC*®. Use nail plates if necessary.

4. Drill the framing members as needed and install 14/2 Type NM cable between the single-gang plastic nail-on box and the three-gang plastic nail-on box. Secure and support the cables with staples as required by the *NEC*®. Use nail plates if necessary.

5. Connect the conductors to the devices in the boxes so that the three duplex receptacles in the charging station are controlled by the 120-volt timer. Review the timer installation manual if it is available. Use Chapters 13 and 18 in the *House Wiring* 3e textbook for help if necessary. When making the conductor connections, review the following procedures:

   – Connecting wires together with a wirenut in Chapter 2 of the *House Wiring* 3e textbook.

   – Using terminal loops to connect circuit conductors to terminal screws on receptacles or switches in Chapter 18 of the *House Wiring* 3e textbook.

   – Installing receptacles (or switches) in a nonmetallic electrical box in Chapter 18 of the *House Wiring* 3e textbook.

   – Installing receptacles (or switches) in a metal electrical box in Chapter 18 of the *House Wiring* 3e textbook.

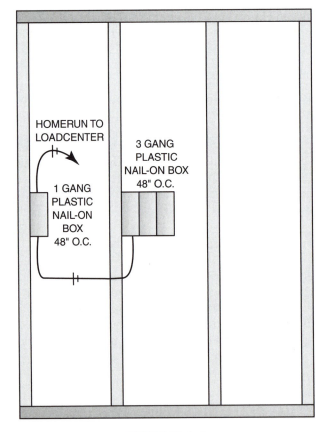

FIGURE 21.1-1

6. Secure the devices to the boxes. Do not install switch or receptacle covers unless your instructor tells you to.

7. Install a 15-amp AFCI circuit breaker in the mock-up loadcenter and make the proper connections for the homerun wiring. Refer to the procedure on installing a single-pole AFCI circuit breaker in Chapter 19 of the *House Wiring* 3e textbook.

8. Test the circuit by:

   • Unlocking the electrical power supply and turning on the circuit breaker.

   • Activate the timer and observe if the charging station receptacles are working correctly.

   • Follow the procedure on testing 120-volt receptacles with a voltage tester in Chapter 20 of the *House Wiring* 3e textbook and check each receptacle for proper voltage, polarity, and grounding.

   • If a problem is found, disconnect the power using a lock-out/tag-out procedure like the one in Chapter 1 of the *House Wiring* 3e textbook, and troubleshoot the circuit.

9. Show all completed work to the instructor.

10. Lock out the power supply, clean up the work area, and return all tools and materials to their proper locations.

11. Complete the Review for this lab exercise.

12. Get the instructor to sign off upon satisfactory completion of this lab exercise.

Review

1. Make out a complete material list for this lab. Include a description of the item used and the quantity. Be sure to include *all* items used.

   _____

   _____

   _____

   _____

   _____

   _____

2. Based on an hourly rate of $50.00 and using the material price list located in the Appendix B, calculate the cost of doing the electrical work in this lab exercise.

   _____

   _____

   _____

   _____

   _____

   _____

   _____

Chapter 22 — Alternative Energy System Installation

Lab 22.1: Wire PV Modules in Series and Parallel

Name: _____ **Date:** _____ **Score:** _____

Introduction

Photovoltaic (PV) modules can be wired together in series, parallel, or series-parallel, to get a desired system voltage and current. Voltage sources connected in series result in the total voltage increasing but the current flow remains the same. With DC sources like PV modules, connect them positive (+) to negative (–) for a series connection. Voltage sources connected in parallel result in the total current increasing but the voltage remains the same. With DC sources like PV modules, connect them positive (+) to positive (+) and negative (–) to negative (–) for a parallel connection. PV systems may use a mix of series and parallel connections to obtain the required voltage and amperage. In this lab exercise you will draw in the wiring to connect PV modules in series and parallel. Before you attempt this lab exercise review the material in Chapter 22 of the *House Wiring* 3e textbook that covers series and parallel connection of PV modules.

Materials and Equipment

House Wiring 3e textbook, pencil

Procedure

1. Using Figure 22.1-1 in the Review section of this lab exercise, draw in the wiring that will connect the PV modules in series. Show the wiring from the modules connected to the combiner box. In the blank spaces provided, indicate the total array voltage and total array current.

2. Using Figure 22.1-2 in the Review section of this lab exercise, draw in the wiring that will connect the PV modules in parallel. Show the wiring from the modules connected to the combiner box. In the blank spaces provided, indicate the total array voltage and total array current.

3. Show the instructor your completed work.

4. Get the instructor to sign off upon satisfactory completion of this lab exercise.

Review

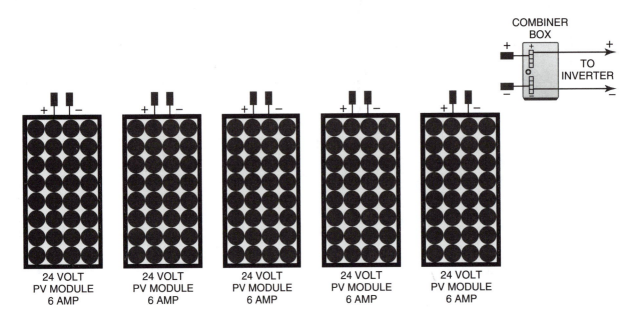

TOTAL ARRAY VOLTAGE = _____
TOTAL ARRAY CURRENT = _____

FIGURE 22.1-1

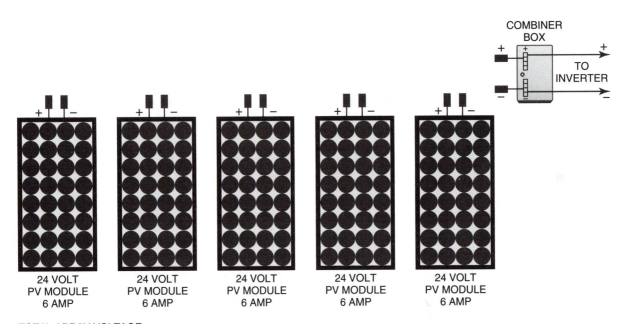

TOTAL ARRAY VOLTAGE = _____
TOTAL ARRAY CURRENT = _____

FIGURE 22.1-2

Lab 22.2: Wire PV System Batteries in Series and Parallel

Name: _____ **Date:** _____ **Score:** _____

Introduction

PV system batteries can be wired together in series, parallel, or series-parallel, to get a desired system voltage and current. Voltage sources connected in series result in the total voltage increasing but the current flow remains the same. With DC sources like batteries, connect them positive (+) to negative (–) for a series connection. Voltage sources connected in parallel result in the total current increasing but the voltage remains the same. With DC sources like batteries, connect them positive (+) to positive (+) and negative (–) to negative (–) for a parallel connection. PV system battery banks may use a mix of series and parallel connections to obtain the required voltage and amperage. In this lab exercise you will draw in the wiring to connect PV system batteries in series and parallel. Before you attempt this lab exercise review the material in Chapter 22 of the *House Wiring* 3e textbook that covers series and parallel connection of PV system batteries.

Materials and Equipment

House Wiring 3e textbook, pencil

Procedure

1. Using Figure 22.2-1 in the Review section of this lab exercise, draw in the wiring that will connect the battery bank in series. Show the wiring from the modules connected to the charge controller. In the blank spaces provided, indicate the total battery bank voltage and total amp-hour capacity.

2. Using Figure 22.2-2 in the Review section of this lab exercise, draw in the wiring that will connect the battery bank in parallel. Show the wiring from the modules connected to the combiner box. In the blank spaces provided, indicate the total battery bank voltage and total amp-hour capacity.

3. Show the instructor your completed work.

4. Get the instructor to sign off upon satisfactory completion of this lab exercise.

Review

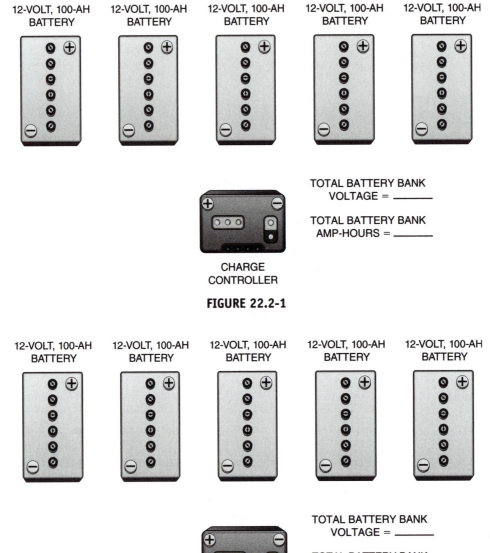

12-VOLT, 100-AH BATTERY   12-VOLT, 100-AH BATTERY   12-VOLT, 100-AH BATTERY   12-VOLT, 100-AH BATTERY   12-VOLT, 100-AH BATTERY

CHARGE CONTROLLER

TOTAL BATTERY BANK VOLTAGE = _____

TOTAL BATTERY BANK AMP-HOURS = _____

FIGURE 22.2-1

12-VOLT, 100-AH BATTERY   12-VOLT, 100-AH BATTERY   12-VOLT, 100-AH BATTERY   12-VOLT, 100-AH BATTERY   12-VOLT, 100-AH BATTERY

CHARGE CONTROLLER

TOTAL BATTERY BANK VOLTAGE = _____

TOTAL BATTERY BANK AMP-HOURS = _____

FIGURE 22.2-2

Description
Long-nose pliers
Lineman pliers
Diagonal cutting pliers
10-inch pump pliers
Keystone Tip Screwdriver, 4-inch heavy duty
Keystone Tip Screwdriver, 6-inch medium duty
Keystone Tip Screwdriver, 8-inch heavy duty
Phillips-head screwdriver, 4-inch number 2 point
Electrician's knife
25-foot tape measure
Electrician's hammer
T-stripper wire stripper
Tool pouch
Solenoid type voltage tester (Wiggy)
Digital multimeter
Clamp-on ammeter
Rotary Cutter (for Type MC cable)
Screw-holding screwdriver
Scratch-awl
Magnetized torpedo level
Hacksaw
Adjustable wrench, 10 inch
EMT Conduit bender, ½ inch and ¾ inch
Metal file or reaming tool
Pipe vise
Manual conduit threader with ½-inch cutting die
Pipe cutter for RMC
Pipe reamer for RMC
Coaxial cable stripper for RG-6 cable
Crimping tool and die for RG-6 coaxial cable
Compression fitting tool for RG-6 coaxial cable
Cable stripper for Cat 5e UTP cable

(Continued)

Description
Punchdown tool with 110 tip
Cable tester for 4 pair UTP cable
Crimping tool and die for RJ-45 modular plugs
Corded or cordless pistol-grip drill
Manual knockout set
Corded or cordless ½-inch right-angle drill
Corded or cordless hammer drill
Corded or cordless reciprocating saw
Various drill bits as needed
Safety glasses

Appendix B

Cost Sheet for Equipment and Material Used in the Lab Exercises

Equipment/Materials Description	Cost per Item
Boxes – Nonmetallic	
1-gang device box, nail-on	$0.72
2-gang device box, nail-on	$1.79
3-gang device box, nail-on	$2.69
4-inch round ceiling box, nail-on	$2.03
Boxes – Handy Boxes	
4 × 2⅛ × 1⅞-inch with ½-inch KOs	$1.24
4 × 2⅛ × 2⅛-inch with ½-inch KOs	$2.69
Boxes – Metal Gangable Device Boxes	
3 × 2 × 2¾-inch with clamps and mounting bracket	$4.04
3 × 2 × 3½-inch with clamps and mounting bracket	$5.08
Boxes – Octagon	
4 × 1½-inch with 1/2 KOs	$1.77
4 × 1½-inch with clamps and 1/2 KOs	$2.15
4 × 1½-inch with clamps and bracket	$4.18
4 × 2⅛-inch with 1/2 KOs	$2.60
4 × 2⅛-inch with clamps and 1/2 KOs	$2.79
4 × 2⅛-inch with clamps and bracket	$5.10
Boxes – Square	
4 × 1½-inch with 1/2 KOs	$1.66
4 × 1½-inch with ½-inch and ¾-inch KOs	$1.60
4 × 1½-inch with clamps and ½-inch KOs	$1.66
4 × 1½-inch with clamps and bracket	$4.07
4 × 2⅛-inch with bracket and ½-inch and ¾-inch KOs	$4.50
4 × 2⅛-inch with clamps and 1/2 KOs	$1.53
4 × 2⅛-inch with clamps and bracket	$4.10
4 × ½-inch raised plaster ring, 1-gang	$1.20
4 × ½-inch raised plaster ring, 2-gang	$1.65
Boxes – Covers	
4-inch square flat blank	$0.42
4-inch octagon flat blank	$0.71
Boxes – Supports	
Adjustable bar hanger	$3.25
Circuit Breakers	
15-amp, 1-pole	$9.51
20-amp, 1-pole	$9.51
15-amp, 2-pole	$21.91
20-amp, 2-pole	$21.91
30-amp, 2-pole	$21.91
40-amp, 2-pole	$21.91

(Continued)

Equipment/Materials Description	Cost per Item
50-amp, 2-pole	$21.91
60-amp, 2-pole	$21.91
15-amp, AFCI, 1-pole	$59.29
15-amp, GFCI, 1-pole	$74.43
20-amp, GFCI, 1-pole	$74.43
Conduit – By the Foot	
½-inch RMC	$1.42
¾-inch RMC	$1.32
½-inch IMC	$0.90
¾-inch IMC	$1.02
2-inch IMC	$4.45
½-inch EMT	$0.24
¾-inch EMT	$0.42
½-inch PVC	$0.13
Conduit – Fittings, RMC, and IMC	
½-inch steel locknut	$0.11
¾-inch steel locknut	$0.15
½-inch RMC, one-hole strap	$0.16
¾-inch RMC, one-hole strap	$0.21
2-inch conduit hanger	$0.96
Conduit – Fittings, EMT, and PVC	
½-inch EMT connector, set screw	$0.21
½-inch EMT connector, compression	$0.31
¾-inch EMT connector, set screw	$0.28
¾-inch EMT connector, compression	$0.61
½-inch EMT coupling, set screw	$0.23
½-inch EMT coupling, compression	$0.54
¾-inch EMT coupling, set screw	$0.33
¾-inch EMT coupling, compression	$0.64
½-inch EMT, one-hole strap	$0.10
¾-inch EMT, one-hole strap	$0.15
½-inch EMT conduit hanger	$0.39
¾-inch EMT conduit hanger	$0.47
½-inch PVC connector	$0.17
Fasteners – Bolts, Screws, Etc.	
#10 × ¾-inch sheet metal screws (box of 100)	$3.10
8-32 × ¾-inch machine screws (box of 100)	$2.34
6-32 × ¾-inch machine screws (box of 100)	$2.10
Grounding	
Grounding screws, 10-32 (box of 100)	$4.31
Ground rod, copper, 8-foot × ⅝-inch	$28.12
Ground rod, galvanized steel, 8-foot × ⅝-inch	$7.74
Acorn type ground rod clamp, ⅝-inch	$1.65
Water pipe grounding clamp, brass	$2.39
Grounding crimp sleeve (box of 100)	$8.20
Grounding bar kit	$9.45
Lamps	
60-watt, 120-volt, med. base, gen. service	$0.33

Equipment/Materials Description	Cost per Item
Loadcenters	
100-amp, 1 PH, 20 circuit, main breaker	$129.51
100-amp, 1 PH, 8 circuit, MLO	$46.77
Miscellaneous	
4-foot electric baseboard heater	$34.24
Thermostat, 2-pole, 240-volt	$26.84
15-amp, 125-volt, attachment plug	$3.52
Chime	$9.50
Chime transformers	$5.00
Smoke detectors, 3-wire, feed-thru	$13.06
Cable ties, 8-inch (per package of 100)	$2.87
Nail plates (steel 1/16-inch)	$0.41
Service Entrance Fittings	
Weatherhead, 2-inch conduit, clamp type	$10.93
Weatherhead for #2 AL SEU	$5.86
Sill plate for #2 AL SEU	$1.62
Duct seal (1-pound brick)	$1.80
Service mast kit, 2-inch	$26.37
Threaded hub, 1¼-inch	$5.12
Threaded hub, 2-inch	$5.12
Meter socket, 1 PH, 100-amp	$35.19
Tape – By the Roll	
Rubber splicing tape (roll)	$11.01
Black vinyl tape (roll)	$3.70
White vinyl tape (roll)	$3.70
Wire and Cable – By the Foot	
2 AWG aluminum XHHW	$1.80
6 AWG bare grounding wire	$0.62
NMSC 14/2	$0.30
NMSC 14/3	$0.40
NMSC 12/2	$0.39
NMSC 12/3	$0.61
NMSC 10/2	$0.69
NMSC 10/3	$0.97
NMSC 8/3	$1.47
SEU, 2 AWG aluminum, 3-wire	$0.83
SER, 6 AWG aluminum, 4-wire	$0.72
SJ Cord, 16/3	$0.58
SJ Cord, 14/3	$0.97
Underground feeder (UF) 14/2	$0.37
Underground feeder (UF) 14/3	$0.46
Thermostat cable 18/2 (chime wire)	$0.10
Thermostat cable 18/3 (chime wire)	$0.15
Wire and Cable Fittings	
Cable connector, ½-inch metal	$0.25
Cable connector, ½-inch plastic	$0.11
Cable connector, ¾-inch metal	$0.60
SEU cable connector, 1¼-inch	$1.49
SEU cable connector, watertight, 1¼-inch	$5.97

(Continued)

Equipment/Materials Description	Cost per Item
SEU 2 AWG AL, 1-hole strap	$0.21
SEU 2 AWG AL, fold-over clips	$0.33
Staples, bell wire (box of 100)	$2.58
Staples, Type NM, insulated, 14 AWG – 10 AWG (box of 100)	$2.14
Staples, larger for 8/3 and 6/3 Type NM (box of 100)	$5.53
Wirenuts	$0.12
Wirenuts, green grounding	$0.15
Antioxidant for AL wire (bottle)	$6.94
Split bolt connector, size 2	$4.71
Wiring Devices	
Duplex receptacle, 15-amp, 125-volt, brown	$0.52
Duplex receptacle, 20-amp, 125-volt, brown	$1.44
Duplex receptacle, 15-amp, 125-volt, brown, tamper-resistant	$1.55
Lamp holder, keyless plastic	$1.69
Receptacle, 30-amp dryer, flush-mount	$7.60
Receptacle, 30-amp dryer, surface-mount	$10.10
Receptacle, 50-amp range, flush-mount	$7.60
Receptacle, 50-amp range, surface	$10.10
Receptacle, GFCI, 15-amp, 125-volt, brown	$8.73
Switch, 1-pole, 15-amp, 125-volt, brown	$0.80
Switch, 3-way, 15-amp, 125-volt, brown	$1.25
Switch, 4-way, 15-amp, 125-volt, brown	$8.64
Switch, doorbell, surface-mount	$4.05
Timer Switch, 15- or 20-amp, 125-volt, box-mount	$22.50

Appendix C — Lab Competency Profile

Student Name _____

I certify that I have completed the lab exercises as shown to the best of my ability.

Student Signature _____ Date _____

I certify that the student named above received training in the topic areas shown and satisfactorily completed the lab exercises indicated.

Instructor Name _____

Instructor Signature _____ Date _____

Directions: Instructors will inspect and grade each lab exercise using the rating scale shown below. Lab exercise grading is based on five factors: 1) Safety; 2) Workmanship; 3) Correct operation; 4) Completion in a timely manner; and 5) Compliance with the *National Electrical Code®*. Circle the appropriate number to indicate the degree of student competency for each lab exercise.

RATING SCALE

4 = The lab exercise was done in a manner that was safe, neat, and workmanlike; operated correctly; was completed in a timely manner; and all applicable *NEC®* rules were met.

3 = The lab exercise was done in a manner that was safe, neat, and workmanlike, however; there were *NEC®* code violations; or the lab did not operate correctly; or it was done in too great a length of time.

2 = The lab exercise was not done following safe work practices; it was not done in a neat and workmanlike manner; it did not operate correctly; and there were numerous *NEC®* violations.

1 = The student did not do this lab exercise.

Chapter 1: Residential Workplace Safety		
Lab 1.1	Demonstrate an understanding of both General Safety and Electrical Safety by scoring 100% on a comprehensive safety exercise.	4 3 2 1
Lab 1.2	Find information in the *National Electrical Code®*.	4 3 2 1
Lab 1.3	Use a Material Safety Data Sheet (MSDS).	4 3 2 1

Chapter 2: Hardware and Materials Used in Residential Wiring

Lab 2.1	Identify the parts of a typical metal device box.	4 3 2 1
Lab 2.2	Identify different types of wire connectors.	4 3 2 1
Lab 2.3	Identify the parts on a duplex receptacle.	4 3 2 1
Lab 2.4	Identify the parts on single-pole, double-pole, three-way, and four-way switches.	4 3 2 1

Chapter 3: Tools Used in Residential Wiring

Lab 3.1	Identify several guidelines for the care and safe use of electrical hand tools.	4 3 2 1
Lab 3.2	Using lineman pliers.	4 3 2 1
Lab 3.3	Using a wire stripper.	4 3 2 1
Lab 3.4	Using an electrician's knife.	4 3 2 1
Lab 3.5	Using an electrician's knife and rotary stripping tool to strip off the outer sheathing off various sizes of Type NM, Type UF, and Type MC cables.	4 3 2 1
Lab 3.6	Using a screwdriver and appropriate fasteners to install electrical boxes.	4 3 2 1
Lab 3.7	Stripping large electrical conductors with an electrician's knife.	4 3 2 1
Lab 3.8	Use various electrical tools to cut, strip, and install an attachment plug to a length of flexible cord.	4 3 2 1
Lab 3.9	Set up and use a pistol-grip drill.	4 3 2 1
Lab 3.10	Using a manual knockout set.	4 3 2 1
Lab 3.11	Set up and use a hacksaw.	4 3 2 1
Lab 3.12	Set up and use a right-angle drill.	4 3 2 1
Lab 3.13	Set up and use a hammer drill.	4 3 2 1
Lab 3.14	Set up and use a reciprocating saw.	4 3 2 1

Chapter 4: Test and Measurement Instruments Used in Residential Wiring

Lab 4.1	Using a voltage tester.	4 3 2 1
Lab 4.2	Using a digital multimeter.	4 3 2 1
Lab 4.3	Using a clamp-on ammeter.	4 3 2 1

Chapter 5: Understanding Residential Building Plans

Lab 5.1	Identify common architectural electrical symbols.	4 3 2 1
Lab 5.2	Identify the structural parts of a house.	4 3 2 1

Chapter 6: Determining Branch Circuit, Feeder Circuit, and Service Entrance Requirements

Lab 6.1	Calculate the minimum number of general lighting circuits.	4 3 2 1
Lab 6.2	Calculate common residential electric cooking loads, branch-circuit conductor sizes, and circuit breaker sizes.	4 3 2 1
Lab 6.3	Determine the ampacity of a conductor.	4 3 2 1
Lab 6.4	Calculate the size of a residential service entrance using the standard method.	4 3 2 1
Lab 6.5	Calculate the size of a residential service entrance using the optional method.	4 3 2 1

Chapter 7: Introduction to Residential Service Entrances

Lab 7.1	Identify the major parts of a residential service entrance.	4 3 2 1

Chapter 8: Service Entrance Equipment and Installation

| Lab 8.1 | Install an overhead service entrance using service entrance cable. | 4 3 2 1 |
| Lab 8.2 | Install an overhead–mast-type service entrance. | 4 3 2 1 |

Chapter 9: General Requirements for Rough-In Wiring

| Lab 9.1 | Indicate the proper locations on a building plan for the minimum number of receptacles and lighting outlets. | 4 3 2 1 |

Chapter 10: Electrical Box Installation

Lab 10.1	Installing device boxes.	4 3 2 1
Lab 10.2	Installing outlet boxes.	4 3 2 1
Lab 10.3	Installing outlet boxes with an adjustable bar hanger.	4 3 2 1

Chapter 11: Cable Installation

| Lab 11.1 | Installing a cable run from a loadcenter. | 4 3 2 1 |

Chapter 12: Raceway Installation

Lab 12.1	Cut, ream, and thread rigid metal conduit (RMC) by using a manual threader, reamer, and pipe cutter.	4 3 2 1
Lab 12.2	Bend a 90° stub-up with a hand bender.	4 3 2 1
Lab 12.3	Bend a back-to-back bend with a hand bender.	4 3 2 1
Lab 12.4	Bend an offset bend with a hand bender.	4 3 2 1
Lab 12.5	Bend a three-point saddle bend with a hand bender.	4 3 2 1
Lab 12.6	Bend box offsets with a hand bender.	4 3 2 1
Lab 12.7	Bend and install EMT as part of a conduit system.	4 3 2 1

Chapter 13: Switching Circuit Installation

Lab 13.1	Install a single-pole switch for a lighting load with the power source feeding the switch.	4 3 2 1
Lab 13.2	Install a single-pole switch for a lighting load with the power source feeding the lighting outlet (switch loop).	4 3 2 1
Lab 13.3	Install a three-way switching circuit with the power source feeding the first three-way switch location.	4 3 2 1
Lab 13.4	Install a three-way switching circuit with the power source feeding the lighting outlet location.	4 3 2 1
Lab 13.5	Install a three-way switching circuit with the power source feeding the lighting outlet location and with three-wire cable run from the lighting outlet to each three-way switch location.	4 3 2 1
Lab 13.6	Install a four-way switching circuit with the power source feeding the first three-way switch location.	4 3 2 1
Lab 13.7	Install a four-way switching circuit with the power source feeding the lighting outlet location.	4 3 2 1
Lab 13.8	Install a split-duplex receptacle so that the bottom half is controlled by a single-pole switch and the top half is "hot" at all times.	4 3 2 1
Lab 13.9	Install a single-pole switch controlling a lighting load with a continuously "hot" receptacle located downstream of the lighting outlet—the power source will feed the switch.	4 3 2 1

Chapter 14: Branch-Circuit Installation

Lab 14.1	Install a general lighting branch circuit.	4 3 2 1
Lab 14.2	Install a small-appliance branch circuit.	4 3 2 1
Lab 14.3	Install an electric range branch circuit.	4 3 2 1
Lab 14.4	Install an electric clothes dryer branch circuit.	4 3 2 1
Lab 14.5	Install an electric baseboard heating branch circuit.	4 3 2 1
Lab 14.6	Install a smoke detector branch circuit.	4 3 2 1
Lab 14.7	Install a low-voltage chime circuit.	4 3 2 1

Chapter 15: Special Residential Wiring Situations

Lab 15.1	Install a garage branch circuit.	4 3 2 1

Chapter 16: Video, Voice, and Data Wiring Installation

Lab 16.1	Install crimp style F-Type connectors on RG-6 coaxial cable.	4 3 2 1
Lab 16.2	Install compression style F-Type connectors on an RG-6, 75-ohm coaxial cable.	4 3 2 1
Lab 16.3	Install RJ-45 jacks on the ends of a four-pair Category 5e UTP cable.	4 3 2 1
Lab 16.4	Install RJ-45 modular plugs on the ends of Category 5e UTP cable.	4 3 2 1

Chapter 17: Lighting Fixture Installation

Lab 17.1	Install a light fixture to an electrical box with a strap.	4 3 2 1
Lab 17.2	Install a light fixture to an electrical box with a stud and strap.	4 3 2 1

Chapter 18: Device Installation

Lab 18.1	Identify common receptacle and plug configurations.	4 3 2 1

Chapter 19: Service Panel Trim-Out

Lab 19.1	Install and trim out a single-phase circuit breaker loadcenter.	4 3 2 1
Lab 19.2	Install and trim out a single-phase main lug only sub-panel.	4 3 2 1

Chapter 20: Checking Out and Troubleshooting Electrical Wiring Systems

Lab 20.1	Complete a successful service call.	4 3 2 1

Chapter 21: Green Wiring Practices

Lab 21.1	Install a charging station.	4 3 2 1

Chapter 22: Alternative Energy System Installation

Lab 22.1	Wire PV modules in series and parallel.	4 3 2 1
Lab 22.2	Wire PV system batteries in series and parallel.	4 3 2 1